Bioinformatics:
genes, proteins
and computers

Bioinformatics:
genes, proteins and computers

Edited by

Christine Orengo
Department of Biochemistry and Molecular Biology,
University College London, London, UK

David Jones
Department of Computer Science,
University College London, London, UK

Janet Thornton
European Bioinformatics Institute, Cambridge, UK

BIOS Scientific Publishers
Taylor & Francis Group

First published 2003
Reprinted 2004

A CIP catalogue record for this book is available from the British Library.

ISBN 1 85996 054 5

Garland Science/BIOS Scientific Publishers
4 Park Square, Milton Park, Abingdon, Oxon, OX14 4RN, UK and
270 Madison Avenue, New York, NY 10016, USA
World Wide Web home page: www.garlandscience.com

Garland Science/BIOS Scientific Publishers is a member of the Taylor & Francis Group.

Distributed exclusively in the United States, its dependent territories, Canada, Mexico, Central and South America, and the Caribbean by Springer-Verlag New York Inc., 175 Fifth Avenue, New York, USA, by arrangement with Garland Science/BIOS Scientific Publishers.

Production Editor: Andrea Bosher
Typeset by Phoenix Photosetting, Chatham, UK
Printed by Cromwell Press, Trowbridge, UK

Contents

Colour plates can be found between pages 50 and 51, 210 and 211, 242 and 243

Abbreviations

AFLP	amplified fragment length polymorphism
ANN	artificial neural networks
BAC	bacterial artificial chromosome
CAPRI	Critical Assessment of Prediction of Interaction
CAPS	cleaved amplified polymorphic DNA
CASP	Critical Assessment of Structure Prediction
cDNA	complementary DNA
CDR	complementary determining regions
COG	Clusters of Orthologous Groups
CORBA	Common Object Request Broker Architecture
CPP	Coupled Perturbation Protocol
CSS	Cascading Style Sheets
DAS	Distributed Annotation System
DBMS	database management system
DDD	DALI Domain Database/Dictionary
DHS	dictionary of homologous superfamilies
DTD	data type definition
EBI	European Bioinformatics Institute
EC	Enzyme Commission
EM	electron microscopy
EM	energy minimization
EMSD	European Macromolecular Structure Database
ESI	electrospray ionization
EST	expressed sequence tag
FOD	frequently occurring domain
GCM	genetic code matrix
GML	Generalized Markup Language
GO	Gene Ontology
GPCR	G-protein-coupled receptor
HMM	Hidden Markov Models
HSP	high-scoring segment pair
HTML	HyperText Markup Language
HTTP	Hyper Text Transport Protocol
ICAT	isotope coded affinity tag
MALDI	matrix-assisted laser desorption/ionization
MD	molecular dynamics
MDM	mutation data matrix
MIAME	Minimum Information About a Microarray Experiment
MOP	Maximum Overlap Protocol
MPP	Minimum Perturbation Protocol
mRNA	messenger RNA
MS	mass spectrometry
MSP	maximal segment pair
ncRNA	non-coding RNA

NMR	nuclear magnetic resonance
OLAP	On-Line Analytic Processing
OMG LSR	Object Management Group Life Sciences Research
ORF	open reading frame
PAC	P1-derived artificial chromosome
PAH	polycyclic aromatic hydrocarbons
PAM	percent/point accepted mutation
PCR	polymerase chain reaction
PDB	Protein Data Bank/Protein Structure Databank
PDF	probability density function
PMF	peptide mass fingerprinting
PSSM	position-specific scoring matrix
RAPD	randomly amplified polymorphic DNA
RCSB	Research Collabatory of Structural Biology
RDF	Resource Description Framework
RFLP	restriction fragment length polymorphism
RMSD	root mean square deviation
RPC	Remote Procedure Calling
rRNA	ribosomal RNA
RT-PCR	reverse transcriptase-polymerase chain reaction
SAGE	Serial Analysis of Gene Expression
SCFG	stochastic context-free grammar
SCR	structurally conserved region
SGML	Standard Generalized Markup Language
SMM	small molecule metabolism
SNP	single nucleotide polymorphism
SOAP	Simple Object Access Protocol
SOM	self-organizing map
SP	sum of pairs
SSLP	single sequence length polymorphism
STS	sequence tagged site
SVM	support vector machines
SVR	structurally variable region
TOF	time of flight
TPA	tissue plasminogen activator
tRNA	transfer RNA
VRML	Virtual Reality Modeling Language
WML	Wireless Markup Language
WSDL	Web Services Description Language
XML	eXtensible Markup Language
XSL	eXtensible Stylesheet Language
XSLT	eXtensible Stylesheet Language Transformations
YAC	yeast artificial chromosome

Contributors

Blackstock, W.P., *Cellzone UK, Elstree, UK*

Jackson, R.M., *Department of Biochemistry and Molecular Biology, University College London, London, UK*

Jones, D.T., *Department of Computer Science, University College London, London, UK*

Kellam, P., *Wohl Virion Centre, University College London, London, UK*

Liu, X., *Department of Information Systems and Computing, Brunel University, Uxbridge, UK*

Martin, A.C.R., *School of Animal and Microbial Sciences, The University of Reading, Reading, UK*

Martin, N.J., *School of Computer Science and Information Systems, Birkbeck College, London, UK*

Nagl, S.B., *Department of Biochemistry and Molecular Biology, University College London, London, UK*

Orengo, C., *Department of Biochemistry and Molecular Biology, University College London, London, UK*

Pearl, F., *Department of Biochemistry and Molecular Biology, University College London, London, UK*

Sgouros, J.G., *Computational Genome Analysis Laboratory, Imperial Cancer Research Fund, London, UK*

Sillitoe, I., *Department of Biochemistry and Molecular Biology, University College London, London, UK*

Teichmann, S.A., *Structural Studies Division, MRC Laboratory of Molecular Biology, Cambridge, UK*

Thornton, J., *European Bioinformatics Institute, Cambridge, UK*

Todd, A.E., *Department of Biochemistry and Molecular Biology, University College London, London, UK*

Twyman, R.M., *Department of Biology, University of York, Heslington, York, UK*

Valdar, W.S.J., *Wellcome Trust Centre for Human Genetics, Oxford, UK*

Weir, M.P., *Impharmatica, London, UK*

Foreword
Professor Janet Thornton

With the flood of biological data, which started in the early 1990s, 'bioinformatics' is gradually becoming an accepted discipline in main stream biology – or 'necessary evil' depending on one's perspective! At the simplest level it has been described as 'mere data curation' but it is the view of the authors of this book that bioinformatics is one of the critical keys needed to unlock the information encoded in genome data, protein structure data and high-throughput transcriptome and proteome information. This view reflects the absolute need to use all the available information to interpret new experimental data. Increasingly this is only possible using computational approaches – and so bioinformatics lies at the very heart of modern biology. The discipline should not be separated from experimental work, but fully integrated so that 'in vitro,' 'in vivo' and 'in silico' approaches can be used in synergy to solve problems and make discoveries. Whilst this is important to increase our academic understanding, the applications of and need for these skills in the pharmaceutical and biotechnology industries cannot be overstated. To succeed these industries are embracing bioinformatics into the core of their research and development programs.

As a subject, bioinformatics is difficult to define. A possible succinct definition is 'the collection, archiving, organization and interpretation of biological data'. This goes beyond the collection and storage of data, to include the elucidation of fundamental principles through classification, organization and interpretation. Therefore although one role of the bioinformatician is to develop and provide tools and databases, we are also in a position to ask and answer fundamental questions about molecular evolution, biological function and the control of biological systems, which cannot be tackled in any other way. The ultimate goal as with most theoretical sciences is to use the information to increase our understanding to the point where we can make reliable predictions. For bioinformatics this has evolved from predicting structure from sequence, to predicting function from structure, networks and complexes from transcriptome and proteome data, and ultimately one would like to design novel genes/proteins and small molecules with specific functions. Such a goal is ambitious but not impossible and would transform the design of novel therapeutics, from vaccines to drugs.

However to be able to address such problems, it is necessary for those who practise bioinformatics to be experts not only in biology, but also in the computational/mathematical approaches needed. From the start bioinformatics has been a highly multidisciplinary subject, recruiting not only biologists, but also mathematicians, physicists, computer scientists, and most recently engineers. This presents a challenge for training. On the one side the physical scientists need to learn the concepts of biology – whose details can be overwhelming. On the other side the biologists need to absorb the concepts and skills of using computers, rigorous statistics and databases. This book seeks to provide an introduction to many – though not all – aspects of modern bioinformatics – concentrating on the principles involved, rather than the technology, which is inevitably evolving at a great rate. It is aimed at the third year undergraduate or masters student with a biological background, or physical scientists who have some knowledge of basic molecular biology and wish to understand some of the basic principles of bioinformatics.

Most of the authors in this book have contributed or taught on an undergraduate bioinformatics course at UCL and this book aims to provide supplementary material for this and similar courses in universities and research institutes. Therefore, this book concentrates more on concepts than detailed descriptions of algorithms and analysis methods. However, throughout the book we have provided some extra details in text boxes for those readers seeking more in-depth information. At the end of each chapter we have provided a short list of selected papers for further study concentrating where possible on reviews. As this is a rapidly evolving field we have decided against extensive lists of specific references.

Probably the core focus of bioinformatics over the last 10 years has been the development of tools to compare nucleotide and amino acid sequences to find evolutionary relatives. This subject is likely to remain central, so the first part of the book (Chapters 2–6) addresses this problem. During the late 1980s it was gradually realized that three-dimensional structure is much better conserved than sequence, and therefore structural information can reveal relationships, which are hidden at the sequence level. The structural data help to classify protein domain families and better understand the evolution of new sequences, structures and function. Therefore the second part of the book (Chapters 7–10) considers the comparison of protein structures, protein families and the modeling of structure from sequence, leading to a consideration of the evolution of function within homologous families. In Chapter 12 this is extended to consideration of genome data and how such information can be used to help in functional assignment to a gene product, which is critical for genome annotation and target identification in the pharmaceutical industry. The following chapter (13) addresses the problem of predicting protein–protein or protein–ligand complexes, which are important for structure-based drug design. With the advent of transcriptome and proteome technologies the need for the development of tools to handle such data, let alone interpret them has been a real challenge. Chapters 14–18 present an introduction to these areas, which are just developing, but will become increasingly important and sophisticated over the next few years. The ontologies and databases are just being established as we go to press and questions of how to generate and store data are still not completely clear.

So, what is the future of bioinformatics? The flood of data is not going to abate in the next 10 years and this will increase the need for better, faster computational approaches. We will need new databases, new concepts and a much closer link between experiment and modeling. Clearly we can compare bioinformatics to the revolution in experimental molecular biology, which was first located in a few specialized laboratories but is now ubiquitous in almost all biological laboratories. If the computational tools are well designed, then gradually all biologists will become 'applied' bioinformaticians at some level. To some extent this is already occurring. In parallel to this the need for research groups, whose main focus is 'theoretical/computational' biology, will surely increase as modeling and prediction of the complexities of life become a tractable proposition. This book serves as a first introduction to this growing and exciting field. We hope you find it a good introduction and enjoy learning about bioinformatics as much as we do practising it!

1

Molecular evolution

Sylvia B. Nagl

Concepts

- Information is a measure of order that can be applied to any structure or system. It quantifies the instructions needed to produce a certain organization and can be expressed in bits. Large biomolecules have very high information content.
- The concept of the gene has undergone many changes. New concepts are emerging that define genes as functional units whose action is dependent on biological context.
- Multigene and multidomain families have arisen by gene duplication in genomes. Complete and partial gene duplication can occur by unequal crossing over, unequal sister chromatid exchange and transposition.
- Sequences or protein structures are homologous if they are related by evolutionary divergence from a common ancestor. Homology cannot be directly observed, but must be inferred from sequence or structural similarity.

1.1 Molecular evolution is a fundamental part of bioinformatics

Genomes are dynamic molecular entities that evolve over time due to the cumulative effects of mutation, recombination, and selection. Before we address bioinformatics techniques for analysis of evolutionary relationships between biological sequences, and between protein structures, in later chapters, we will first survey these mechanisms that form the basis of genome evolution.

1.1.1 A brief history of the gene

The systematic study of the laws of heredity began with the work of Gregor Mendel (1822–1884). In 1865, Mendel who was an Augustinian monk living in Brno, then part of the Austro–Hungarian Empire, published a paper describing the results of plant-breeding experiments that he had begun in the gardens of his monastery almost a decade earlier. Mendel's work received little attention during his lifetime, but when his paper was rediscovered by biologists in 1900, a scientific revolution ensued. The key concept of the Mendelian revolution is that heredity is mediated by discrete units that can combine and dissociate in mathematically predictable ways.

Mendel had studied physics and plant physiology at the University of Vienna, where he was also introduced to the new science of statistics. His experimental methods show the mental habits of a physicist, and in looking for mathematical patterns of inheritance, he became the first mathematical biologist. He set out to arrange in a statistically accurate way the results of deliberate crosses between varieties of sweet pea plants with characteristics that could be easily identified. His elegant experiments led to a clear distinction between *genotype* (the hereditary make-up of an organism) and *phenotype* (the organism's physical and behavioral characteristics), and his results on the pattern of inheritance have become known as *Mendel's law*.

Mendel was the first to refer to hypothetical 'factors' that act as discrete units of heredity and are responsible for particular phenotypic traits. The rediscovery of his work at the beginning of the 20th century prompted a search for the cellular and molecular basis of heredity. The existence of 'germ plasm' was postulated, a material substance in eggs and sperm that in some way carried heritable traits from parent to offspring. While the molecular basis of hereditary factors – protein versus nucleic acid – remained in dispute until the mid-20th century, their cellular basis in chromosomes was soon discovered. In 1909, W. Johannsen coined the word *gene* to denote hypothetical particles that are carried on chromosomes and mediate inheritance.

In sexually reproducing *diploid* organisms, such as Mendel's pea plants and the fruit fly *Drosophila melanogaster* used in the breeding experiments of early Mendelians, the pattern of inheritance of some phenotypic traits could be explained by postulating a pair of genes underlying each trait – a pair of *alleles* occupying a *locus* on a chromosome. It was recognized early on that a single trait might be caused by several genes (*polygenic traits*) and that a single gene may have several effects (*pleiotropy*).

It is important to realize that 'gene' was an abstract concept to Mendel, and the Mendelian biologists of the first half of the 20th century. The founders of genetics, not having any knowledge of the biochemical basis of heredity, had to infer the characteristics of genes by observing the phenotypic outcomes of their breeding experiments. They developed a theoretical framework based on sound mathematics, now called *classical genetics*, that worked extremely well and is still useful today for the interpretation of the data obtained by the new molecular genetics.

Molecular genetics seeks to elucidate the chemical nature of the hereditary material and its cellular environment. At the beginning of the 1950s, it finally became clear that DNA was the critical ingredient of the genes. Rosalind Franklin (1920–1958), a crystallographer working at King's College in London, conducted a careful analysis of DNA using X-ray diffraction which indicated that the macromolecule possessed a double-stranded helical geometry. In 1953, James Watson and Francis Crick produced a successful model of the molecular structure of DNA based on her data by employing the molecular-model building method pioneered by Linus Pauling (1901–1994).

It was clear as soon as the structure of DNA was elucidated that its central role depends on the fact that it can be both replicated and read. This has given rise to the concept of *genetic information* being encoded in DNA (see section 1.1.2). This idea is often expressed by metaphors that conceive of a gene as a 'word' and a genome, the total genetic material of a species, as a linguistic 'text' written in DNA code.

Over the ensuing five decades, rapidly accumulating knowledge on the fine structure of DNA and its functional organization has led to many critical adjustments and refinements in our understanding of the complex roles of genetic material in the cell. These discoveries all highlight the crucial contributions of the cellular environment in regulating the effects of DNA sequences on an organism's phenotype. The causal chain between DNA and phenotype is indirect, different cellular environments link identical DNA sequences to quite different phenotypic outcomes. One of the outcomes of the Human Genome Project is a significantly expanded view of the role of the genome within the integrated functioning of cellular systems.

The concept of the gene has undergone a profound transformation in recent years (*Figure 1.1*). For molecular biology, the traditional definition of gene action, originating from George Beadle's *one gene–one enzyme hypothesis* (1941) led to the concept of the gene as a stretch of DNA that codes for a single polypeptide chain. But even ignoring the fact that open reading frames might overlap, the relationship between DNA sequences and protein chains is many-to-many, not one-to-one. Now, the essential contributions of alternative splicing, RNA editing and post-translational modifications to the synthesis of the actual gene product have become recognized, and with them the limitations of the classic gene concept.

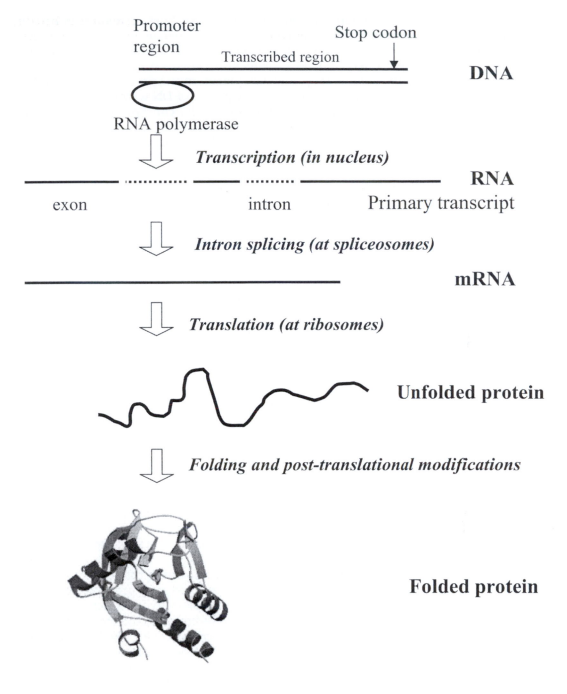

Figure 1.1

From gene to folded protein.

The gene as a unit of function can no longer be taken to be identical with the gene as a unit of intergenerational transmission of molecular information. New concepts are emerging that place emphasis on a *functional perspective* and define genes as 'segments of DNA that function as functional units', 'loci of cotranscribed exons' or, along similar lines, 'distinct transcription units or parts of transcription units that can be translated to generate one or a set of related amino acid sequences'. Another less widely adopted, yet, from a functional perspective, valid definition has recently been proposed by Eva Neumann-Held which states that a gene 'is a process that regularly results, at some stage in development, in the production of a particular protein.' This process centrally involves a linear sequence of DNA, some parts of which correspond to the protein via the genetic code. However, the concept of gene-as-process could also include such entities as coding regions for transcription factors that bind to its regulatory sequences, coding regions for RNA editing and splicing factors, the regulatory dynamics of the cell as a whole and the signals determining the specific nature of the final transcript, and beyond that, the final protein product. In conclusion, diverse interpretations of the concept of the gene exist and the meaning in which the term is applied can only be made clear by careful definition. At the very least, inclusion or exclusion of introns, regulatory regions and promoters need to be made explicit when we speak of a gene from the perspective of molecular biology.

1.1.2 What is information?

Biological, or genetic, information is a fundamental concept of bioinformatics. Yet what exactly is information? In physics, it is understood as a measure of order that can be applied to any structure or system. The term 'information' is derived from the Latin *informare,* which means to 'form', to 'shape', to 'organize'. The word 'order' has its roots in textile weaving; it stems from the Latin *ordiri,* to 'lay the warp'. *Information theory*, pioneered by Claude Shannon, is concerned with information as a universal measure that can be applied equally to the order contained in a hand of playing cards, a musical score, a DNA or protein sequence, or a galaxy.

Information quantifies the instructions needed to produce a certain organization. Several ways to achieve this can be envisaged, but a particularly parsimonious one is in terms of binary choices. Following this approach, we compute information inherent in any given arrangement of matter from the number of 'yes' and 'no' choices that must be made to arrive at a particular arrangement among all equally possible ones (*Figure 1.2*). In his book *The Touchstone of Life* (1999), Werner Loewenstein illustrates this with the following thought experiment. Suppose you are playing bridge and are dealt a hand of 13 cards. There are about 635×10^9 different hands of 13 cards that can occur in this case, so the probability that you would be dealt this hand of cards is about 1 in 635×10^9 – a large number of choices would need to be made to produce this exact hand. In other words, a particular hand of 13 cards contains a large amount of information.

Order refers to the structural arrangement of a system, something that is easy to understand in the case of a warp for weaving cloth but is much harder to grasp in the case of macromolecular structures. We can often immediately see whether an everyday structure is orderly or disorderly – but this intuitive notion does not go beyond simple architectural or periodic features. The intuition breaks down when we deal with macromolecules, like DNA, RNA and proteins. The probability for spontaneous assembly of such molecules is extremely low, and their structural specifications require enormous amounts of information since the number of ways they can be assembled as linear array of their constituent building blocks, nucleotides and amino acids, is astronomical. Like being dealt a particular hand during a bridge game, the synthesis of a particular biological sequence is very unlikely – its information content is therefore very high.

The large molecules in living organisms offer the most striking example of information density in the universe. In human DNA, roughly 3×10^9 nucleotides are strung together on the scaffold of the phosphate backbone in an aperiodic, yet perfectly determined, sequence. Disregarding spontaneous somatic mutations, all the DNA molecules of an individual display the same sequence. We are not yet able to precisely calculate the information inherent in the

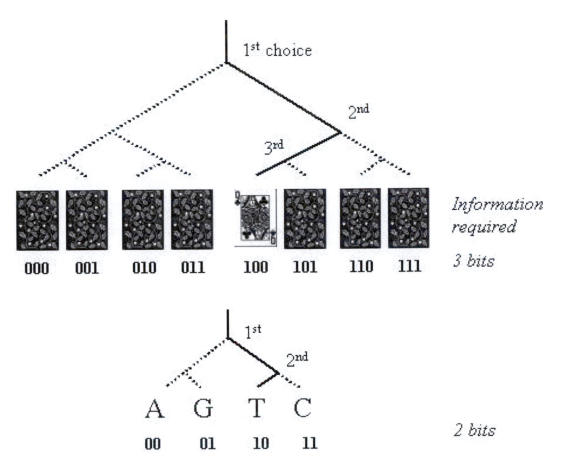

Figure 1.2

What is information? Picking a particular playing card from a pack of eight cards requires 3 yes/no choices (binary choices of 0 or 1). The information required can be quantified as '3 bits'. Likewise, picking one nucleotide among all four equally likely ones (A, G, T, C) requires 2 choices (2 bits).

human genome, but we can get an idea of its huge information storing capacity with a simple calculation. The positions along the linear DNA sequence, that can be occupied by one of the four types of DNA bases, represent the elements of stored information (*Figure 1.2*). So, with 3×10^9 positions and four possible choices for each position, there are $4^{3,000,000,000}$ possible states. The number of possibilities is greater than the estimated number of particles in the universe.

Charles Darwin pointed out in *Origin of Species* how natural selection could gradually accumulate information about biological structures through the processes of random genotypic variation, natural selection and differential reproduction. What Darwin did not know was exactly how this information is stored, passed on to offspring and modified during evolution – this is the subject of the study of *molecular evolution*.

1.1.3 Molecular evolution

This section provides an overview of the processes giving rise to the dynamics of evolutionary change at the molecular level. It primarily focuses on biological mechanisms relevant to the evolution of genomic sequences encoding polypeptide chains.

1.1.3.1   The algorithmic nature of molecular evolution

In his theory of evolution, Darwin identified three major features of the process that occur in an endlessly repeating cycle: generation of heritable variation by random mutation at the level of the genotype, natural selection acting on the phenotype, and differential reproductive success. Darwin discovered the power of cumulative algorithmic selection, although he lacked the terminology to describe it as such.

Algorithm is a computer science term and denotes a certain kind of formal process consisting of simple steps that are executed repetitively in a defined sequential order and will reliably produce a *definite kind of result* whenever the algorithm is run. Knitting is an algorithmic process – following the instructions 'knit one-purl one' will result in a particular ribbing pattern; other examples include braiding hair, building a car on an assembly line, or protein synthesis. Evolution by means of natural selection can also be thought of as algorithmic (mutate-select-replicate) (*Figure 1.3*).

One important feature of an algorithm is its substrate neutrality: The power of the procedure is due to its logical structure, not the materials used in its instantiation, as long as the materials permit the prescribed steps to be followed exactly. Because of substrate neutrality, we can simulate simplified versions of the algorithmic process of evolution in the computer using binary strings of 1s and 0s as the information carriers. These procedures are known as *genetic algorithms* or *evolutionary computation* in computer science.

Cumulative selection will work on almost anything that can yield similar, but nonidentical, copies of itself through some replication process. It depends on the presence of a medium that stores information and can be passed on to the next generation; this medium is DNA or RNA (as in certain types of viruses) in terrestrial life forms. Whenever selection acts, it can be thought of as selecting DNA or RNA copies with particular phenotypic effects over others with different effects. During cumulative *natural selection*, each cycle begins with the results of selection in the previous generation due to the heritability of genetic information.

Most genetic mutations are thought to be either neutral in their phenotypic effects or deleterious, in which case they are removed by *negative selection*. Because of the nonadaptive nature of many mutations, there has also been strong selection for proofreading and error correction mechanisms that reduce the rate at which mutations occur. Rarely, advantageous germline mutations may confer a survival and reproductive advantage on individuals in the next generation who will then, on average, pass on more copies of their genetic material because they will tend to have a larger number of offspring (*positive selection*).

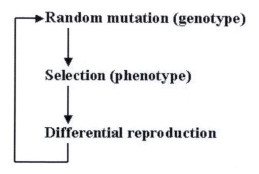

Figure 1.3

Evolution is an algorithmic process.

Over many generations, the accumulation of small changes can result in the evolution of DNA (or RNA) sequences with new associated phenotypic effects. This process also gives rise to the evolution of entirely new biological functions. Yet, such advantageous changes are believed to be less frequent than neutral or nearly neutral sequence substitutions at the level of protein evolution. The *neutralist hypothesis* states that therefore most genetic change is not subject to selection, and mutation-driven molecular changes far exceed selection-driven phenotypic changes (see also Chapter 4).

1.1.3.2 Causes of genetic variation

The analysis of evolution at the molecular level must consider the processes which alter DNA sequences. Such alterations are brought about by *mutations* which are errors in DNA replication or DNA repair. These changes provide the genetic variation upon which natural selection can act.

When considered purely as a chemical reaction, complementary base-pairing is not very accurate. It has been estimated that, without the aid of any enzyme, a copied polynucleotide would probably have point mutations at 5–10 positions out of every hundred. Such an error rate of up to 10% would very quickly corrupt a nucleotide sequence. The polymerase enzymes that carry out replication can improve the error rate in two ways – highly accurate nucleotide selection during the polymerization reaction and proofreading carried out by their exonuclease activity.

The genetic material of RNA viruses, including influenza, polio, and HIV viruses, consists of RNA not DNA. These viruses copy themselves by various enzymatic procedures, but they all use a copying polymerase that lacks the more advanced error-correction apparatus. The action of this enzyme is a huge improvement on an enzyme-free copying procedure, but it still has an estimated copying error rate ranging from 1 in 1000 to 1 in 100,000 (10^{-3} to 10^{-5}). A conserved set of proofreading and repair enzymes, absent from RNA viruses, is present in bacteria and all other cellular life, including humans. These new functions, resulting in dramatically reduced copying error rates, made possible the evolution of larger genomes because they could now be copied without incurring a lethal accumulation of mistakes.

Bacteria seem to make copying mistakes at a rate roughly 10,000–1,000,000 times lower than RNA viruses. The bacterium *Escherichia coli* is able to synthesize DNA at an error rate of only 1 in 10^7 nucleotide additions. The overall error rate for replication of the entire *E. coli* genome is only 1 in 10^{10} to 1 in 10^{11} (10^{-10} to 10^{-11}) due to the bacterial mismatch repair system that corrects the errors that the replication enzymes make. The implication is that there will be, on average, only one uncorrected replication error for every 1000 times the *E. coli* genome is copied. In humans, the estimated mutation rate per nucleotide position per generation is 3×10^{-8}, and the total mutation rate per generation is ~200. Various estimates of harmful mutations per generation fall between 2–20. These estimates, approximate as they are, suggest that the error rate is relatively constant between bacteria and humans.

The error-reducing machinery was an extremely valuable evolutionary innovation, but once acquired, it appears to have been either prohibitively difficult or nonadvantageous to optimize it further. In fact, the cellular machinery seems to have evolved to maintain a finely tuned balance between minimizing errors and permitting a certain degree of random mutational change on which natural selection can act. In sexually reproducing organisms, DNA recombination mechanisms for the active generation of genetic variability in gametes exist alongside the error correction machinery (see section 1.2.3). The process of sex cell maturation (meiosis) not only shuffles the maternal and paternal chromosomes by independent assortment, but also promotes the swapping of large pieces of homologous chromosomal stretches.

Mutations do not occur randomly throughout genomes. Some regions are more prone to mutation and are called *hotspots*. In higher eukaryotes, examples of hotspots are short tandem repeats (for nucleotide insertions and deletions) and the dinucleotide 5'-CpG-3' in which the cytosine is frequently methylated and replicated with error, changing it to 5'-TpG-3'. Hotspots in prokaryotes include the dinucleotide 5'-TpT-3' and short palindromes (sequences that read the same on the coding and the complementary strand).

The results of genotypic changes provide the molecular record of evolution. The more closely related two species are, the more similar are their genome sequences and their gene products, i.e. proteins. In the case of proteins, similarity relationships between sequences closely mirror those of their corresponding DNA sequences, but the three-dimensional structures provide more information. In cases where the sequences have changed beyond recognition the protein structures may retain a strong enough resemblance that a potential evolutionary relationship can still be discerned. By systematic analysis of sequences and structures *evolutionary trees* can be constructed which help to reveal the *phylogenetic relationships* between organisms.

1.1.3.3   Classification of mutations

Mutations can be classified by the length of the DNA sequence affected by the mutational event. *Point mutations* (single-nucleotide changes) represent the minimal alteration in the genetic material, but even such a small change can have profound consequences. Some human genetic diseases, such as sickle cell anemia or cystic fibrosis, arise as a result of point mutations in coding regions that cause single amino-acid substitutions in key proteins. Mutations may also involve any larger DNA unit, including pieces of DNA large enough to be visible under the light microscope. Major *chromosomal rearrangements* thus also constitute a form of mutation, though drastic changes of this kind are normally lethal during embryonic development or lead to decreased (zero) fertility. They do therefore contribute only rarely to evolutionary change. In contrast, *silent mutations* result in nucleotide changes that have no effect on the functioning of the genome. Silent mutations include almost all mutations occurring in extragenic DNA and in the noncoding components of genes.

Mutations may also be classified by the type of change caused by the mutational event into *substitutions* (the replacement of one nucleotide by another), *deletions* (the removal of one or more nucleotides from the DNA sequence), *insertions* (addition of one or more nucleotides to the DNA), *inversions* (the rotation by 180 degrees of a double-stranded DNA segment consisting of two or more base pairs), and *recombination* (see section 1.2.3).

1.1.3.4   Substitutional mutation

Some mutations are spontaneous errors in replication which escape the proofreading function of the DNA polymerases that synthesize new polynucleotide chains at the replication fork. This causes a mismatch mutation at positions where the nucleotide that is inserted into the daughter strand is not complementary (by base-pairing) to the corresponding nucleotide in the template DNA. Other mutations arise because of exposure to excessive UV light, high-energy radiation from sources such as X-rays, cosmic rays or radioisotopes, or because a chemical mutagen has reacted with the parent DNA, causing a structural alteration that affects the base-pairing capability of the affected nucleotide. Mutagens include, for example, aflatoxin found in moldy peanuts and grains, the poison gas nitrogen mustard, and polycyclic aromatic hydrocarbons (PAHs) which arise from incomplete combustion of organic matter such as coal, oil, and tobacco. These chemical interactions usually affect only a single strand of the DNA helix, so only one of the daughter molecules carries the mutation. In the next round of replication, it will be passed on to two of the 'granddaughter' molecules. In single-celled organisms, such as bacteria and yeasts, all nonlethal mutations are inherited by daughter cells and become permanently integrated in the lineage that originates from the mutated cell. In multicellular organisms, only mutations that occur in the germ cells are relevant to genome evolution.

Nucleotide substitutions can be classified into *transitions* and *transversions*. Transitions are substitutions between purines (A and G) or pyrimidines (C and T). Transversions are substitutions between a purine and a pyrimidine, e.g., G to C or G to T. The effects of nucleotide substitutions on the translated protein sequence can also form the basis for classification. A substitution is termed *synonymous* if no amino acid change results (e.g., TCT to TCC, both code for serine). *Nonsynonymous*, or amino-acid altering, mutations can be classified into missense or nonsense mutations. A *missense mutation* changes the affected codon into a codon for a different amino acid (e.g., CTT for leucine to CCT for proline); a *nonsense mutation* changes a codon into one of the termination codons (e.g., AAG for lysine to the stop codon TAG), thereby prematurely terminating translation and resulting in the production of a truncated protein. The degeneracy of the genetic code provides some protection against damaging point mutations since many changes in coding DNA do not lead to a change in the amino acid sequence of the encoded protein. Due to the nature of the genetic code, synonymous substitutions occur mainly at the third codon position; nearly 70% of all possible substitutions at the third position are synonymous. In contrast, all the mutations at the second position, and 96% of all possible nucleotide changes at the first position, are nonsynonymous.

1.1.3.5   Insertion and deletion

Replication errors can also lead to insertion and deletion mutations (collectively termed indels). Most commonly, a small number of nucleotides is inserted into the nascent polynucleotide or some nucleotides on the template are not copied. However, the number of nucleotides involved may range from only a few to contiguous stretches of thousands of nucleotides. Replication slippage or slipped-strand mispairing can occur in DNA regions containing contiguous short repeats because of mispairing of neighboring repeats, and results in deletion or duplication of a short DNA segment. Two important mechanisms that can give rise to long indels are unequal crossing-over and DNA transposition (see section 1.2.3).

A *frameshift mutation* is a specific type of indel in a coding region which involves a number of nucleotides that is not a multiple of three, and will result in a shift in the reading frame used for translation of the DNA sequence into a polypeptide. A termination codon may thereby be cancelled out or a new stop codon may come into phase with either event resulting in a protein of abnormal length.

1.2   Evolution of protein families

1.2.1   Protein families in eukaryotic genomes

With the increasing number of fully sequenced genomes, it has become apparent that *multigene families*, groups of genes of identical or similar sequence, are common features of genomes. These are more commonly described as *protein families* by the bioinformatics community. In some cases, the repetition of identical sequences is correlated with the synthesis of increased quantities of a gene product (*dose repetitions*). For example, each of the eukaryotic, and all but the simplest prokaryotic, genomes studied to date contain multiple copies of ribosomal RNAs (rRNAs), components of the protein-synthesizing molecular complexes called ribosomes. The human genome contains approximately 2000 genes for the 5S rRNA (named after its sedimentation coefficient of 5S) in a single cluster on chromosome 1, and roughly 280 copies of a repeat unit made up of 28S, 5.8S and 18S rRNA genes grouped into five clusters on chromosomes 13, 14, 15, 21 and 22.

It is thought that amplification of rRNA genes evolved because of the heavy demand for rRNA synthesis during cell division when thousands of new ribosomes need to be assembled. These rRNA genes are examples of protein families in which all members have identical or nearly identical sequences. Another example of dosage repetition is the amplification of an esterase gene in a Californian *Culex* mosquito resulting in insecticide resistance.

Other protein families, more commonly in higher eukaryotes, contain individual members that are sufficiently different in sequence for their gene products to have distinct properties. Members of some protein families are clustered (*tandemly repeated*), for example the globin genes, but genes belonging to other families are dispersed throughout the genome. Even when dispersed, genes of these families retain sequence similarities that indicate a common evolutionary origin. When sequence comparisons are carried out, it is in certain cases possible to see relationships not only within a single family but also between different families. For example, all of the genes in the α- and β-globin families have discernible sequence similarity and are believed to have evolved from a single ancestral globin gene.

The term *superfamily* was coined by Margaret Dayhoff (1978) in order to distinguish between closely related and distantly related proteins. When duplicate genes diverge strongly from each other in either sequence and/or functional properties, it may not be appropriate to classify them as a single protein family. According to the Dayhoff classification, proteins that exhibit at least 50% amino acid sequence similarity (based on physico-chemical amino acid features) are considered members of a family, while related proteins showing less similarity are classed as belonging to a superfamily and may have quite diverse functions. More recent classification schemes tend to define the cut-off at 35% amino acid identity (rather than using a similarity measure) at the amino acid sequence level for protein families. Following either definition, the α globins and the β globins are classified as two separate families, and together with the myoglobins form the globin superfamily. Other examples of superfamilies in higher eukaryotes include the collagens, the immunoglobulins and the serine proteases.

One of the most striking examples of the evolution of gene superfamilies by duplication is provided by the homeotic (*Hox*) genes. These genes all share common sequence elements of 180 nucleotides in length, called the homeobox. *Hox* genes play a crucial role in the development of higher organisms as they specify the body plan. Mutations in these genes can transform a body segment into another segment; for example, in *Antennapedia* mutants of *Drosophila* the antennae on the fly's head are transformed into a pair of second legs.

More than 350 homeobox-related genes from animals, plants, and fungi have been sequenced, and their evolution has been studied in great detail as it gives very important clues for understanding the evolution of body plans. *Figure 1.4* illustrates the reconstructed evolution of the Hox gene clusters. *Drosophila* has a single cluster of homeotic genes which consists of eight genes each containing the homeobox element. These eight genes, and other homeobox-containing genes in the *Drosophila* genome, are believed to have arisen by a series of gene duplications that began with an ancestral gene that existed about 1000 million years ago. The modern genes each specify the identity of a different body segment in the fruit fly – a poignant illustration of how gene duplication and diversification can lead to the evolution of complex morphological features in animals.

1.2.2   Gene duplication

Protein families arise mainly through the mechanism of *gene duplication*. In the 1930s, J.B.S. Haldane and Hermann Muller were the first to recognize the evolutionary significance of this process. They hypothesized that selective constraints will ensure that one of the duplicated DNA segments retains the original, or a very similar, nucleotide sequence, whilst the duplicated (and therefore redundant) DNA segment may acquire divergent mutations as long as these mutations do not adversely affect the organism carrying them. If this is the case, new functional genes and new biological processes may evolve by gene duplication and enable the evolution of more complex organisms from simpler ones. As we have seen in the case of ribosomal RNA, gene duplication is also an important mechanism for generating multiple copies of identical genes to increase the amount of gene product that can be synthesized.

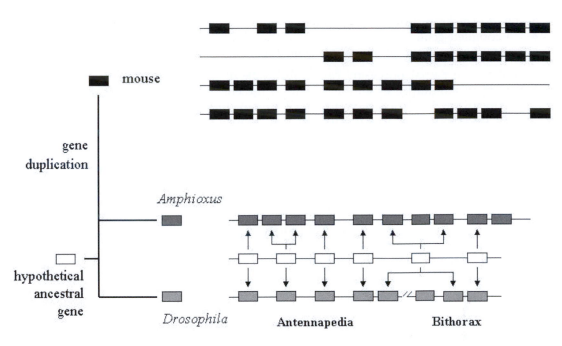

Figure 1.4

Evolution of Hox gene clusters. The genomic organization of Antennapedia-related homeobox genes in mouse (*Mus musculus*), amphioxus (*Branchiostoma floridae*) and the fruit fly (*Drosophila melanogaster*) is shown. Amphioxus is thought to be a sister group of the vertebrates, it has been called 'the most vertebrate-like invertebrate'. Moving in the 3'-to-5' direction in each species cluster, each successive gene expresses later in development and more posterior along the anterior–posterior axis of the animal embryo.

The evolution of new functional genes by duplication is a rare evolutionary event, since the evidence indicates that the great majority of new genes that arise by duplication acquire deleterious mutations. Negative selection reduces the number of deleterious mutations in a population's gene pool. Alternatively, mutations may lead to a version of the copied sequence that is not expressed and does not exert any effect on the organism's phenotype (a silent *pseudogene*). The most common inactivating mutations are thought to be frameshift and nonsense mutations within the gene-coding region.

Partial gene duplication plays a crucial role in increasing the complexity of genes – segments encoding functional and structural protein domains are frequently duplicated. Domain duplication within a gene may lead to an increased number of active-sites or, analogous to whole gene duplication, the acquisition of a new function by mutation of the redundant segment. In the genomes of eukaryotes, internal duplications of gene segments have occurred frequently. Many complex genes might have evolved from small primordial genes through internal duplication and subsequent modification. The shuffling of domains between different gene sequences can further increase genome complexity (see section 1.2.5).

1.2.3 Mechanisms of gene duplication

A duplication event may involve

- a whole genome (polyploidy)
- an entire chromosome (aneuploidy or polysomy)

- part of a chromosome (partial polysomy)
- a complete gene
- part of a gene (partial or internal gene duplication).

Complete and partial polysomy can probably be discounted as a major cause for gene number expansion during evolution, based on the effects of chromosome duplications in modern organisms. Duplication of individual human chromosomes is either lethal or results in genetic diseases such as Down syndrome, and similar effects have been observed experimentally in *Drosophila*. Whole genome duplication, caused by the formation of diploid rather than haploid gametes during meiosis, provides the most rapid means for increasing gene number. Subsequent fusion of two diploid gametes results in a polyploid genotype, which is not uncommon in plants and may lead to speciation. A genome duplication event has also been identified in the evolutionary history of the yeast *Saccharomyces cerevisiae* and has been dated to have taken place approximately 100 million years ago. Genome duplication provides the potential for the evolution of new genes because the extra genes can undergo mutational changes without harming the viability of the organism.

Complete or partial gene (domain) *duplication* can occur by several mechanisms (*Figure 1.5*). *Unequal crossing-over* during meiosis is a reciprocal *recombination* event that creates a sequence duplication in one chromosome and a corresponding deletion in the other. It is initiated by similar nucleotide sequences that are not at identical locations in a pair of homologous chromosomes and cause the chromosomes to align to each other out of register. *Figure 1.5* shows an example in which unequal crossing-over has led to the duplication of one repeat in one daughter chromosome and deletion of one repeat in the other. *Unequal sister chromatid exchange* occurs by the same mechanism, but involves a pair of chromatids from the same chromosome. *Transposition* is defined as the movement of genetic material (*transposons*) from one chromosomal location to another. Transposons carry one or more genes that encode other functions in addition to those related to transposition.

There are two types of transposition events. In conservative transposition the DNA element itself moves from one site to another, whilst in replicative transposition, the element is copied and one copy remains at its original site while the other inserts at a new site (*Figure 1.5*). Replicative transposition therefore involves an increase in the copy number of the transposon. Some transposable elements transpose via a RNA intermediate. All such elements contain a reverse transcriptase gene whose product is responsible for the relatively error-prone reverse transcription of the DNA element.

Transposable elements can promote gene duplications, and genomic rearrangements including inversions, translocations, duplications and large deletions and insertions. For example, a duplication of the entire growth hormone gene early in human evolution might have occurred via *Alu-Alu* recombination (*Alu* repeats represent one of the most abundant families of retrosequences in the human genome).

1.2.4 The concept of homology

Evolutionary relationships between members of protein families can be classified according to the concept of *homology* (*Figure 1.6*). Sequences or structures are homologous if they are related by evolutionary divergence from a common ancestor. From this follows that homology cannot be directly observed, but must be inferred from calculated levels of sequence or structural similarity. Reliable threshold values of similarity are dependent on the mathematical methods used for analysis (this will be discussed in detail in later chapters). In contrast, analogy refers to the acquisition of a shared feature (protein fold or function) by convergent evolution from unrelated ancestors.

Among homologous sequences or structures, it is possible to distinguish between those that have resulted from gene duplication events within a species genome and perform different but related functions within the same organism (*paralogs*) and those that perform the

(a)

Unequal sister chromatid exchange

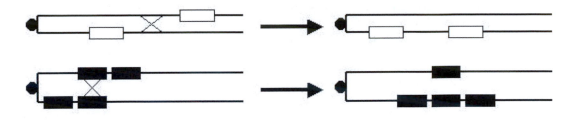

Unequal crossing over during meiosis

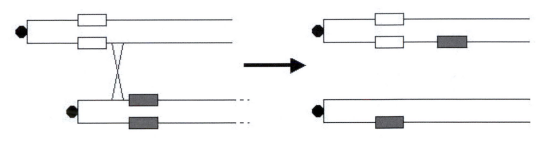

(b)

Transposition via an RNA intermediate

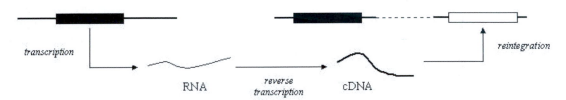

DNA transposons

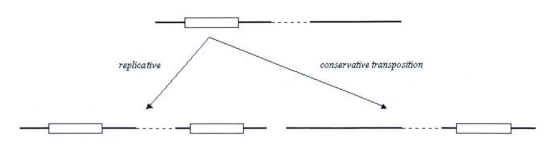

Figure 1.5

Mechanisms of gene duplication.

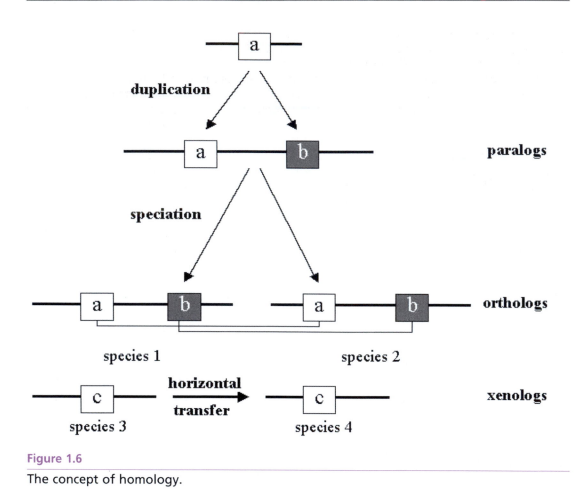

Figure 1.6

The concept of homology.

same, or a highly similar, function in different species (*orthologs*). Horizontal gene transfer is defined as the transfer of genetic material from one genome to another, specifically between two species. Sequences acquired via horizontal gene transfer are called *xenologs*.

A high degree of gene conservation across organisms, correlated with conservation of function, has been well established by whole genome sequencing. For example, comparison between two complete eukaryotic genomes, the budding yeast *Saccharomyces cerevisiae* and the nematode worm *Caenorhabditis elegans*, suggested *orthologous relationships* between a substantial number of genes in the two organisms. About 27% of the yeast genome (~5700 genes) encode proteins with significant similarity to ~12% of nematode genes (~18,000). Furthermore, the same set of yeast genes also has putative orthologs in the *Drosophila* genome. Most of these shared proteins perform conserved roles in core biological processes common to all eukaryotes, such as metabolism, gene transcription, and DNA replication.

1.2.5   The modularity of proteins

Domain shuffling is a term applied to both domain duplication within the same gene sequence (see section 1.2.3) and domain insertion. *Domain insertion* refers to the exchange of structural or functional domains between coding regions of genes or insertion of domains from one region into another. Domain insertion results in the generation of *mosaic (or chimeric) proteins*. Domain insertion may be preceded by domain duplication events.

One of the first mosaic proteins to be studied in detail was *tissue plasminogen activator (TPA)*. TPA converts plasminogen into its active form, plasmin, which dissolves fibrin in blood clots. The TPA gene contains four well-characterized DNA segments from other genes: a fibronectin finger domain from fibronectin or a related gene; a growth-factor module that is homologous to that present in epidermal growth factor and blood clotting enzymes; a car-boxy-terminal region homologous to the proteinase modules of trypsin-like serine pro-teinases; and two segments similar to the kringle domain of plasminogen (*Figure 1.7*). A special feature of the organization of the TPA gene is that the junctions between acquired domains coincide precisely with exon–intron borders. The evolution of this composite gene thus appears to be a likely result of exon shuffling. The domains present in TPA can also be identified in various arrangements in other proteins.

The combination of multiple identical domains, or of different domains, to form mosaics gives rise to enormous functional and structural complexity in proteins. Because domains in mosaics perform context-dependent roles and related modules in different proteins do not always perform exactly the same function, the domain composition of a protein may tell us only a limited amount about the integrated function of the whole molecule.

1.3   Outlook: Evolution takes place at all levels of biological organization

Contemporary evolutionary thought is complementary to the study of gene and protein evolution through bioinformatics. This intellectually engaging and lively area of research continues to pose fundamental challenges to our understanding of the mechanisms by which natural selection exerts its effects on genomes. Whilst all evolutionary change can

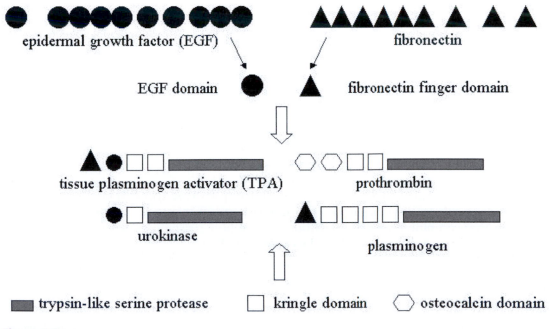

Figure 1.7

Domain shuffling: Mosaic proteins. The domain compositions of tissue plasminogen activator (TPA) and several other proteins are shown schematically.

be represented as a change in the frequencies of particular DNA segments within genomes, an ongoing debate is concerned with the question of *why* these frequencies change. The received view conceives of natural selection as the result of competition between individual organisms in a population, and by default, their individual copies of the species genome.

Gene selectionism takes the view that only relatively short segments of DNA, able to confer certain phenotypic effects on the organism, satisfy a crucial requirement for units of selection. According to this view, the natural selection of individual phenotypes cannot by itself produce cumulative change, because phenotypes are extremely temporary manifestations, and genomes associated with particular phenotypes are broken up when gametes are formed. In contrast, genes in germline cell lineages are passed on intact through many generations of organisms and are thus exposed to cumulative selection.

This argument can even be taken a step further. Extensive sequence comparisons, made possible by the large-scale sequencing of prokaryotic, archeal and eukaryotic genomes, have consistently revealed the modular composition of genes and proteins, and have enabled detailed evolutionary studies of their constituent smaller units, protein *domains*. Domains are discrete molecular entities that form the stable components of larger genes and proteins and are passed on intact between generations. Thus, DNA segments encoding protein domains can arguably be seen as stable units of selection.

One preliminary conclusion that can be drawn from this debate is that a range of mechanisms is likely to be operative in genome evolution. Bioinformatics can analyze the resulting patterns of evolutionary change present in biological sequences and three-dimensional protein structures. In elucidating the complex underlying processes, other fields of biology, notably molecular biology and evolutionary biology, are bioinformatics' indispensable partners.

Emergence of multi-drug resistance in malignant tumors

Gene amplification mechanisms contribute to one of the major obstacles to the success of cancer chemotherapy, that is, the ability of malignant cells to develop *resistance to cytotoxic drugs*. In tumor cells, gene amplification can occur through unscheduled duplication of already replicated DNA. Resistance to a single drug, methotrexate, is conferred by DHFR gene amplification and, even more importantly, amplification of the *mdr1* gene, encoding a membrane-associated drug efflux pump, leads to multi-drug resistance which enables malignant cells to become resistant to a wide range of unrelated drugs. Drug-resistant mutations are predicted to occur at a rate of 1 in every 10^6 cell divisions, so a clinically detectable tumor ($>10^9$ cells) will probably contain many resistant cells. Most anti-cancer drugs are potent mutagens, and therefore, the treatment of a large tumor with single agents constitutes a very strong selection pressure likely to result in the emergence of drug resistance. In contrast, combination treatments that combine several drugs without overlapping resistances delay the emergence of drug resistance and can therefore be expected to be more effective.

Acknowledgments

I wish to thank my colleagues Gail Patt and Scott Mohr, Core Curriculum, College of the Arts and Sciences, Boston University, for sharing their ideas and their enthusiasm about evolutionary theory when I was teaching there.

References and further reading

Depew, D.J., and Weber, B.H. (1996) *Darwinism Evolving: Systems Dynamics and the Genealogy of Natural Selection.* Cambridge, MA, MIT Press, 588 pp.

Greaves, M. (2000) *Cancer: The Evolutionary Legacy.* Oxford, Oxford University Press, 276 pp.

Keller, E.F. (2002) *The Century of the Gene.* Cambridge, MA, Harvard University Press, 192 pp.

Li, W.-H. (1997) *Molecular Evolution.* Sunderland, MA, Sinauer Associates, 487 pp.

Loewenstein, W.R. (1999) *The Touchstone of Life: Molecular Information, Cell Communication and the Foundations of Life.* London, Penguin Books, 476 pp.

Gene finding

John G. Sgouros and Richard M. Twyman

- Genes are the functional elements of the genome and represent an important goal in any mapping or sequencing project.
- The aim of some projects may be the isolation of a single gene, whereas others may seek to identify every gene in the genome. This depends on the size and complexity of the genome, and the availability of genetic and physical maps.
- Whatever the scope of the project, the problem of gene finding essentially boils down to the identification of genes in large anonymous DNA clones.
- Contemporary gene-finding in bacteria generally involves the scanning of raw sequence data for long open reading frames. Bacterial genomes are small, gene-dense and the genes lack introns. Problems may be encountered identifying small genes, genes with unusual organization or genes using rare variations of the genetic code.
- The genes of higher eukaryotes account for less than 5% of the genome and are divided into small exons and large introns. Gene finding in eukaryotes therefore involves more complex analysis, based on specific motifs (signals) differences in base composition compared to surrounding DNA (content) and relationships with known genes (homology). No gene prediction algorithm in eukaryotes is 100% reliable. Problems may include the failure to detect exons, the detection of phantom exons, mis-specification of exon boundaries and exon fusion.
- Particular problems apply to the detection of non-coding RNA genes as these lack an open reading frame and do not show compositional bias. Homology searching, in combination with algorithms that detect potential secondary structures, can be a useful approach.

2.1 Concepts

An important goal in any mapping or sequencing project is the identification of genes, since these represent the functional elements of the genome. The problems associated with gene finding are very different in prokaryotes and eukaryotes. Prokaryotic genomes are small (most under 10 Mb) and extremely gene dense (>85% of the sequence codes for proteins). There is little repetitive DNA. Taken together, this means that it is possible to sequence a prokaryotic genome in its entirety using the shotgun method within a matter of months. Very few prokaryotic genes have introns. Therefore, a suitable strategy for finding protein-encoding genes in prokaryotes is to search for long open reading frames (uninterrupted strings of sense codons). Special methods are required to identify genes that do not encode proteins, i.e. genes for *non-coding RNAs*, and these are discussed later.

In contrast, eukaryotic genomes are large, ranging from 13 Mb for the simplest fungi to over 10,000 Mb in the case of some flowering plants. Gene and genome architecture increases in complexity as the organism itself becomes more complex. In the yeast *Saccharomyces cerevisiae* for example, about 70% of the genome is made up of genes, while this falls to 25% in the fruit fly *Drosophila melanogaster* and to just 1–3% in the genomes of vertebrates and higher plants. Much of the remainder of the genome is made up of repetitive

DNA. This increased size and complexity means that sequencing an entire genome requires massive effort and can only be achieved if high-resolution genetic and physical maps are available on which to arrange and orient DNA clones. Of the eukaryotic genome sequences that have been published thus far, only six represent multicellular organisms. In most higher eukaryotes, it is therefore still common practice to seek and characterize individual genes.

In *S. cerevisiae*, only 5% of genes possess introns, and in most cases only a single (small) intron is present. In *Drosophila*, 80% of genes have introns, and there are usually one to four per gene. In humans, over 95% of genes have introns. Most have between one and 12 but at least 10% have more than 20 introns, and some have more than 60. The largest human gene, the Duchenne muscular dystrophy locus, spans more than 2 Mb of DNA. This means that as much effort can be expended in sequencing a single human gene as is required for an entire bacterial genome. The typical exon size in the human genome is just 150 bp, with introns ranging vastly in size from a few hundred base pairs to many kilobase pairs. Introns may interrupt the open reading frame (ORF) at any position, even within a codon. For these reasons, searching for ORFs is generally not sufficient to identify genes in eukaryotic genomes and other, more sophisticated, methods are required.

2.2   Finding genes in bacterial genomes

2.2.1   Mapping and sequencing bacterial genomes

The first bacterial genome sequence, that of *Haemophilus influenzae*, was published in 1995. This achievement was remarkable in that very little was known about the organism, and there were no pre-existing genetic or physical maps. The approach was simply to divide the genome into smaller parts, sequence these individually, and use computer algorithms to reassemble the sequence in the correct order by searching for overlaps. For other bacteria, such as the well-characterized species *Escherichia coli* and *Bacillus subtilis*, both genetic and physical maps were available prior to the sequencing projects. Although these were useful references, the success of the shotgun approach as applied to *H. influenzae* showed that they were in fact unnecessary for re-assembling the genome sequence. The maps were useful, however, when it came to identifying genes.

Essentially, contemporary gene finding in bacteria is a question of *whole genome annotation*, i.e. systematic analysis of the entire genome sequence and the identification of genes. Even where a whole genome sequence is not available, however, the problem remains the same – finding long open reading frames in genomic DNA. In the case of uncharacterized species such as *H. influenzae*, it has been necessary to carry out sequence annotation *de novo*, while in the case of *E. coli* and *B. subtilis*, a large amount of information about known genes could be integrated with the sequence data.

2.2.2   Detecting open reading frames in bacteria

Due to the absence of introns, protein-encoding genes in bacteria almost always possess a long and uninterrupted *open reading frame*. This is defined as a series of sense codons beginning with an *initiation codon* (usually ATG) and ending with a *termination codon* (TAA, TAG or TGA). The simplest way to detect a long ORF is to carry out a *six-frame translation* of a query sequence using a program such as *ORF Finder*, which is available at the NCBI web site (http://www.ncbi.nih.gov). The genomic sequence is translated in all six possible reading frames, three forwards and three backwards. Long ORFs tend not to occur by chance so in practice almost always correspond to a gene.

The majority of bacterial genes can be identified in this manner but very short genes tend to be missed because programs such as ORF finder require the user to specify the minimum size of the expected protein. Generally, the value is set at 300 nucleotides (100 amino acids) so proteins shorter than this will be ignored. Also, *shadow genes* (overlapping open reading frames on opposite DNA strands) can be difficult to detect. Content sensing algorithms, which use hidden Markov models to look at nucleotide frequency and dependency data, are useful for the identification of shadow genes. However, as the number of completely sequenced bacterial genomes increases, it is becoming a more common practice to find such genes by comparing genomic sequences with known genes in other bacteria, using databank search algorithms such as BLAST or FASTA (see Chapter 3).

ORF Finder and similar programs also give the user a choice of standard or variant genetic codes, as minor variations are found among the prokaryotes and in mitochondria. The principles of searching for genes in mitochondrial and chloroplast genomes are much the same as in bacteria. Caution should be exercised however because some genes use quirky variations of the genetic code and may therefore be overlooked or incorrectly delimited. One example of this phenomenon is the use of *non-standard initiation codons*. In the standard genetic code, the initiation codon is ATG, but a significant number of bacterial genes begin with GUG, and there are examples where UUG, AUA, UUA and CUG are also used. In the initiator position these codons specify *N*-formylmethionine whereas internally they specify different amino acids. Since ATG is also used as an internal codon, the misidentification of an internal ATG as the initiation codon in cases where GUG, etc. is the genuine initiator may lead to gene truncation from the 5' end. The termination codon TGA is another example of such ambiguity. In genes encoding selenoproteins, TGA occurs in a special context and specifies incorporation of the unusual amino acid *selenocysteine*. If this variation is not recognized, the predicted gene may be truncated at the 3' end. This also occurs if there is a suppressor tRNA which reads through one of the normal termination codons.

2.3 Finding genes in higher eukaryotes

2.3.1 Approaches to gene finding in eukaryotes

A large number of eukaryotic genomes currently are being sequenced, many of these representing unicellular species. Six higher (multicellular) eukaryotic genomes have been sequenced thus far, at least up to the draft stage. The first was the nematode *Caenorhabditis elegans*, followed by the fruit fly *Drosophila melanogaster*, a model plant (*Arabidopsis thaliana*), the human (*Homo sapiens*), and most recently, rice (*Oryza sativa*) and the Japanese pufferfish (*Fugu rubripes*). In each case, the sequencing projects have benefited from and indeed depended on the availability of comprehensive genetic and physical maps. These maps provide a framework upon which DNA clones can be positioned and oriented, and the clones themselves are then used for the reconstruction of the complete genome sequence. Once the sequence is available, the data can be annotated in much the same way as bacterial genomes, although more sophisticated algorithms are required to identify the more complex architecture of eukaryotic genes (see below). However, any eukaryote genome-sequencing project requires a great deal of organization, funding and effort and such a project cannot yet be described as 'routine'. In organisms without a completed sequence, it is still common practice for researchers to map and identify genes on an individual basis. High-resolution maps of the genome are required for this purpose too, so before discussing how genes are identified in eukaryotic genomic DNA, we consider methods for constructing maps and identifying the positions of individual genes.

2.3.1.1 Genetic and physical maps

There are two types of map – genetic and physical. A *genetic map* is based on recombination frequencies, the principle being that the further apart two loci are on a chromosome, the more likely they are to be separated by recombination during meiosis. Genetic maps in model eukaryotes such as *Drosophila*, and to a certain extent the mouse, have been based on morphological phenotypes reflecting underlying differences in genes. They are constructed by setting up crosses between different mutant strains and seeing how often recombination occurs in the offspring. The recombination frequency is a measure of linkage, i.e. how close one gene is to another. Such gene–gene mapping is appropriate for genetically amenable species with large numbers of known mutations, but this is not the case for most domestic animals and plants. In the case of humans, many mutations are known (these cause genetic diseases) but it is not possible to set up crosses between people with different diseases to test for linkage! Therefore, genetic mapping in these species has relied on the use of *polymorphic DNA markers* (*Box 2.1*). New genes can be mapped against a panel of markers or the markers can be mapped against each other to generate dense marker frameworks. In the case of humans, markers must be used in combination with pedigree data to establish linkage.

Box 2.1 Polymorphic DNA markers used for genetic mapping

Any marker used for genetic mapping must be *polymorphic*, i.e. occur in two or more common forms in the population, so that recombination events can be recognized. There are several problems with traditional markers such as morphological phenotypes or protein variants (e.g. blood groups, HLA types or proteins with different electrophoretic properties) including:

- They are not abundant enough in most species for the construction of useful genetic maps.
- They often affect the fitness of the individual (this is the case for human diseases).
- It is often not possible to accurately map the markers themselves.

These problems have been addressed by the use of *DNA markers*, i.e. polymorphic DNA sequences that are abundant in the genome, easily detected in the laboratory but generally have no overt phenotypic effect.

The first polymorphic DNA markers were *restriction fragment length polymorphisms* (RFLPs). These are single nucleotide variations that create or abolish a restriction enzyme site. The first genome-wide human genetic map was based on about 400 RFLPs and RFLP maps have been constructed for a number of other animals. Although a useful starting point, RFLP maps are not very dense and, since most RFLPs are diallelic, some markers tend to be uninformative. A higher-resolution map of the human genome was constructed using *simple sequence length polymorphisms* (SSLPs), also known as *microsatellites*. These are variations in the number of tandem repeats of relatively simple sequences such as the dinucleotide CA. In humans and other mammals, microsatellites are highly abundant, multi-allelic, and easy to detect. RFLPs and microsatellites are not very useful in plants so different types of markers have been used. These include *randomly amplified polymorphic DNA markers* (RAPDs) which are generated by polymerase chain reaction (PCR) amplification with arbitrary primers and *cleaved amplified polymorphic sequences* (CAPS). The most widely used markers in plants are *amplified fragment length polymorphisms* (AFLPs), which are easily integrated into physical maps.

There is much current interest in the use of *single-nucleotide polymorphisms* (SNPs) as markers in all organisms, since these are very abundant (one every kilobase pair in the human genome) and easy to type in a high-throughput manner. A high-resolution SNP map of the human genome is under construction by several research organizations and private companies.

While the distances shown on a genetic map are defined arbitrarily on the basis of recombination frequencies, a *physical map* involves real physical distances measured in kilobase pairs or megabase pairs. The lowest resolution physical maps are based on chromosome banding, a technique applicable to mammals and insects but not other species. Other types of physical map include restriction maps (based on the positions of sites for rare-cutting restriction enzymes like *Not*I) and radiation hybrid maps, which are generated by typing human chromosome fragments in hybrid cell panels. These have a maximum resolution of several hundred kilobase pairs. The most useful markers for radiation hybrid maps are sequence tagged sites (STSs) and this principle is described in *Box 2.2*. The highest resolution physical map, with a resolution of 1 bp, is the DNA sequence itself.

Box 2.2   Sequence tagged sites as physical markers

Sequence tagged sites are DNA sequences, 1–200 bp in length, that are unique to a particular region of the genome and easily detected by PCR. A reference map with STSs distributed every 100 kbp or so provides the framework onto which particular DNA fragments, such as BAC and PAC clones, can be assembled. These clones are tested by PCR for the presence of a panel of STSs and, should one be found, the clone can be mapped onto the genomic framework. An STS reference map of the human genome was created by typing a panel of radiation hybrids. This is a physical mapping technique in which human cells are lethally irradiated, causing chromosome fragmentation, and the chromosome fragments are rescued by fusion of the human cells to rodent cells. Panels of cells containing different human chromosome fragments can be tested by PCR for the presence of STS markers. As in genetic mapping, the closer two markers are to each other, the more likely they are to be present on the same chromosome fragment. Analysis of these data allows the order of markers along the chromosome to be deduced and a marker map to be constructed. Most STSs are invariant but some incorporate microsatellite polymorphisms and can therefore be used to integrate physical and genetic maps. STSs can also be used as probes for fluorescent *in situ* hybridization to integrate clone contig and cytogenetic maps.

2.3.1.2   Whole genome sequencing

Once genetic and physical maps are available, they can be used to find individual genes (discussed in the next section) or as a framework for whole genome sequencing. In the latter case, the next stage is to assemble a *clone contig map*, in which real DNA clones containing genomic DNA are ordered and aligned to generate a contiguous sequence. Clones can be assembled into contigs by searching for overlapping sequences and mapped onto the genomic scaffold using STSs and other DNA markers. For large eukaryotic genomes, the first generation of clone contig maps is assembled from large-capacity vectors. *Yeast artificial chromosomes* (YACs) were originally used for this purpose but the high degree of chimerism and rearrangement among YAC clones makes such maps unreliable. *Bacterial artificial chromosomes* (BACs) and *P1-derived artificial chromosomes* (PACs), although a lower capacity, have a much greater stability and are the favored templates for genome sequencing projects. Once a clone contig map is in place, the individual clones can be subcloned into cosmids and a higher-resolution map generated. The cosmids can then be subcloned into sequencing vectors and the sequences reconstructed. The sequence can then be scanned for genes as discussed later.

2.3.1.3 Finding individual genes

Genetic and physical maps are also required to isolate individual genes from eukaryotic genomes if no functional information is available about the gene product. A scheme for this positional cloning strategy is shown in *Figure 2.1*. A good example is the identification of the human cystic fibrosis gene. In this case, no supporting information was available from cytogenetic analysis (in the form of deletions or translocation breakpoints) but an increasing gradient of linkage disequilibrium was used to identify the first exon of the gene.

Initially, linkage analysis in large family pedigrees was used to map the cystic fibrosis gene to 7q31–q32. Two flanking physical markers were identified, one of which was a DNA clone

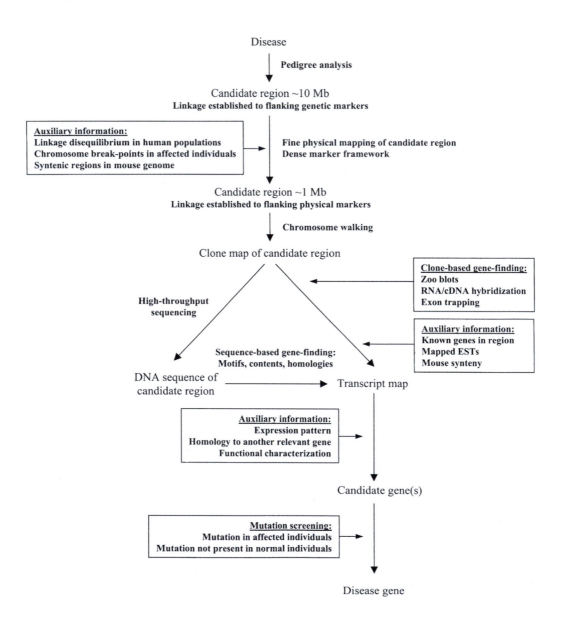

Figure 2.1

Simplified scheme for the positional cloning of human genes.

and the other the *MET* oncogene. This defined an initial candidate region of some 500 kbp. In the absence of any biochemical information about the gene product, the only viable approach was to clone the entire candidate region and search it systematically for genes. This was achieved by chromosome walking, a technique in which clones are used as hybridization probes to identify overlapping clones, and the process repeated until the whole candidate region is covered. At the time this project was undertaken (1985–1989), there were no high-capacity vectors and huge numbers of cosmid and phage λ clones had to be processed. A technique called chromosome jumping, now rendered obsolete by the availability of BACs and PACs, was used to avoid a large segment of unclonable DNA. Cloned regions were then aligned with the available physical map of the cystic fibrosis region, a restriction map generated using rare-cutter enzymes.

This study also predated the widespread use of computers for gene finding. A number of different strategies were employed to locate the cystic fibrosis gene including zoo blotting, cDNA blotting and the detection of CpG islands. These techniques and others are briefly discussed in *Box 2.3*.

Box 2.3   Experimental transcript mapping

The techniques discussed here exploit three unique features of genes that are not shared with non-coding DNA: their conservation between species, their structure and their expression as RNA.

Zoo blots. Coding sequences are strongly conserved during evolution, whereas non-coding DNA is generally under less selective pressure and is less conserved. DNA from a genomic clone that may contain a gene is hybridized to a Southern blot containing whole genomic DNA from a variety of species (a zoo blot). At reduced stringency, clones containing genes from one species will generate strong hybridization signals on genomic DNA from other species.

RNA and cDNA blots. Genes are expressed as RNA. If a genomic clone hybridizes to a blot of RNA or cDNA, or to a cDNA library, it is likely that the genomic clone contains a gene. Problems with this approach include low sensitivity to weakly expressed genes and a tendency to pick up pseudogenes and nonfunctional gene fragments in genomic clones.

Enzymatic identification of CpG islands. CpG islands are short stretches of hypomethylated GC-rich DNA often found associated with the promoters of vertebrate genes. They can be detected by computer (see main text) but enzymatic methods involving restriction enzymes with CG-rich recognition sites are also available. An alternative PCR-based technique has also been used.

cDNA capture. A hybridization approach in which genomic clones are used to capture related cDNAs in solution. This is more sensitive than RNA or cDNA blots but still relies on the gene being adequately expressed in the cells from which the cDNA was derived.

Exon trapping (exon amplification). An artificial splicing assay which does not rely on gene expression. The general principle is that a genomic clone is inserted into an 'intron' flanked by two artificial exons within an expression vector. The vector is then introduced into mammalian cells and the recombinant expression cassette is transcribed and spliced to yield an artificial mRNA that can be amplified by RT-PCR. If the genomic clone does not contain an exon, the RT-PCR product will contain the two artificial exons in the vector and will be of a defined size. If the genomic clone does contain an exon, it will be spliced into the mature transcript and the reverse transcriptase PCR (RT-PCR) product will be larger than expected.

2.3.2   Computational gene finding in higher eukaryotes

2.3.2.1   The role of bioinformatics

Today's mapping and sequencing projects generate enormous amounts of sequence data very rapidly. The manual annotation methods discussed in *Box 2.3*, while highly discriminatory and applicable to unsequenced DNA fragments, do not have the throughput to cope with the amount of data produced. Bioinformatic methods are necessary in large-scale sequencing projects to temporarily fill the information gap between sequence and function. A number of algorithms have been developed to scan and annotate raw sequence data, but it should be appreciated that none of the methods discussed below is 100% reliable and any computer-based gene predictions should be confirmed using other experimental methods.

2.3.2.2   Recognition of eukaryotic genes

All gene-finding algorithms must discriminate gene DNA from non-gene DNA, and this is achieved by the recognition of particular gene-specific features. Where simple detection of open reading frames is not sufficient, there are three types of feature that are recognized: signals, contents and homologies.

Signals are discrete motifs with a recognizable consensus sequence. Features that might be recognized as signals include promoter elements, polyadenylation sites and splice donor and acceptor sites. Algorithms that look for gene signals are known as *signal sensors* and can either search for signals in isolation or in the context of other local signals. Weight matrices are often employed to assign a score to each position in a putative signal and the total score over the entire motif is determined and used to predict whether or not the signal is genuine. Signal sensing is not sufficient on its own to identify eukaryotic genes because they are so diverse in structure. The TATA box, for example, is a transcriptional control element generally found about 30 bp upstream of the transcriptional start site of a gene. This would appear to be an excellent signal. However, a TATA box is only found in about 70% of genes. A polyadenylation signal is found in all genes but in many cases the sequence differs greatly from the consensus (AATAAA). Splice sites are very highly conserved in structure, but are only found in genes containing introns.

The *content* of a gene is not a specific motif but a general characteristic of an extended, variable-length sequence that differs from the surrounding DNA. For example, coding DNA and non-coding DNA differ significantly in terms of *hexamer frequency* (the frequency with which specific strings of six nucleotides are used). There are differences not only in terms of nucleotide frequency, but also in terms of *nucleotide dependency* (whether two particular nucleotides occur at different positions at the same time). These statistics are calculated over a *specific window size*, and the frequencies and dependencies are compared between windows. Hexamer frequencies and other base composition parameters depend on the codon usage of the organism. Different species prefer different codons for the same amino acid, where choice is allowed by degeneracy in the genetic code (this reflects the availability of different tRNAs). For this reason, computer algorithms that incorporate content sensing must be trained using the codon preference tables of the organism from which the sequence is derived. Another example of how content sensing can be used is to look for unusually high frequencies of the dinucleotide CG (in vertebrate genomes) or the trinucleotide CNG (in plant genomes) as these define *CpG islands*, which are found upstream of many housekeeping-type genes.

Gene-finding algorithms that employ signal and/or content sensing alone are known as *ab initio* programs because they do not incorporate any information about known genes. The third type of feature leading to the recognition of eukaryotic genes is *homology*, i.e. matches to previously identified genes or ESTs (expressed sequence tags). The most effective gene-finding programs incorporate elements of signal and content sensing as well as database homology searching using BLAST or FASTA (see Chapter 3) to produce an integrated gene-finding package. However, a word of caution is required because eukaryotic genomes tend to be littered with *pseudogenes* (nonfunctional gene remnants and reverse transcripts of messenger RNAs) and *gene fragments* (lone exons or multiple exons resulting from nonproductive recombination events). These may contain some of the signals and content of genuine genes and show homology to them, but they are not expressed.

2.3.2.3   Detection of exons and introns

The division of genes into small exons and large introns means that gene finding can be broken into a two-tier process of detecting exons (based on content and the position of splice signals) and then building these into a *whole gene model*. The use of auxiliary information, particularly in the form of cDNA sequences and regulatory elements, is much more important for the correct identification and delimitation of eukaryotic genes than it is for prokaryote genes. This is because many eukaryotic genes undergo *alternative splicing*, so gene predictions based on cDNA sequences can miss out or incorrectly delimit particular exons.

Many annotation errors involve exons, and the errors fall into a number of categories:

- Failure to detect genuine exons (low sensitivity)
- Prediction of false exons (low specificity)
- Incorrect specification of intron/exon boundaries
- Exon fusion

The failure to predict genuine exons is particularly relevant when the exons are very small (some human genes, for example, contain exons of 9 bp) or where the exons are at the 5′ or 3′ ends of the gene, since these may contain the untranslated region of the mRNA.

2.4   Detecting non-coding RNA genes

The methods discussed above are suitable for the detection of genes that are translated into protein. However, an increasing number of genes have been found to function only at the RNA level. These include *ribosomal RNA* (rRNA) and *transfer RNA* (tRNA) genes, as well as a host of others with specialized functions grouped under the label *non-coding RNAs* (ncRNAs). The most straightforward way to identify such genes is by database homology searching, but it should be noted that sequence conservation is not as important as the conservation of secondary structure for many ncRNAs, so *pattern-matching algorithms* that detect regions with the potential to form secondary structures may be more useful. Mathematical models known as *stochastic context-free grammars (SCFGs)* have been applied to statistically model ncRNAs in which variations in secondary structure are tolerated. An example is the tRNAscan-SE package that is used to find tDNA (genes for tRNA). The detection of novel ncRNA genes, i.e. where no precedents exist, remains a problem, since it is not clear what features to search for. Unlike protein-coding genes there is no open reading frame to detect and no codon bias to exploit in content sensing. In bacterial genomes, and those of simple eukaryotes such as *S. cerevisiae*, the existence of large genomic regions with no open reading frames (dubbed *gray holes*) suggests the presence of ncRNA genes, and these can be analyzed in more detail for the production of non-coding transcripts. These methods cannot be applied in higher eukaryote genomes and the identification of novel ncRNAs, particularly those whose thermodynamic structures are not significantly more stable than would be expected from a random DNA sequence, remains a difficult issue.

References and further reading

Burge, C.B., and Karlin, S. (1998) Finding the genes in genomic DNA. *Curr Opin Struct Biol* **8**: 346–354.
Eddy, S. (1999) Non-coding RNA genes. *Curr Opin Genet & Dev* **9**: 695–699.

Gaasterland, T., and Oprea, M. (2001) Whole-genome analysis: annotations and updates. *Curr Opin Struct Biol* **11**: 377–381.

Haussler, D. (1998) Computational genefinding. *Bioinformatics: A Trends Guide* **5**: 12–15.

Lewis, S., Ashburner, M., and Reese, M.G. (2000) Annotating eukaryote genomes. *Curr Opin Struct Biol* **10**: 349–354.

Rogic, S., Macksworth, A.K., and Oulette, F.B.F. (2001) Evaluation of gene-finding programs on mammalian sequences. *Genome Res* **11**: 817–832.

Stein, L. (2001) Genome annotation: from sequence to biology. *Nature Reviews Genet* **2**: 493–503.

Strachan, T., and Reid, A.P. (1999) Identifying human disease genes. In: *Human Molecular Genetics 2*, BIOS Scientific Publishers Ltd, Oxford UK, pp 351–375.

Sequence comparison methods

3

Christine Orengo

- The evolutionary mechanisms giving rise to changes in protein sequences and the challenges faced when aligning protein sequences are discussed.
- The use of dot plots for illustrating matching residue segments between protein sequences is described.
- Accurate methods for automatically aligning protein sequences, based on dynamic programming are considered.
- Rapid but more approximate methods for automatically searching sequence databases, based on segment matching are described. These employ hashing techniques for rapidly comparing residue segments between protein sequences.
- Statistical methods for assessing the significance of a sequence similarity are described, including Z-scores, *P*-values and E-values derived by comparing the match score to distributions of scores for unrelated sequences.
- Intermediate sequence search methods which exploit known relationships between sequences in protein families can be used to increase the sensitivity of sequence search methods.
- The performance of the existing database search methods in recognizing proteins which are homologs, i.e. evolutionarily related sequences, is assessed.
- The challenges and strategies involved in multiple sequence alignment are considered

3.1 Concepts

Proteins are essential agents for controlling, effecting and modulating cellular functions and phenotypic behavior. The international genome projects have brought a wealth of protein sequence information from all kingdoms of life and a major goal, now, is to understand how these proteins function. At present the largest international databases contain more than half a million non-identical sequences. Although physical characterization of the biochemical properties of all these proteins is impractical, proteins with similar sequences have diverged from a common ancestral gene and usually have similar structures and functions, with certain caveats (see Chapter 10). Therefore, development of reliable sequence comparison methods has been one of the major focuses of bioinformatics.

Although evolutionary relationships can often be more reliably detected by comparing the three-dimensional structures of proteins (see Chapter 6), methods for comparing sequences are generally much faster. Furthermore, since it is much easier to determine a protein's sequence than the 3-D structure, there are at least 30 times more sequences known than structures. Therefore, a bioinformatics analysis of any newly determined protein sequence typically involves comparing that sequence against libraries or databases of sequences to find related proteins with known functional properties, which can be used to annotate the unknown protein.

In order to understand the computational methods used for sequence alignment it is helpful to consider the changes which can occur between two homologous sequences during evolution. As discussed in Chapter 1, mutations in the nucleic acid bases of the DNA coding for the protein can give rise to different amino acid residues. Insertions and deletions of residues can also occur, though less frequently than mutations, and these can be quite extensive, for example, more than 50 residues. Usually insertions occur in the loops connecting secondary structures, where they are less likely to disrupt hydrogen-bonding patterns. Furthermore, gene shuffling and circular permutations can give rise to insertions of domains from other genes and can also lead to rearrangements in the order of the residues in the protein sequence. Analysis of protein families has revealed that significant changes in the sequences of relatives can occur, and are tolerated if they do not affect the folding or stability of the protein. In some protein families as few as five identical residues in 100 are retained between some relatives.

In this chapter, the concepts underlying automated methods for comparing protein sequences will be described and the reliability of these methods discussed. The accuracy with which structural and functional attributes can be inherited between relatives depends on the evolutionary distance between those relatives. Sander and Schneider have shown empirically that sequences of 100 residues or more, sharing at least 35% identical residues, are definitely homologs; a result confirmed recently by Rost with a much larger dataset (see *Figure 3.1*).

At these levels of sequence identity, Chothia and Lesk demonstrated that protein structures are very similar. Furthermore, analysis of protein families has shown that functional properties can also be inherited, although more confidently for single domain proteins. For multidomain proteins higher levels of sequence identity (>60%) are generally required otherwise similarity may only extend to one domain and differences in the other domains may modulate function (see Chapter 10).

However, even at lower levels of sequence identity, where functional annotation must be performed cautiously, sequence alignment enables the identification of equivalent residues that may be functionally important; a hypothesis which could be tested experimentally through mutation studies. In particular, multiple sequence alignment which optimizes the alignment of several homologs can be used to search for patterns of highly conserved residue positions (see Chapters 4 and 5). These may be associated with active-sites or patches of residues on the surface of the protein, important for mediating interactions with other proteins (see Chapter 13).

Pair-wise sequence alignment methods are generally used to detect close homologs (≥ 35% identity) and even to reveal evolutionary relationships in what Doolittle has referred to as the twilight zone of sequence similarity, i.e. down to as low as 25% identity. Below that, in the midnight zone, coined by Rost, multiple alignment methods and sequence profiles must be used to infer homology (see Chapter 5). In very distant homologs sequences can diverge beyond recognition by such methods, and a new breed of algorithms, such as threading, has arisen to detect these very remote relationships (see Chapter 9).

3.2   Data resources

3.2.1   Databases

There are several repositories of protein and nucleotide sequences. The largest is the GenBank database at the NCBI in the US, which contained over 12 million nucleotide sequences in 2002. The associated GenPept resource holds the translated amino acid sequences for these. Currently these can be clustered into half a million non-identical or non-redundant sequences. GenBank contains many partial sequences and it may be preferable to use other resources containing complete sequences known to be protein-coding regions.

The SWISS-PROT and TrEMBL databases at the EBI contain only complete sequences from genes known to code for proteins. The SWISS-PROT database also contains functional annotations for these sequences obtained either experimentally or through inheritance. See Chapter 5 for descriptions of secondary protein databases, containing sequences clustered into evolutionary families.

The DNA databank of Japan (DDBJ) is maintained by the National Institute of Genetics, Japan.

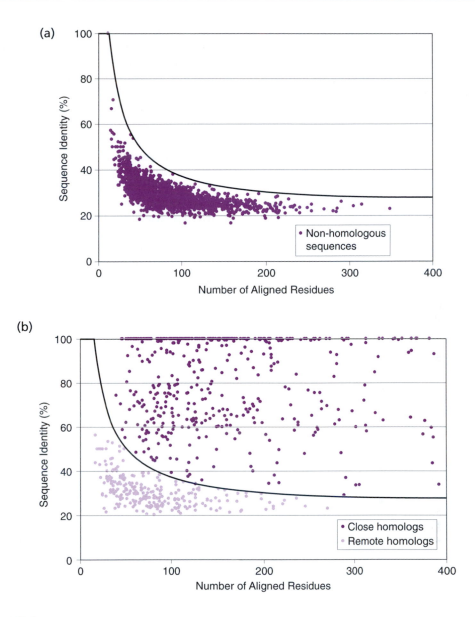

Figure 3.1

Plots showing sequence identities measured between pairs of (a) unrelated sequences (b) homologs. The black line shows the threshold sequence identity for assigning homology for a given number of aligned residues.

3.2.2   Substitution matrices

Substitution matrices provide information on residues possessing similar properties (e.g. physicochemical) or residues exchanging frequently in known protein families. Such matrices can therefore be used to score the substitution of one residue for another in a protein alignment. The simplest matrix is the unitary matrix, which scores 1 for identical residues and 0 for non-identical residues. The most commonly used resources are derived by analysis of residues exchanging in close relatives and include the PAM matrices derived by Dayhoff and co-workers in the 1970s and more recently the BLOSUM matrices of Henikoff and Henikoff, derived from multiple sequence alignments. The Joy matrices of Overington and co-workers are derived from structural alignments of related proteins and capture those residues frequently exchanging in structurally equivalent positions. Chapter 4 describes these matrices and their methods of construction in more detail.

3.3 Algorithms for pairwise sequence comparison

There is not scope in this chapter to provide detailed descriptions of all the algorithms used for sequence alignment, nor to provide information on all their implementations. The concepts behind the algorithms will be explained and the most commonly used implementations, accessible over the web, will be mentioned. There are now several reviews and books describing sequence alignment methods. These are listed at the end of this chapter and the reader is encouraged to read these for additional insights.

The basic concepts underlying all methods are similar. First, it is necessary to have a scheme for scoring the similarities of residues in order to be able to line up potentially equivalent residues. Secondly an optimization strategy must be used to explore alternative ways of lining up these residues in order to maximize the number of similar residues aligned.

3.3.1 Challenges faced when aligning sequences

Since there are only 20 amino acid residues from which polypeptide chains are constructed there is a finite possibility, particularly for short proteins, that two proteins will have similar sequences of residues, purely by chance. For random unrelated sequences, there is a finite possibility that 5–10% of their sequences will be identical. This can increase to 10–20% if gaps are allowed when aligning the sequences (see Doolittle's review). It can be seen from the plots in *Figure 3.1*, that there are many random, unrelated sequences with sequence identities of 20%.

If we consider similarities in the nucleic acid sequences of proteins, the number of unrelated proteins exhibiting high sequence identities is even more pronounced as here the nucleic acid alphabet is smaller, at only four characters. Therefore, where possible, it is better to compare amino acid rather than nucleic acid sequences, in order to minimize the number of false homologies identified.

Methods developed for comparing protein sequences range from fast approximate approaches, which seek to infer homology on the basis of substantive matching of residue segments between proteins to those which attempt to accurately determine all the equivalent residue positions. Fast approximate methods are generally used for scanning a database with a sequence of unknown function in order to find a putative homolog of known function. Any putative relatives identified in the database can then be realigned using the accurate, but slower, methods. Both approaches are described below.

One of the hardest tasks in aligning two protein sequences is recognizing the residue insertions or deletions, also described as indels. It can be seen from *Figure 3.2* that the matching of identical residues between two distantly related globin sequences can be significantly improved by placing gaps, corresponding to insertions or deletions, in the correct positions. Putting gaps in the best places, in order to line up as many equivalent positions as possible, is an extremely difficult task to perform by eye, especially for large sequences. Furthermore, in very distant homologs these residue insertions or deletions can be very extensive. Therefore the most accurate methods available use quite sophisticated optimization strategies which are able to cope with these indels and thereby identify the best alignment of the sequences.

Most robust algorithms for aligning sequences employ an optimization strategy known as dynamic programming to handle insertions/deletions. This will be described for a simple example later in the chapter.

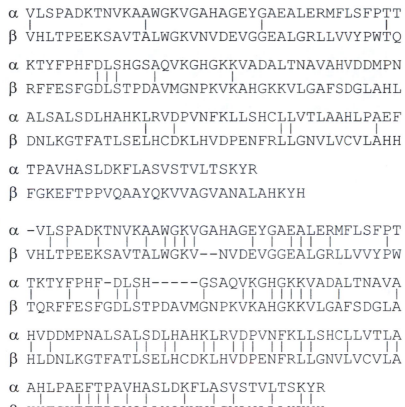

α VLSPADKTNVKAAWGKVGAHAGEYGAEALERMFLSFPTT
β VHLTPEEKSAVTALWGKVNVDEVGGEALGRLLVVYPWTQ

α KTYFPHFDLSHGSAQVKGHGKKVADALTNAVAHVDDMPN
β RFFESFGDLSTPDAVMGNPKVKAHGKKVLGAFSDGLAHL

α ALSALSDLHAHKLRVDPVNFKLLSHCLLVTLAAHLPAEF
β DNLKGTFATLSELHCDKLHVDPENFRLLGNVLVCVLAHH

α TPAVHASLDKFLASVSTVLTSKYR
β FGKEFTPPVQAAYQKVVAGVANALAHKYH

(a)

α -VLSPADKTNVKAAWGKVGAHAGEYGAEALERMFLSFPT
β VHLTPEEKSAVTALWGKV--NVDEVGGEALGRLLVVYPW

α TKTYFPHF-DLSH-----GSAQVKGHGKKVADALTNAVA
β TQRFFESFGDLSTPDAVMGNPKVKAHGKKVLGAFSDGLA

α HVDDMPNALSALSDLHAHKLRVDPVNFKLLSHCLLVTLA
β HLDNLKGTFATLSELHCDKLHVDPENFRLLGNVLVCVLA

α AHLPAEFTPAVHASLDKFLASVSTVLTSKYR
(b) β HHFGKEFTPPVQAAYQKVVAGVANALAHKYH

Figure 3.2

Example alignment of two short globin sequences showing the improvement in number of identical residues matched which can be obtained by allowing gaps in either sequence. (a) No gaps permitted; (b) introduction of gaps enables more identical residues to be aligned.

3.3.1.1 Dot plots and similarity matrices for comparing protein sequences

The simplest way to visualize the similarity between two protein sequences is to use a similarity matrix, known as a dot plot (see *Figure 3.3*). These were introduced by Philips in the 1970s and are two-dimensional matrices which have the sequences of the proteins being compared along the vertical and horizontal axes, respectively. For a simple visual representation of the similarity between two sequences, individual cells in the matrix can be shaded black if residues are identical, so that matching sequence segments appear as runs of diagonal lines across the matrix (see *Figure 3.3a*).

Some idea of the similarity of the two sequences can be gleaned from the number and length of matching segments shown in the matrix. Identical proteins will obviously have a diagonal line in the center of the matrix. Insertions and deletions between sequences give rise to disruptions in this diagonal. Regions of local similarity or repetitive sequences give rise to further diagonal matches in addition to the central diagonal.

Because of the limited protein alphabet, many matching sequence segments may simply have arisen by chance. One way of reducing this *noise* is to only shade runs or '*tuples*' of

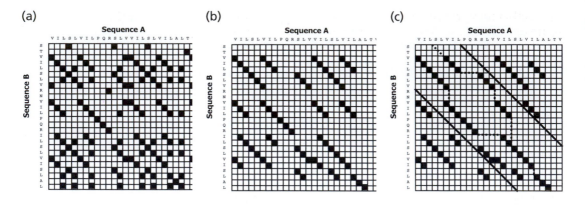

Figure 3.3

Example of a dot plot for comparing two simple protein sequences. (a) All cells associated with identical residue pairs between the sequences are shaded black; (b) only those cells associated with identical tuples of two residues are shaded black; (c) only cells associated with tuples of three are shaded and the optimal path through the matrix has been drawn. This is constrained to be within the window given by the two black lines parallel to the central diagonal. An alternative high-scoring path is also shown.

residues, e.g. a tuple of 3 corresponds to three residues in a row. This is effective because the probability of matching three residues in a row by chance is much lower than single-residue matches. It can be seen from *Figures 3.3b,c* that the number of diagonal runs in the matrix has been considerably reduced by looking for 2-tuples (*Figure 3.3b*) or 3-tuples (*Figure 3.3c*).

However, to obtain a more accurate estimate of the similarity between two protein sequences, it is better to determine which residues are most likely to be equivalent in the two sequences and score the similarity of these equivalent residues. This corresponds to finding a single pathway through the dot plot. In other words, one that crosses any given row or column in the matrix, only once. Pathways which track back on themselves or multiple pathways across the same row or column would correspond to matching a single residue in one protein with several residues in the other protein.

Therefore, the aim is to find a single path through the dot plot which has the most biological significance, i.e. the path which aligns the most identical residues or residues which are expected to be tolerated as mutations. In 1970, Needleman and Wunsch devised a method for finding this path. The first step is to convert the dot plot to a *score matrix* or *path matrix*, in which cells are scored according to the similarity of the residue pairs associated with them. The optimal path will then be the one having the highest score.

Figure 3.3c shows two possible pathways through the score matrix being used to compare two simple protein sequences. The most biologically significant pathway, with the highest score, is that linking 21 identical residue pairs and having only two *gaps* in the alignment. The gaps are associated with vertical or horizontal lines in the matrix and correspond to insertions in one or other sequence.

As well as matching identical residues, methods for scoring the path matrix can reflect similarities in the properties of residues. In distant homologs, the number of identical residues may be very low, even as low as 5%, giving weak signals in the matrix. However, residue properties are often conserved when identities are not. Therefore, to recognize similarities between the sequences of distantly related proteins, it is better to score cells in the path matrix, according to whether residues are likely to have been substituted for one another, rather than by whether they are identical. The residue substitution matrices, described above and in Chapter 4, are used to do this. They contain information on the probability of one residue mutating to another, within a protein sequence. These probabilities may be derived from similarities in residue properties or from the frequencies with which residues are found to exchange in protein families.

3.3.1.2 Dynamic programming

When comparing large sequences, the multiplicity of paths means that it is not feasible to analyze these matrices manually. The automatic algorithm developed by Needleman and Wunsch in 1970 employed an optimization technique for analyzing this score matrix and finding the highest scoring pathway. This technique, known as dynamic programming, has also been widely used in other applications in computer science; for example in solving the traveling salesman problem for which the optimal sequence of cities visited is the sequence involving the least traveling time.

In sequence alignment the algorithm is attempting to match the maximum number of identical or similar residue pairs whilst tolerating the minimum number of insertions or deletions in the two sequences. This latter constraint is desirable because residue insertions or deletions are likely to be unfavorable, particularly if they disrupt hydrogen bonding networks in secondary structures or residue arrangements in an active-site. Analysis of protein families has shown that residue positions within these secondary structures are generally well conserved and that insertions and deletions of residues occur much more frequently in coil regions or 'loops' connecting the secondary structures. An average sequence of 250 residues may comprise fewer than 15 secondary structures and therefore one may expect the best alignment to open 14 or fewer gaps. Most methods applying dynamic programming algorithms to align sequences have adjustable parameters for constraining or penalizing gaps in protein sequences, discussed in more detail below. So, in more distant sequences, you can relax the parameter to allow more gaps for the additional insertions/deletions expected.

The best way of understanding how dynamic programming works is to consider a simple computer implementation of it applied to sequence alignment. There are effectively three major steps involved in dynamic programming:

- Score the 2-D score matrix according to the identities or similarities of residues being compared.
- Accumulate the scores in the 2-D matrix penalizing for insertions/deletions, as required.
- Identify the highest scoring path through the matrix.

(1) Scoring the Matrix. The 2-D score matrix is first populated with scores according to the identities or similarities of residues associated with each cell. In *Figure 3.4a*, the unitary substitution matrix has been used to score the path matrix by assigning 1 for an identical residue pair or match and 0 for a mismatch.

(2) Accumulating the Matrix. In the next step, scores in this matrix are accumulated from the bottom right corner of the matrix to the top left corner of the matrix. This is done by calculating the scores a row at a time, starting at the furthest right cell of any given row. Then for each cell (i,j) in the row, the best score from all the possible paths leading up to that cell is added to the score already in the cell.

Since the matrix is being accumulated from the bottom right corner, all possible paths to this (i,j) cell must pass through one of the cells in the adjacent $(j + 1th)$ row or $(i + 1th)$ column to the right and below the current cell, shown highlighted in *Figure 3.4b*. The best score from a path leading to cell (i,j) must therefore be selected from either the cell diagonal to the current (i,j) cell, that is the $(i + 1, j + 1)$ cell, or from the highest scoring cell in the $j + 1th$ row to the right or in the $i + 1th$ column below it.

Thus for the simple example shown in *Figure 3.4*, the i + 2th cell in the j + 1th row has the highest score of 5 which is added to the score of 1 in cell (i, j) giving a total score of 6. For situations like this, where the diagonal (i + 1, j + 1) cell does not contain the highest score and a score is added from another cell along the (j + 1th) row or (i + 1th) column, this is equivalent to allowing a gap in the sequence alignment. This can be penalized by charging a gap penalty, discussed further below, the value of which will depend on the scoring scheme used. In this example, where the identity matrix is being used, a suitable penalty may be 3.

The accumulation of scores at each position in the matrix can be expressed mathematically as follows:

$$S_{i,j} = S_{i,j} + \max \begin{bmatrix} S_{i+1,j+1} \\ S_{i+m,j+1} - g \\ S_{i+1,j+m} - g \end{bmatrix}$$

Where $S_{i+m,j+1}$ is the maximum score observed in the j + 1 row. $S_{i+1, j+m}$ is the maximum score observed in the i + 1 column and g is the gap penalty imposed.

This simple accumulation operation can be repeated for each cell in the score matrix, giving the final accumulated score matrix (*Figure 3.4b*). As mentioned above, the accumulation process is started in the cell at the bottom right corner of the matrix. However, for all the cells in the bottom row or furthest right column there are obviously no scores to accumulate as there are no paths leading into these cells. Therefore, these cells will only contain scores reflecting the similarity of the associated residue pairs.

(3) Tracing Back Along the Highest Scoring Path. Once the matrix has been accumulated, the highest scoring path through the matrix can be very easily found. Since the matrix has been accumulated from bottom right to top left, the best path is traced back from top left to bottom right. The path will start in the highest scoring cell in the top row or furthest left column of the matrix and will be traced from there in a similar manner to the procedure for accumulating. That is the best path from each (i,j) cell will proceed through the highest scoring of the cells, in the (j + 1th) row or (i + 1th) column to the right and below it in the matrix. This is illustrated in *Figure 3.4c*.

The alignment of the two sequences obtained by processing the matrix in this way is shown below. It can be seen from *Figure 3.4c,d* that eight identical residues are matched and four gaps are introduced in the path to enable these residues to be aligned.

Dynamic programming gives a very elegant computational solution to the combinatorial problem of evaluating all possible pathways through the matrix. By accumulating the matrix in this manner, it is very easy to find the next cell along the path when tracing back across the matrix. This is because the trace back procedure is constrained by the requirement that residues can only be paired with one other residue so the path cannot track back on itself. Thus from any given cell the path can only be traced through cells below and to the right of the cell. Again, if the path does not proceed through the (i + 1, j + 1) cell diagonal to the current (i,j) cell but traces through one of the cells in the column or row to the right and below it, this is biologically equivalent to allowing insertions/deletions in the sequences.

In this simple implementation, cells were scored 1 or 0 depending on whether residues matched or not. As discussed already, cells can alternatively be scored using more sophisticated residue substitution matrices (e.g. PAM, BLOSUM, see Chapter 4). Since the score will reflect the tolerance for a particular mutation at any residue in the alignment, this is a better way of modeling the evolutionary changes. This is particularly important for distantly related sequences where use of the unitary substitution matrix would give rise to very sparse score matrix making it hard to find the optimal alignment between the sequences.

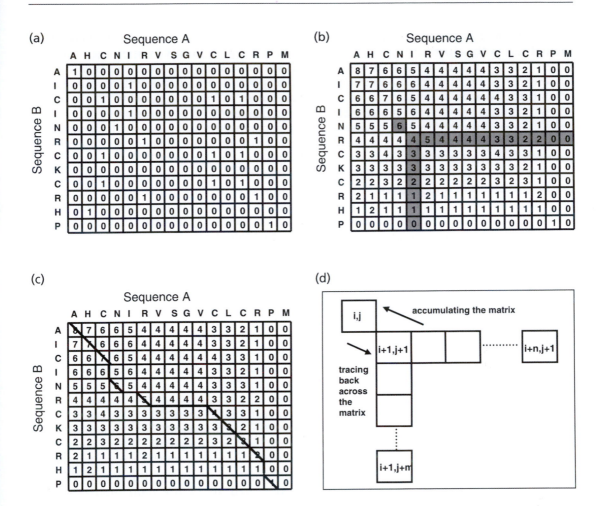

Figure 3.4

Simple example showing the steps in dynamic programming. (a) Scoring the path matrix; (b) accumulating the matrix; (c) mechanism for accumulating the matrix and tracing back across the matrix; (d) tracing back through the matrix to obtain the highest-scoring path giving the alignment of the two sequences.

3.3.1.3 Gap penalties

Another important modification to the scoring process involves penalizing for insertions/deletions to prevent excessive gaps in the alignment that may have little biological meaning. There are various mechanisms for implementing gap penalties. Some algorithms use a length-independent penalty whilst others charge a fixed penalty for opening a gap and then increment this penalty according to the length of the insertion. As discussed already, analysis of protein families for which structures are known has shown that insertions and deletions occur more frequently in loops connecting secondary structures. Therefore if the structures of one or both of the proteins are known, some implementations penalize gaps more heavily when they occur in secondary structure regions.

In summary, dynamic programming finds the optimal path by searching for optimal sub-alignments from each position and this removes the need to explore all paths through the matrix; the final path follows a route through all these locally optimal paths. Several implementations of the algorithm speed up the analysis of the path matrix by applying a '*window*' to the matrix and only seeking paths within this window (see *Figure 3.4c*). This is equivalent to restricting the number of insertions/deletions permitted as no insertions can extend beyond the window boundary. Window sizes are often adjusted depending on the differences in lengths between the proteins and also by considering whether the proteins are likely to be close or distant homologs having more extensive insertions. In *Figure 3.3c* the diagonal lines drawn across the matrix illustrate the position of the window and constrain the optimal pathway identified by the algorithm to be within these boundaries.

Some implementations of the algorithm also determine sub-optimal alignments (see for example Waterman and Eggert). This can sometimes be helpful in the analysis of distant homologs for which there may be extensive changes or insertions in the loops so that the signal in the score matrix becomes weak in these regions. Several local paths may be suggested by the data and should perhaps be considered by the biologist.

3.3.1.4 Local implementation of dynamic programming

The Needleman and Wunsch algorithm employs a global implemention of dynamic programming, in that it seeks to identify equivalent positions over the entire lengths of the proteins being compared. However, as discussed in Chapter 1, many proteins are modular and comprise more than one domain. Domain recruitment and domain shuffling are now established as very common evolutionary mechanisms by which organisms expand their functional repertoire. Because of this proteins which share one or more homologous protein domains may not be homologous over their entire lengths.

Therefore, in 1981, Smith and Waterman devised a local implementation of the dynamic programming algorithm which seeks a local region of similarity by reducing the accumulated path score once the path starts tracing through regions with insignificant similarity. This is implemented simply by imposing negative scores for residue mismatches. When tracing back through, the matrix paths can start anywhere and are terminated once the score falls below zero. Several paths aligning putative homologous domains or identifying regions of local similarity can thus be obtained from the iterative analysis of the path matrix using a Smith–Waterman algorithm (see *Figure 3.5*). The Smith–Waterman algorithm is employed in the widely used database search algorithm *SSEARCH* because of its ability to detect equivalent domains between multidomain proteins.

3.3.1.5 Calculating pair-wise sequence identity

Sequence similarity between two proteins is frequently measured by calculating the number of identical residues aligned between the proteins and then normalizing by the size of the smaller protein (see equation (1) below). This corresponds to normalizing by the maximum number of identical residues which could be matched between the two proteins. Thus the sequence identity measured for the simple example described above and illustrated in *Figure 3.4d* is 8/12 = 67%. Alternative, statistical approaches for assessing the significance of sequence similarity are described in section 3.5 below.

$$\text{Sequence identity (\%)} = \frac{\text{number of identical residues}}{\text{number of residues in the shortest sequence}} \times 100 \qquad (1)$$

3.4   Fast database search methods

Methods employing dynamic programming algorithms are slow but yield the most reliable alignments of protein sequences. Although it is possible to speed up the algorithm using a window to reduce the amount of the path matrix that must be searched, these methods still

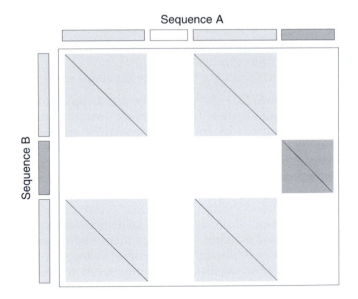

Figure 3.5

Schematic illustration of local paths identified in the score matrix used to compare two large multidomain proteins. Each local path is associated with the alignment of corresponding domains in the two sequences.

remain computationally very expensive. They are therefore impractical for searching large sequence databases such as GenBank. Some recent initiatives reduce the time by parallelizing the code and running the algorithm on several machines simultaneously. However, alternative strategies for database searching have been to develop faster, more heuristic approaches to identify probable homologs, which can then be realigned using the more accurate Needleman–Wunsch or Smith–Waterman algorithms.

These heuristic methods are not guaranteed to find the optimal alignment but can give a good indication of sequence similarity, usually by identifying matching sequence segments between the proteins being compared. There are two main approaches which have been deployed, the FASTA suite of programs first developed by Lipman and Pearson in the mid-1980s and the BLAST programs first introduced by the Altschul group in 1990.

Both methods first match segments of identical residues (k-tuples or k residues in a row) between sequences, where k is the length of the tuple. Typical tuple lengths are 1 or 2 for amino acid sequences, longer for nucleic acid sequences and the user can generally adjust the tuple length employed. A computational technique, known as *hashing*, is used for rapid matching. That is for each k-tuple in a sequence, information on the location of that segment in the sequence is stored at a specific computer address or *'hash bin'* determined by the composition and order of the residues in the segment. Two sequences can thus be very rapidly compared by hashing their sequences in the same way and then identifying the occurrences of common segments from their presence in corresponding hash bins. Matching segments identified in this way can also be visualized using a dot plot, as described above.

3.4.1 FASTA

The major steps employed in FASTA can be summarized as follows:

- Identify all matching K-tuples, comprising identical residues, between the sequences.

- Extend these segments to include similar residues, using an appropriate substitution matrix.
- Impose a window on the matrix to restrict insertions/deletions and select the 10 highest scoring segments within this window.
- Use limited dynamic programming to join the segments within this window.

The various implementations of FASTA typically use k-tuples of 1 and 2 for protein sequences. Having identified all identical tuples between two proteins (see *Figure 3.6*), these are extended into longer segments by including potentially equivalent residue pairs, i.e. those having similar residue properties or found to exchange frequently in protein families. This is done by re-scoring the path matrix using a substitution matrix. The PAM matrices of Dayhoff are typically used and the matrix can be chosen to reflect the expected evolutionary distance between the proteins (e.g. PAM120 for close homologs and PAM 250 for more distant homologs, see Chapter 4).

In the original implementation of the algorithm, sequence similarity was estimated from the total number and length of matching segments identified in the path matrix. Recent implementations employ a more biologically meaningful strategy by attempting to identify a single path through the matrix, whilst maintaining speed by performing only limited amounts of dynamic programming. This is done by imposing a window on the path matrix, and selecting the 10 highest scoring fragments within this window, above an allowed threshold. Dynamic programming is then used to join these segments, giving a single path through the matrix which can be used to calculate a pairwise similarity score as described above (see *Figure 3.6*). The significance of the similarity is assessed by calculating how frequently this score is observed when scanning the query sequence against a database of unrelated sequences (see section 3.5 below).

3.4.2 BLAST

In the original BLAST method developed by Altschul and co-workers, the algorithm identifies the high-scoring segment pair (HSP) between the two sequences as a measure of their similarity. BLAST scores are also assessed statistically from a database search by considering how frequently unrelated sequences return similar scores. The initial step seeks matching tuples, or words, of three residues between the protein sequences being compared. Rare words, that is words occurring less frequently in protein sequences are considered more significant, so that for a new query sequence the occurrence of up to 50 more rarely occurring words is identified, hashed and subsequently searched for in all the database sequences being scanned.

Matching words are then extended in a similar manner to that described for FASTA, i.e. using a substitution matrix to identify similar residues. The score of the highest-scoring segment pair, also described as the maximal segment pair (MSP), gives an indication of the similarity of the two proteins.

In more recent implementations, the algorithm has been speeded up by various heuristics which reflect the biological requirement that the alignment of the two sequences is given by a single path through the matrix. Since in many related proteins, several matching segments fall on the same diagonal across the matrix, the algorithm only extends those segments for which another high-scoring segment is found within a threshold distance on the same diagonal.

Often several high-scoring segments are matched between protein sequences and a more reliable estimation of the degree of similarity between the proteins can only be obtained by somehow considering the segments together, again reflecting the biological requirement to find a single path linking these segments. Therefore, as in recent versions of FASTA, in the powerful gapped-BLAST or WU-BLAST implementation, limited dynamic programming is performed to link together these high-scoring segments and obtain a more biologically meaningful measure of sequence similarity.

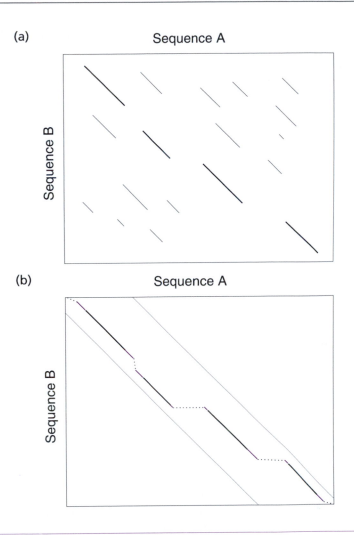

Figure 3.6

Schematic illustration of the steps involved in sequence comparison using FASTA. (a) Identification of matching residue segments between the sequences. Identical 2-tuples are first identified. (b) These are extended using a Dayhoff substitution matrix. The highest scoring segments within a window are selected and joined using dynamic programming.

The major steps employed in gapped BLAST can therefore be summarized as follows:

- Identify all matching words, comprising identical residues, between the sequences.
- Identify high-scoring segments, above a given threshold, occurring within an allowed distance from another high-scoring segment on the same diagonal.
- Extend these segments to include similar residues, using an appropriate substitution matrix.
- Use limited dynamic programming to join the highest scoring segments.

Both FASTA and BLAST are very widely used for database searching even though they are not guaranteed to find the optimal alignment between protein sequences. Their popularity is largely due to the increases in speed gained by using these simplified approaches. Searching a large database of about half a million sequences can be up to 100 times faster using BLAST than Smith–Waterman. In addition to the advantages of speed, both methods use robust statistical approaches for assessing the likely significance of any sequences matched during a database scan, described below.

3.5   Assessing the statistical significance of sequence similarity

As discussed already, there is a finite probability that proteins will have similarities in their sequences by chance alone. The plots in *Figure 3.1*, above, confirms this, showing that the number of unrelated sequences with identities above a given threshold explodes in the twilight zone (<25% sequence identity). However, the genome initiatives are determining the sequences of many diverse relatives in different organisms and it is clear that the identities of many of these relatives are below 25% identity. Frequently, the relationship of a query protein to a relative of interest, perhaps of known functional characterization, falls in this ambiguous region of sequence similarity. In these cases, statistical methods for assessing significance have been found to be more reliable than sequence identity.

Statistical approaches compare the score for aligning a protein sequence with a putative relative, to the distribution of scores obtained by aligning that protein with unrelated sequences. The larger the database of unrelated sequences, the more reliable are the statistics. The simplest approach is to calculate a Z-score (see also Glossary). This is the number of standard deviations (s.d) from the mean score (m) for the database search, given by the score returned from a putative match (S) (see *Figure 3.7*). The mean score and standard deviation are determined from the distribution of scores for unrelated sequences.

A high Z-score of 15 is considered very significant and the sequences can confidently be described as homologs. For Z-scores in the range 5–15, the proteins are probably homologous, though more distant relatives and there may be significant variations in their structures, although the structural core would probably be conserved between them. For lower Z-scores, there is less certainty in the nature of the relationship.

A more reliable method for assessing significance, now implemented in several database search protocols, involves mathematically fitting the tail of the distribution of scores returned by random, unrelated sequences (see *Figure 3.8*). This approach was adopted because it was observed that the distributions of scores obtained by scanning sequence databases, could best be described as extreme value distributions in which the tail decays more slowly than for a normal distribution (see *Figure 3.8a* and Glossary).

This type of distribution is often observed when recording maximal values. For example, if you were plotting the frequency of people's heights, by recording the height of the tallest person in each house in the country. Similarly, when aligning protein sequences in a data-

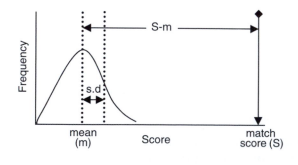

Figure 3.7

Example of a distribution of scores returned from comparing unrelated sequences used to calculate the Z-score for a potential match to a query sequence.

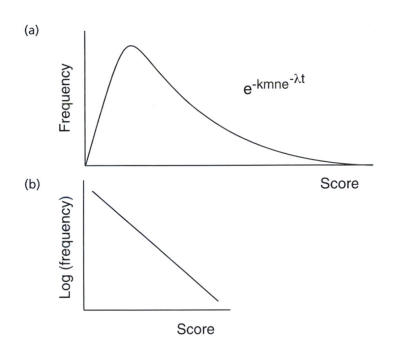

(a)

Frequency

$e^{-kmne^{-\lambda t}}$

Score

(b)

Log (frequency)

Score

Figure 3.8

(a) Extreme value distribution of scores for unrelated protein sequences. (b) Linear relationship between the scores and the log frequency of the scores in the tail of the frequency vs score distribution. This is used to calculate probability values (*p*-values) for matches to a query sequence.

base scan, you are seeking the maximum score that can possibly be achieved, amongst many alternative alignments. The tail of this extreme value distribution exhibits a linear relationship between the score value and log of the frequency at which this score is observed (see *Figure 3.8*). Parameters can easily be estimated to fit this relationship and it is then possible to calculate the frequency with which any score is returned by unrelated sequences.

More importantly, this frequency information can be used to derive a probability value (*p*-value) that any given match is to an unrelated sequence. For example a *p*-value of 0.01 means that 1 in 100 matches giving this score are to unrelated sequences. The related measure, expectation value (E-value), takes into account the size of the database being scanned (see Glossary). The larger the database, obviously, the more likely that the query sequence will match an unrelated sequence by chance and therefore the higher the probability that an unrelated sequence will return a score above a given threshold. Therefore, most database searches usually calculate expectation values. These are obtained by simply multiplying the *p*-value by the size of the database. Obviously, the lower the *p*-value or E-value, the greater the probability that the query sequence has matched a homologous protein.

Searching large databases, e.g. GenBank using gapped BLAST, E-values of less than 10^{-10} to 10^{-40} are typically obtained for very close relatives. Usually any matches with E-values below 0.01 suggest an evolutionary relationship. However, in the absence of any matches at this threshold, it is usual to consider matches up to an E-value of 1, seeking additional evidence to confirm these very tentative relationships. Values above 1 suggest that no relative can be detected using the algorithms employed and thus any homologs present in the database are extremely distant and below the limits of reliability of the current search methods.

In the absence of a database search the statistical significance of a pair-wise similarity score between two sequences can be estimated by simply jumbling the sequence of one of the proteins and plotting the distribution of scores for alignments to these jumbled random sequences. One of the most robust implementations of the Smith–Waterman algorithm is the SSEARCH program which can be run from the SRS server at the EBI or from the NCBI website. This typically jumbles sequences 1000 times to estimate significance. For a more detailed discussion on statistical approaches in bioinformatics, see the glossary and references listed at the end of this chapter.

3.6 Intermediate sequence searching

The percentage of relatives identified by any of the sequence search methods described in this chapter can also be increased by scanning against protein family libraries or *intermediate sequence libraries* rather than scanning the basic sequence repositories like GenBank, EMBL or Trembl. Protein family resources like Pfam, PRINTS, SCOP or CATH cluster sequences into families according to clear sequence, structural and/or functional similarity (see Chapters 5 and 7). Query sequences can be scanned against representatives from each protein family to identify a putative family to which the protein belongs. A sequence need only match one representative from that family to infer relationships to all other relatives in the family.

This technique has been described as *intermediate sequence searching*. Protein sequence A is found to match sequence B in a given protein family and this allows relationships between A and other relatives in the same protein family (e.g. C, D, E) to be also inferred. Other implementations of this approach which do not rely on scanning protein family databases, effectively take the first match from a database search, e.g. using GenBank (i.e. sequence B matching sequence A) and then scan the database again with B to obtain all the matches to B. This process can be repeated to identify successive intermediate matches (e.g. B→C, C→F, etc.), though it is obviously important to employ reliable statistics and safe thresholds to identify relatives and to check matches thoroughly. For example, a multiple alignment (see below) of all the relatives identified in this way could be used to check for conserved residue patterns allowing removal of any erroneous outlying sequences not exhibiting these patterns.

3.7 Validation of sequence alignment methods by structural data

In order to assess the reliability of sequence alignment methods and the accuracy of alignments generated, several benchmarking methods have been developed which effectively use structural data to test whether sequences matched are genuinely homologous and also to determine the quality of the alignments. There are now several classifications of protein structures (see Chapter 7), which can be used to select sets of very distant homologs having low sequence identity (e.g. <30%) but known to be homologs because of their high structural and functional similarity. These data-sets contain relatives which are diverse enough to provide a real test of the sensitivity and accuracy of the sequence comparison methods.

Most benchmarking studies reported by groups working with the SCOP, CATH or DALI domain structure databases have yielded similar results. Rigorous pairwise alignment methods based on Needleman–Wunsch or Smith–Waterman algorithms which employ full dynamic programming are more sensitive than the more approximate FASTA or BLAST methods, although the latter are much faster. The SSEARCH algorithm is found to identify twice as many relatives as gapped BLAST. However, the profile-based methods (see Chapter 5), such as PSI-BLAST are even more powerful, identifying three times as many relatives as gapped BLAST. Intermediate sequence search methods typically perform as well as PSI-BLAST, depending on the database used. Hidden Markov models, described in Chapter 5, are typically as powerful as profile-based methods, particularly SAM-T99, devised by the Karplus group, which is currently the most powerful method available.

Currently none of the sequence search methods described in this chapter are able to recognize more than about 60–70% of the very distant homologs which can be identified from functionally validated structural similarities, though undoubtedly the technologies will improve over the next decade. Performance will also be enhanced as more sequence databases expand giving improved substitution matrices and family specific profiles.

3.8   Multiple sequence alignment

In many cases distant relatives belonging to the same homologous superfamily, that is deriving from the same common ancestor, share few identical residues. A recent analysis of relatives in the CATH protein family database (see Chapter 7) shows that most homologs in currently known superfamilies share less than 25% sequence identity with each other. At these levels of sequence identity, pair-wise algorithms become unreliable as the signal in the 2-D path matrix is often very weak and most algorithms have difficulty in identifying the optimal path.

One powerful solution is to exploit the observation that some residue positions are much more highly conserved across the superfamily than others, even in very distant relatives. These positions may be associated with functional sites or surface patches important for mediating interactions with other proteins. Therefore, these residues, or residues with similar properties, would be expected to be present in all members of the family. Residue preferences at these conserved positions are diagnostic for the family and can be encoded as a regular pattern or a 1-D-profile for the family (see Chapter 5).

Multiple sequence alignment methods were introduced in the early 1980s in order to identify these patterns. Besides their value in deriving patterns for homolog detection, multiple alignments are important for identifying positions that are essential for the stability or function of the protein. The reader is encouraged to read Chapter 4 for a more detailed discussion of methods used to analyze residue conservation and an illustration of a multiple sequence alignment, colored according to conserved positions.

Unfortunately, dynamic programming cannot easily be extended to more than three protein sequences as it becomes prohibitively expensive in computing time. Therefore, more heuristic methods have been developed which adopt an evolutionary approach to the problem by performing a progressive alignment of all the sequences. Thus sequences are aligned in the order of their evolutionary proximity, with the closest homologs being aligned first and the next closest at each subsequent step.

In order to determine these phylogenetic relationships, sequences are first compared pair-wise, using a standard dynamic programming method such as Needleman–Wunsch. This gives a matrix of pair-wise similarity scores, which can be used to derive the phylogenetic tree, detailing the evolutionary relationships between the sequences.

Most protocols use hierarchical clustering methods to generate the tree (see Chapter 15 for a description of hierarchical clustering). In other words the most similar proteins are clustered first and next most similar proteins or clusters successively merged. The same order is used for the multiple alignment. However, in some approaches individual sequences are always added to a growing multiple alignment, whilst other methods first align all closely related sequences pair-wise and then align the consensus sequences derived from these pair-wise comparisons (see *Figure 3.9*).

As each relative is added to the multiple alignment, a consensus sequence can be derived from analysis of residues aligned at each position. This can be achieved in various ways, discussed in more detail below and in Chapters 4 and 5. Usually the consensus sequence is a description of residue preferences at each position, based on the frequency with which certain residue types are found at that position.

Different approaches use slightly different methods for clustering the sequences and deriving the trees. In the CLUSTAL method, developed by Higgins and co-workers in 1987, the

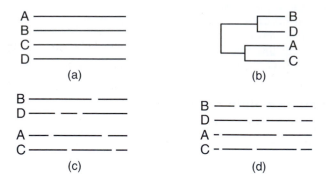

Figure 3.9

Schematic representation showing how phylogenetic analysis determines the order in which sequences are added to a multiple alignment. For a given set of sequences (a), pairwise comparisons are performed to derive a phylogentic tree (b), which governs the order in which sequences are aligned. First B is aligned against D, then A against C (c). Finally the pairwise alignments are aligned (d), preserving gaps, to give the complete multiple alignment of the sequences.

tree-drawing method of Saitou is employed. Phylogenetic methods used in the tree-drawing program PHYLIP, of Felsenstein and co-workers can also be used. Other approaches, such as MULTAL developed by Taylor in 1988, use hierarchical clustering to group the sequences initially but this clustering is revised after the addition of each sequence. All the remaining sequences are compared to the new consensus sequence and the sequence having highest similarity is aligned next. Multiple alignment methods have been modified in numerous ways since their inception. Some of the more common improvements are described below.

3.8.1   Sequence weighting

In some protein families or sequence data-sets, close relatives from a particular subfamily may be over-represented. This can result in the residue preferences encapsulated in the consensus sequence, becoming biased towards this subfamily. To address this problem recent implementations of CLUSTAL (CLUSTALW, see reference at the end of the chapter), and related algorithms, determine weights for individual sequences depending on how many close relatives they have in the data-set (see *Figure 3.10*). In other words the more close relatives a sequence has, the lower the weight of the sequence. Again, different weighting schemes have been implemented by the various alignment methods.

3.8.2   Deriving the consensus sequence and aligning other sequences against it

Methods developed by different groups use a variety of approaches to derive a consensus sequence from a progressive multiple alignment, though similar concepts are usually applied. Generally the methods analyze the frequency with which each of the 20 amino acid residues appear at every position in the alignment. The consensus sequence is effectively a profile describing these frequencies. Frequencies can be adjusted according to the weight of each sequence, so that sequences which are over-represented in the data-set are down-weighted, as described above.

   When aligning new sequences against this consensus sequence or profile, scores can be generated in several ways. Residues aligning at a given position can be scored according to how frequently such residues occur at this position in the aligned relatives, using the fre-

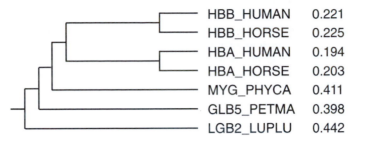

HBB_HUMAN	0.221
HBB_HORSE	0.225
HBA_HUMAN	0.194
HBA_HORSE	0.203
MYG_PHYCA	0.411
GLB5_PETMA	0.398
LGB2_LUPLU	0.442

Figure 3.10

Sequence weighting exploited in the multiple alignment method of CLUSTALW by Higgins. Weights are calculated from the pairwise similarities between all the sequences and are shown alongside the codes associated with each sequence.

quency information encoded in the consensus sequence. For example, a tyrosine residue aligned against a position at which only tyrosines have previously been aligned will score more highly than a tyrosine aligned against a position at which a variety of residues appear.

Alternatively, substitution matrices can be used to score residues depending, for example, on whether they align against positions with similar residue properties. Some methods use a more Bayesian approach, combining both scoring schemes. Substitution matrices are used early in the multiple alignment when there are insufficient sequences aligned to build up a reliable picture of residue preferences. As the alignment proceeds and more relatives are added, residue frequency information can also be included in the scoring method. The contributions of the two approaches can be adjusted as the alignment progresses. In recent implementations of CLUSTALW different substitution matrices are employed depending on the pair-wise similarity of the sequences being aligned. For example a PAM120 matrix can be used at the start of the multiple alignment when close relatives are being aligned and this is replaced by the PAM250 matrix as the alignment progresses as more distant relatives are added.

3.8.3 Gap penalties

As with pair-wise sequence alignment, penalties can be charged for introducing gaps in the alignment. Penalties are often adjusted according to information accruing in the multiple alignment. In the CLUSTALW method, penalties are relaxed at positions in the alignment where there are deletions of residues in relatives already aligned and made more stringent in positions conserved amongst relatives. If structural data are available they can also be used to restrict gaps in positions which correspond to secondary structure regions in the protein.

References and further reading

Attwood, T.K., and Parry-Smith, D. (1999) Pairwise alignment techniques, pp. 108–132 and Multiple sequence alignment, pp. 133–144 in *Introduction to Bioinformatics*. Addison Wesley Longman Ltd.

Barton, G. (1996) Protein sequence alignment and database scanning in *Protein Structure Prediction*. Ed M.J. Sternberg. The Practical Approach Series. pp. 31–63.

Barton, G. (1998) Creation and analysis of protein multiple sequence alignments in Bioinformatics: a Practical Guide to the Analysis of Genes and Proteins. Ed A.D. Baxevanis and B.F. Ouellette. *Wiley-Liss*, Inc. pp. 215–232.

Bourne, P. (2002) Structural Bioinformatics. Wiley.

Duret, L., and Abdeddaim, S. (2000) Multiple alignments for structural, functional or phylogenetic analyses of homologous sequences in *Bioinformatics: Sequence, Structure and Databanks*. Ed O. Higgins and W. Taylor. *Oxford University Press*.

Higgins, D.G., and Taylor, W.R. (2000) Multiple sequence alignment. *Methods Mol. Biol.* **143**; 1–18.

Kanehisa, M. (2000) Sequence analysis of nucleic acids and proteins in *Post-Genome Informatics*. Ed M. Kanehisa. Oxford University Press. pp. 64–85.

Schuler, G. (1998) Sequence alignment and database searching in *Bioinformatics: a Practical Guide to the Analysis of Genes and Proteins*. Ed A.D. Baxevanis and B.F. Ouellette. *Wiley-Liss*, Inc. pp. 187–214.

Yona, G., and Brenner, S.E. (2000) Comparison of protein sequences and practical database searching in *Bioinformatics. Sequence, Structure and Databanks*. Eds O. Higgins and W. Taylor. *Oxford University Press*.

Amino acid residue conservation

William S.J. Valdar and David T. Jones

- Related protein sequences differ at one or many positions over their length. What rules govern where they differ and by how much? Molecular evolution offers two contrasting perspectives embodied in the neutralist and selectionist models.
- Some amino acid substitutions are better tolerated by a protein than others. This has implications for how we compare and align related sequences. Substitution matrices are a way of quantifying how often we expect to see the mutation of one amino acid to another without disruption of protein function. The Dayhoff and BLOSUM matrices are examples.
- A multiple alignment reveals patterns of variation and conservation. Examining these patterns provides useful clues as to which positions are crucial for function, and thus constrained in evolution, and which are not. A conservation score attempts to infer the importance of a position in an alignment automatically by assessing how much its amino acids are allowed to vary.

4.1 Concepts

Proteins are products of evolution. Their sequences are encoded histories of mutation and selection over millions of years. Align two closely related protein sequences, say myoglobin from human and from tuna (*Plate 1a*), and you will see they are the same at many positions, but differ at others. Align many such sequences (to form a *multiple sequence alignment, Plate 1b*) you will see rich patterns of variation and consistency.

What does this variation mean and how does it arise? More specifically, why do some positions vary whereas others remain constant, and what can this tell us about the protein? This chapter explores these questions.

4.2 Models of molecular evolution

4.2.1 Neutralist model

According to the *neutral theory of molecular evolution*, once a protein has evolved to a useful level of functionality, most new mutations are either deleterious or neutral. Deleterious mutations are removed from the population by negative selection. Neutral mutations, those that have a slight or negligible effect on protein function, are kept. Most of the substitutions you observe in a multiple sequence alignment are therefore neutral; rather than representing improvements in a protein, these indicate how tolerant the protein is to change at that position. In an already optimized protein, regions that are more constrained by function will show slower rates of substitution and *vice versa*.

Fibrinopeptides, spacers that prevent the sticky surfaces of fibrinogen aggregating until a blood clot forms, are under fewer functional constraints than ubiquitin, which tags proteins for destruction; they also evolve about 900 times faster. The most functionally important residues of hemoglobin, those that secure the heme group, show a much lower rate of substitution than do others in the protein.

4.2.2   Selectionist model

The selectionist model offers a different view. It agrees with the neutralist model that most mutations are deleterious and quickly lost, but it disagrees about those mutations that are kept. According to this model, the majority of accepted mutations confer a selective advantage whereas neutral mutations are rare. Most of the substitutions you observe in a multiple alignment therefore represent evolutionary adaptations to different environments and have arisen through natural selection rather than through random neutral drift.

Evidence exists for both models, and which is more appropriate depends largely on context. Here we assume a neutralist model because that model is by far the more commonly applied in structure and sequence bioinformatics.

4.3   Substitution matrices

The alignment of two biological sequences is the cornerstone of bioinformatics. This is because the degree to which two sequences differ estimates how distantly they are related in evolution. This in turn tells us how different (or similar) they might be in terms of their structure and function. How can we measure this useful evolutionary distance?

4.3.1   Conservative substitutions

First we must consider why sequences differ. The section above described how most of the differences between related sequences we see today represent the accumulation of neutral mutations. This suggests that in the 'true' alignment for a pair of sequences the type of mutations we expect to see most often are those that only slightly disturb structure and function. This implies that some types of mutations are more likely to cause disruption than others, and therefore (if we ignore insertions and deletions for now) that some substitutions are more readily accepted than others. From our knowledge of how proteins work, we know that disruptive substitutions tend to be those between very different amino acids (e.g., aspartate to tryptophan), whereas harmless ones tend to be between similar ones (e.g., leucine to isoleucine). In order to say what set of substitutions is most likely to have occurred, and hence to find the best alignment between the two sequences, we need a way to measure how similar different amino acids are. This is where substitution matrices are useful.

Alignment methods use scoring schemes that tell them whether the current alignment is any good or not. Most scoring schemes model the similarities between different amino acids, and hence the plausibility of a particular substitution, in the form of a 20×20 matrix, where the value of each element represents the similarity of the row and column amino acids. These matrices are symmetric, meaning the value for any two given amino acids is the same regardless of how they are looked up, and so contain 210 distinct elements (190 elements representing all pairs of distinct amino acids + 20 elements along the diagonal representing the similarity of each amino acid to itself). Many similarity matrices have been devised, but only a few are commonly used today.

The simplest matrix is the *identity matrix*, which gives identical amino acids a similarity of 1 and non-identical amino acids a score of 0. Aligning two sequences with such a matrix amounts to maximizing the number of identical residues between the two. A simple metric based on this scoring scheme is referred to as the *percentage identity*. For example, if two proteins have 43 residues out of a total of 144 aligned residues in common then they can be quoted as being 30% identical – or that have 30% sequence identity. Percentage is a very widely used measure of sequence similarity because it is easily understood. However, there is no obvious way to directly relate percentage identity with evolutionary distance because so many important considerations are ignored. Nonetheless, it is often reasonable to assume

a)

```
HUMAN   GLSDGEWQLVLNVWGKVEADIPGHGQEVLIRLFKGHPETLEKFDRFKHLKSEDEMKASED 60
TUNA    ----ADFDAVLKCWGPVEADYTTMGGLVLTRLFKEHPETQKLFPKFAGIAQAD-IAGNAA 55
        . :..:::  **: ** ****  .  ** **** *** : :*  :.  *.: ..

HUMAN   LKKHGATVLTALGGILKKKGHHEAEIKPLAQSHATKHKIPVKYLEFISEAIIQVLQSKHP 120
TUNA    ISAHGATVLKKLGELLKAKGSHAAILKPLANSHATKHKIPINNFKLISEVLVKVMHEKAG 115
        :. ******. ** :** ** * * :****;**********:: :::**.:::*::.*

HUMAN   GDFGADAQGAMNKALELFRKDMASNYKELGFQG 153
TUNA    --LDAGGQTALRNVMGIIIADLEANYKELGFSG 146
        ..:.".." ":.:.: ::  ": :""""".
```

b)

```
d1dwta_   GLSDGEWQQVLNVWGKVEADIAGHGQEVLIRLFTGHPETLEKFDKFKHLKTEAEMKASED 60
d1emy__   GLSDGEWELVLKTWGKVEADIPGHGETVFVRLFTGHPETLEKFDKFKHLKTEGEMKASED 60
d1mwca_   GLSDGEWQLVLNVWGKVEADVAGHGQEVLIRLFKGHPETLEKFDKFKHLKSEDEMKASED 60
d2mm1__   GLSDGEWQLVLNVWGKVEADIPGHGQEVLIRLFKGHPETLEKFDRFKHLKSEDEMKASED 60
d1a6m__   VLSEGEWQLVLHVWAKVEADVAGHGQDILIRLFKSHPETLEKFDRFKHLKTEAEMKASED 60
d1mbs__   GLSDGEWHLVLNVWGKVETDLAGHGQEVLIRLFKSHPETLEKFDKFKHLKSEDDMRRSED 60
d1lht__   GLSDDEWNHVLGIWAKVEPDLSAHGQEVIIRLFQLHPETQERFAKFKNLTTIDALKSSEE 60
d1myt__   ----ADFDAVLKCWGPVEADYTTMGGLVLTRLFKEHPETQKLFPKFAGIAQAD-IAGNAA 55
d1mba__   SLSAAEADLAGKSWAPVFANKNANGLDFLVALFEKFPDSANFFADFKGK-SVADIKASPK 59
          :  : ..   *. * .:   * .:  **  .*:: *  *         :  .

d1dwta_   LKKHGTVVLTALGGILKKKGHHEAELKPLAQSHATKHKIPIKYLEFISDAIIHVLHSKHP 120
d1emy__   LKKQGVTVLTALGGILKKKGHHEAEIQPLAQSHATKHKIPIKYLEFISDAIIHVLQSKHP 120
d1mwca_   LKKHGNTVLTALGGILKKKGHHEAELTPLAQSHATKHKIPVKYLEFISEAIIQVLQSKHP 120
d2mm1__   LKKHGATVLTALGGILKKKGHHEAEIKPLAQSHATKHKIPVKYLEFISEAIIQVLQSKHP 120
d1a6m__   LKKHGVTVLTALGAILKKKGHHEAELKPLAQSHATKHKIPIKYLEFISEAIIHVLHSRHP 120
d1mbs__   LRKHGNTVLTALGGILKKKGHHEAELKPLAQSHATKHKIPIKYLEFISEAIIHVLHSKHP 120
d1lht__   VKKHGTTVLTALGRILKQKNNHEQELKPLAESHATKHKIPVKYLEFICEIIVKVIAEKHP 120
d1myt__   ISAHGATVLKKLGELLKAKGSHAAILKPLANSHATKHKIPINNFKLISEVLVKVMHEK-- 113
          :    .  ::. ". ::*        ":: .   :  : :   ::        :

d1dwta_   GDFGADAQGAMTKALELFRNDIAAKYKELGFQ- 152
d1emy__   AEFGADAQGAMKKALELFRNDIAAKYKELGFQG 153
d1mwca_   GDFGADAQGAMSKALELFRNDMAAKYKELGFQG 153
d2mm1__   GDFGADAQGAMNKALELFRKDMASNYKELGFQG 153
d1a6m__   GDFGADAQGAMNKALELFRKDIAAKYKELGY-- 151
d1mbs__   AEFGADAQAAMKKALELFRNDIAAKYKELGFHG 153
d1lht__   SDFGADSQAAMKKALELFRNDMASKYKEFGFQG 153
d1myt__   AGLDAGGQTALRNVMGIIIADLEANYKELGFSG 146
          .    "..:  "  :  :  ::   ::  .  ..
```

Plate 1

(a) A pairwise alignment of two myoglobin sequences. (b) A multiple sequence alignment of eight myoglobin sequences. (See Chapter 4, Section 4.1). Colour coding of the residues is as follows: hydrophobic (red), positively charged (pink), negatively charged (blue), other (black).

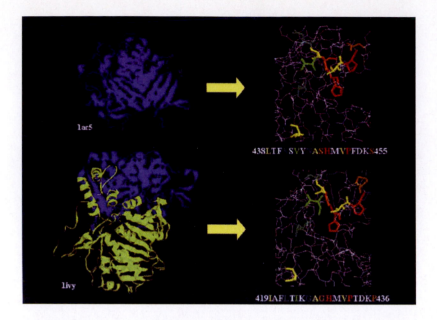

Three-dimensional arrangement of active site residues matching a shared PROSITE pattern which is diagnostic for serine histidine carboxypeptidases. (See Chapter 5, Section 5.3.3)

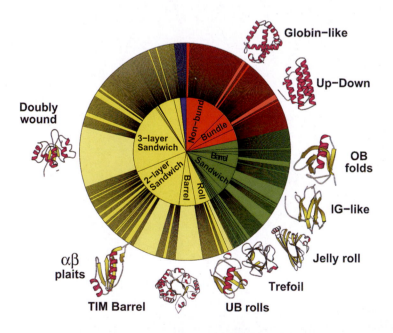

CATHerine wheel showing the population of different levels in the CATH classification hierarchy. The inner circle is divided into architectures, whilst the outer circle is divided into fold groups. The angle subtended by a given segment indicates the population of that particular architecture or fold group. It can be seen that the doubly wound fold and the TIM barrel fold are the most highly populated fold groups in the classification. (See Chapter 7, Section 7.4.4)

that two proteins with a high percentage identity are more closely related than two with a low percentage identity.

To model substitutions more realistically, we would like a matrix that scored the similarity between amino acids in a more sophisticated way, one that recognizes so-called *conservative substitutions*, that is, substitutions that conserve the physical and chemical properties of the amino acids and limit disruption.

4.3.2   Scoring amino acid similarity

The identity matrix for scoring amino acid similarity is clearly naïve. A slightly more realistic one, which is still sometimes used, is *genetic code matrix* (GCM), which contains values based on so-called genetic code scoring. Genetic code scoring considers the minimum number of DNA/RNA base changes (0, 1, 2 or 3) that would be required to interconvert the codons of one amino acid to those of the other amino acid. It is known that codons which are separated by only a single base change, often encode amino acids which are chemically similar, at least according to some criteria. This scoring scheme is also interesting because it tries to relate the actual process of mutation (which of course happens to the nucleic acids in the gene itself) to the observed amino acid substitutions in the encoded protein. Despite these nice features, however, the GCM has not proven to offer a good measure of amino acid similarity, perhaps because although mutation happens at the genetic level, natural selection (at least in coding sequences) tends to occur at the protein level and beyond.

The breakthrough in measuring amino acid similarity came when in the late 1960s and early 1970s, Margaret Dayhoff and co-workers began examining alignments of highly similar sequences and started counting how many times each amino acid was replaced by every other. The result of this type of analysis is a matrix of mutation frequencies. The numbers in matrices like this are described as *empirical* or *empirically derived* because they are based on observed measurements or frequencies. In contrast, numbers based on theoretical reasoning about what we might expect to occur, such as in the GCM and identity matrix, are described as *a priori*.

The mathematical model Dayhoff *et al.* developed for protein evolution resulted in a set of widely used substitution matrices, which are commonly used even today. These are known as Dayhoff, MDM (Mutation Data Matrix), or PAM (Percent Accepted Mutation) matrices. Because they are based on observed substitutions, these matrices, and ones that followed in their footsteps, are called *substitution matrices* (*Figure 4.1*).

4.3.3   Elements of a substitution matrix

To understand what the numbers in these matrices mean, and how they relate to amino acid similarity, we must first acquaint ourselves with so-called *logarithm of odds* (or *log odds*) ratios.

4.3.3.1   Log odds ratio

An *odds ratio* (also known as a likelihood ratio or a relative probability) is the probability of an event happening under one hypothesis divided by the probability of the same event happening under an alternative hypothesis. For instance, suppose we throw a dice and hope for a six. If the dice is fair, the probability of getting a six is $1/6 = 0.167$. If the dice is loaded, the probability of getting a six might be, say, 0.5. The odds ratio for getting a six under the loaded versus fair hypotheses can be expressed as:

$$R(\text{get a six}) = \frac{P(\text{get a six} \mid \text{die is loaded})}{P(\text{get a six} \mid \text{die is fair})} = \frac{0.5}{0.167} = 3,$$

	A	R	N	D	C	Q	E	G	H	I	L	K	M	F	P	S	T	W	Y	V	B	Z	X	*
A	4	-1	-2	-2	0	-1	-1	0	-2	-1	-1	-1	-1	-2	-1	1	0	-3	-2	0	-2	-1	0	-4
R	-1	5	0	-2	-3	1	0	-2	0	-3	-2	2	-1	-3	-2	-1	-1	-3	-2	-3	-1	0	-1	-4
N	-2	0	6	1	-3	0	0	0	1	-3	-3	0	-2	-3	-2	1	0	-4	-2	-3	3	0	-1	-4
D	-2	-2	1	6	-3	0	2	-1	-1	-3	-4	-1	-3	-3	-1	0	-1	-4	-3	-3	4	1	-1	-4
C	0	-3	-3	-3	9	-3	-4	-3	-3	-1	-1	-3	-1	-2	-3	-1	-1	-2	-2	-1	-3	-3	-2	-4
Q	-1	1	0	0	-3	5	2	-2	0	-3	-2	1	0	-3	-1	0	-1	-2	-1	-2	0	3	-1	-4
E	-1	0	0	2	-4	2	5	-2	0	-3	-3	1	-2	-3	-1	0	-1	-3	-2	-2	1	4	-1	-4
G	0	-2	0	-1	-3	-2	-2	6	-2	-4	-4	-2	-3	-3	-2	0	-2	-2	-3	-3	-1	-2	-1	-4
H	-2	0	1	-1	-3	0	0	-2	8	-3	-3	-1	-2	-1	-2	-1	-2	-2	2	-3	0	0	-1	-4
I	-1	-3	-3	-3	-1	-3	-3	-4	-3	4	2	-3	1	0	-3	-2	-1	-3	-1	3	-3	-3	-1	-4
L	-1	-2	-3	-4	-1	-2	-3	-4	-3	2	4	-2	2	0	-3	-2	-1	-2	-1	1	-4	-3	-1	-4
K	-1	2	0	-1	-3	1	1	-2	-1	-3	-2	5	-1	-3	-1	0	-1	-3	-2	-2	0	1	-1	-4
M	-1	-1	-2	-3	-1	0	-2	-3	-2	1	2	-1	5	0	-2	-1	-1	-1	-1	1	-3	-1	-1	-4
F	-2	-3	-3	-3	-2	-3	-3	-3	-1	0	0	-3	0	6	-4	-2	-2	1	3	-1	-3	-3	-1	-4
P	-1	-2	-2	-1	-3	-1	-1	-2	-2	-3	-3	-1	-2	-4	7	-1	-1	-4	-3	-2	-2	-1	-2	-4
S	1	-1	1	0	-1	0	0	0	-1	-2	-2	0	-1	-2	-1	4	1	-3	-2	-2	0	0	0	-4
T	0	-1	0	-1	-1	-1	-1	-2	-2	-1	-1	-1	-1	-2	-1	1	5	-2	-2	0	-1	-1	0	-4
W	-3	-3	-4	-4	-2	-2	-3	-2	-2	-3	-2	-3	-1	1	-4	-3	-2	11	2	-3	-4	-3	-2	-4
Y	-2	-2	-2	-3	-2	-1	-2	-3	2	-1	-1	-2	-1	3	-3	-2	-2	2	7	-1	-3	-2	-1	-4
V	0	-3	-3	-3	-1	-2	-2	-3	-3	3	1	-2	1	-1	-2	-2	0	-3	-1	4	-3	-2	-1	-4
B	-2	-1	3	4	-3	0	1	-1	0	-3	-4	0	-3	-3	-2	0	-1	-4	-3	-3	4	1	-1	-4
Z	-1	0	0	1	-3	3	4	-2	0	-3	-3	1	-1	-3	-1	0	-1	-3	-2	-2	1	4	-1	-4
X	0	-1	-1	-1	-2	-1	-1	-1	-1	-1	-1	-1	-1	-1	-2	0	0	-2	-1	-1	-1	-1	-1	-4
*	-4	-4	-4	-4	-4	-4	-4	-4	-4	-4	-4	-4	-4	-4	-4	-4	-4	-4	-4	-4	-4	-4	-4	1

Figure 4.1

Henikoff and Henikoff's BLOSUM62 substitution matrix

where $P(\text{get a six} \mid \text{die is loaded})$ is shorthand for 'the probability of "get a six" given that "die is loaded"'. The odds ratio tells us that the probability of a six is three times greater from the loaded die than from the fair die. Had the probabilities been the same, R would have been 1; if they were swapped, R would be $1/3 = 0.33$.

Odds ratios are bothersome to work with: they have an awkward range (0 to infinity, with 1 in the middle) and combining them requires multiplying, which is computationally more error prone than summing (see *Box 4.1*). Instead, R is often expressed as a logarithm, i.e., the *log odds ratio*. LogR has a more convenient range: it is positive if the greater probability is on top, negative (by an equivalent amount) if it is on the bottom, and 0 if both hypotheses explain the observation with equal probability. For example, in the example above, $\log R = \log(3/1) = 0.48$. If the probabilities were swapped then $\log R = \log(1/3) = -0.48$.

4.3.3.2 Log odds ratio of substitution

When aligning a position in two sequences, we want to reward a substitution that is common and penalize one that is rare. To do this, we first consider a case where two residues, let us call them *a* and *b*, happen to be aligned to each other, either correctly or incorrectly. We then use a log odds ratio that compares how likely these residues coincide in this way under two different hypotheses. The first hypothesis states that *a* and *b* evolved from each other or from a common ancestor. For example, the original ancestral sequence may have had a leucine at that position which, when the sequences diverged (through gene duplication, speciation or whatever), mutated to a valine in one sequence and an isoleucine in the other. We call this the *match* hypothesis and if it is correct then the two residues should be aligned. The second hypothesis states that the two residues have come together in an alignment by chance. We call this the *random* hypothesis, and if it correct then the two residues should not

---

Box 4.1 Combining probabilities using log odds

Section 4.3 mentioned that log odds ratios are easier to combine than raw relative probabilities, but why should we want to combine probabilities in the first place? Suppose you have a dice and a coin. The probability of getting a six on the dice and tails on the coin is given by the probability of getting a six multiplied by the probability of getting tails:

$$P(\text{get a six, get tails}) = P(\text{get a six})P(\text{get tails}) = \frac{1}{6} \times \frac{1}{2} = 0.083$$

More generally, the probability of several independent events occurring in a row is equal to the product of the probabilities of the each individual event. Suppose we align two sequences with three residues each and want to know the log odds ratio for the alignment:

AFE

GKD

The probability of these residues coming together under the match hypothesis is

$$P(alignment\,|\,match) = P(A,G\,|\,match)P(F,K\,|\,match)P(E,D\,|\,match)$$

The probability of the alignment under the random hypothesis is

$$P(alignment\,|\,random) = P(A,G\,|\,random)P(F,K\,|\,random)P(E,D\,|\,random)$$

which means the relative probability of the alignment being correct is

$$R(alignment) = \frac{P(A,G\,|\,match)P(F,K\,|\,match)P(E,D\,|\,match)}{P(A,G\,|\,random)P(F,K\,|\,random)P(E,D\,|\,random)}$$

$$R(alignment) = R(A,G)R(F,K)R(E,D)$$

Computers can add faster than they can multiply. Also, they tend to give rounding errors with numbers that are not integers. Therefore, it is preferable to sum rounded log odds rather than multiply raw probabilities. Since,

$$\log xy = \log x + \log y,$$

$$\log R(alignment) = \log R(A,G) + \log R(F,K) + \log R(E,D)$$

If each $\log R(i,j)$ has first been scaled and rounded to the nearest integer (e.g., log odds of 3.59 is scaled to 35.9 and rounded to 36), the calculation is even faster.

---

be aligned. For any particular pair of residues a and b, both hypotheses will be possible, but often one will be more likely than the other. We measure this balance of probabilities with the log odds ratio for the match versus the random model:

$$\log R(a,b) = \log\left(\frac{P(a,b\,|\,match)}{P(a,b\,|\,random)}\right)$$

When *a* and *b* being aligned is more likely under the match hypothesis than the random hypothesis, the log odds ratio is positive. When the random hypothesis is more probable, the ratio is negative. When both are equally probable, the ratio is zero. A final point is that the log odds ratios in a substitution are typically scaled and rounded to the nearest integer (for computational convenience), and for this reason the values of elements are only indirectly interpretable as log *R*.

4.3.3.3 Diagonal elements

The log odds ratios of the diagonal elements of a substitution matrix, i.e., those that pitch an amino acid against itself, have a slightly different meaning from above, although they can be calculated the same way. Rather than measuring how likely a residue is to mutate to itself, they indicate how likely residue is to resist mutation at all.

4.3.4 Constructing a Dayhoff matrix

Having explained the basic form of substitution matrices, we will now look in more detail at how the classic Dayhoff mutation data matrix is calculated.

4.3.4.1 Construction of the raw PAM matrix

The basic unit of molecular evolution expressed in an MDM is the *Point Accepted Mutation* (PAM). One PAM represents the amount of evolution that will change, on average, 1% of the residues in a protein. For a given number of PAMs, that is, for a given amount of evolutionary time, we would like to know how likely one amino acid is to mutate to another.

The raw PAM substitution matrix is created by considering the possible mutational events that could have occurred between two closely related sequences. Ideally we would like to compare every present day sequence with its own immediate predecessor and thus accurately map the evolutionary history of each sequence position. Although this is impossible, two main courses of action may be taken to approximate this information. Dayhoff *et al.* used the so-called *common ancestor* method. They took closely related pairs of present day sequences and inferred the sequence of the common ancestor. Given only a pair of present day sequences, you cannot infer a common ancestor exactly. But you can generate a complete phylogenetic tree and from this infer the *most probable* ancestor at each tree node. This is harder than it sounds because whenever you infer a one common ancestor you must also consider how your inference affects the overall topology of the tree. An iterative method is typically used for this.

An alternative to the common ancestor method is to calculate the alignment distance between all pairs of sequences from a set of related present-day sequences. You can then use this all-by-all set of distances, typically stored in a *distance matrix*, to estimate a possible phylogenetic tree. Though construction of the distance matrix is a trivial exercise, the generation of an optimal phylogenetic tree from these data again requires an exhaustive iterative analysis such that the total number of mutations required to produce the present day set of sequences is minimized. Though both of the above methods have advantages and disadvantages, matrix methods are now most widely used.

No matter which method is finally used to infer the phylogenetic tree, construction of the PAM matrix is the same. The raw matrix is generated by taking pairs of sequences, either a present-day sequence and its inferred ancestor or two present-day sequences, and tallying the amino acid exchanges that have apparently occurred. Given the following alignment:

MACDEFLVSD

MAGDEALVSD

we can count four PAMs (C→G, G→C, F→A and A→F). Clearly the raw PAM matrix must be symmetric because we cannot know whether C mutated to G or G mutated to C and so on. There is no harm in this. After all, we want to establish the extent of similarity, and 'similarity' is generally thought of as symmetric concept.

Gaps, which represent insertions or deletions in the alignment, tell us nothing about amino acid similarity and so these are modeled separately, outside the substitution matrix, and discussed in later chapters.

4.3.4.2 Calculation of relative mutabilities

As described earlier (section 4.3.3) the diagonal elements of a substitution matrix describe how much a given amino acid resists change. We do not measure this resistance directly. Rather we first calculate how much the amino acid is susceptible to change, that is, its so-called *relative mutability*, and then invert it. It is easy to calculate the relative mutability of an amino acid from an alignment. Simply count the number of times the amino acid is observed to change, and then divide this by the number of times the amino acids occur in the aligned sequences. From the alignment shown earlier, A (alanine) is seen to change once, but occurs three times. The relative mutability of A from this alignment alone is therefore ⅓. To produce an overall mutability for A, the mutabilities from different alignments must be combined. However, different pairs of sequences, and therefore different alignments, typically represent the passage of different amounts of evolutionary time. In other words, each alignment gives us information about substitutions but usually at a different PAM from our other alignments. So when combining observed mutabilities from different alignments, we must first normalize each mutability by the PAM of the alignment it came from. This normalization is done so that the overall mutability of A will be the amount you expect A to mutate during one PAM.

4.3.4.3 The mutation probability matrix

The middle step in building a Dayhoff-type matrix is generating the *mutation probability matrix*. Elements of this matrix give the probability that a residue in column *j* will mutate to the residue in row *i* in a specified unit of evolutionary time. *Box 4.2* describes how these probabilities are calculated. In the simplest case, the unit of evolutionary time is 1 PAM and the matrix is called the 1 PAM (or PAM1) matrix.

Matrices representing larger evolutionary distances may be derived from the PAM1 matrix by matrix multiplication. This means squaring all values in the PAM1 matrix gives a PAM2 matrix, cubing them gives a PAM3 matrix, and so on. Often we wish to align sequences that are only distantly related. In this case, it is sensible to use matrices with PAM numbers such as 120 or even 250 (known as PAM120 and PAM250). These matrices are more sensitive than matrices with lower PAM numbers, because they give less priority to identical amino acid matches and more to conservative substitutions.

4.3.4.4 Calculating the log-odds matrix

More useful than the mutation probability matrix in the alignment of protein sequences is the *relatedness odds matrix*. This is the matrix of likelihood ratios of the type described in section 4.3.3, where the value of each element is relative probability of *i* and *j* being aligned under different hypotheses, i.e.,

$$R(i,j) = \frac{P(i,j \mid match)}{P(i,j \mid random)}$$

where, if f_i and f_j are the fractional frequencies of *i* and *j* respectively, then

---

Box 4.2   Calculating the mutation probability matrix

A diagonal element of this matrix represents the probability of residue i remaining unchanged. We calculate this as

$$m(i,i) = 1 - \lambda u_i$$

Here, $m(i,i)$ is the matrix element on the diagonal (i.e., row i, column i), u_i is the overall relative mutability of residue i, and λ is a normalization term that is constant for a given matrix.

Non-diagonal elements, those that compare one amino acid with another, are given by

$$m(i,j) = \frac{\lambda u_j A(i,j)}{\sum\limits_i A(i,j)}$$

where $A(i,j)$ is the observed number of $i \leftrightarrow j$ changes as read from the raw PAM matrix. Note this equation treats i and j differently. This is because the non-diagonal value is the probability that j mutates to i, given that j mutates in the first place.

To understand what λ does in the first equation, we must remember that a given matrix describes what we expect to happen during a given amount of evolution. Let the amount of evolution represented by the matrix be p, measured in PAMs. The probability of a mutation at a single site is $p/100$ (for example, this is 1% when p is 1 PAM). This means the probability of no mutations at a given site must be $1 - p/100$. Of course, a mutation can spring from any one of the 20 amino acids. So to ensure the matrix represents no more evolution than p would suggest, this $1 - p/100$ 'resistance' to mutation must be shared among the 20 amino acids. More formally, the diagonal elements must be calculated so that the following relationship holds:

$$\sum_i f_i m(i,i) = 1 - \frac{p}{100}.$$

where f_i is the relative frequency of amino acid i. (Note that this relationship breaks down when p is much greater than 5. This is because we start to observe multiple mutations at the same site: for instance, you would not expect two sequences at a PAM distance of 100 to be 0% identical!) Combining our two equations, we find that

$$\lambda = \frac{p}{100 \times \sum\limits_i f_i u_i}.$$

The evolutionary distance p usually has value 1 so that the basic mutation probability matrix corresponds to an evolutionary distance of 1 PAM. Matrices representing the passage of more evolutionary time are calculated from the 1 PAM matrix rather than by repeating the steps above.

$$P(i,j \mid random) = f_i \times f_j$$

To avoid slow computation (see *Box 4.1*, floating point multiplications), the relatedness odds matrix is usually converted to the log odds-matrix (or *Mutation Data Matrix*) thus:

$$M(i,j) = \text{nint}(10\log_{10}R(i,j))$$

where $R(i, j)$ is an element of relatedness odds matrix and the function nint(x) rounds x to the nearest integer.

This final MDM matrix is the classic 'Dayhoff matrix' as used in many different sequence comparison and analysis programs over the past 20 years or more.

4.3.5 BLOSUM matrices

Although the original Dayhoff matrices are still sometimes used, particularly in phylogenetic analyses, many groups have tried to develop better empirical scoring matrices. One criticism of the original Dayhoff matrices was that they were based on only a small number of observed substitutions (~1500). The PET matrix of Jones *et al.*, attempted to stick closely to the approach employed by Dayhoff *et al.* but to consider a much larger dataset (10 to 20 times larger than the original). Soon after the PET matrix, came another family of matrices from Henikoff and Henikoff. These are known as the *BLOSUM matrices*, and have become the standard comparison matrix in modern sequence comparison applications.

Unlike the Dayhoff matrices, BLOSUM matrices are not based on an explicit evolutionary model. Instead, every possible substitution is counted within conserved blocks of aligned protein sequences taken from many different protein families. No attempt is made to count substitutions along the branches of an evolutionary tree. Without considering an evolutionary tree there can be biases through over and under counting some substitutions.

Empirical studies have shown that the BLOSUM matrices provide a more useful measure of amino acid similarity for the purposes of comparing amino acid sequences or databank searching (see later). It is unclear exactly why this should be, but it may be because the observed substitutions used in constructing the matrices are restricted to those regions that are most highly conserved across each protein family. These regions are not only more likely to be accurately aligned, but also are likely to be more representative of the functionally important regions of proteins.

Like Dayhoff matrices, BLOSUM matrices can be derived at different evolutionary distances. The measure of evolutionary distance for BLOSUM matrices is simply the percentage identity of the aligned blocks in which the substitutions are counted. For example, the widely used BLOSUM62 matrix (*Figure 4.1*) is based on aligned blocks of proteins where on average 62% of the amino acids are identical. This matrix is roughly equivalent to a PAM150 matrix. Conversely, the well-known PAM250 matrix is roughly equivalent to a BLOSUM50 matrix. Because BLOSUM matrices are not based on a definition of unit evolutionary time, unlike Dayhoff matrices, it is not possible to calculate BLOSUM matrices at different evolutionary distances simply by extrapolation.

4.4 Scoring residue conservation

In a multiple sequence alignment, a *variable* position describes one at which diverse substitutions are common, a *conserved* position is one that appears to resist change and a *strictly conserved* or *invariant* position is one that remains constant. But often words are not enough and we would like to put a number on how conserved a position is. This is the job of a *conservation score*.

Assuming the neutralist model of molecular evolution (section 4.2), if the degree of functional constraint dictates how conserved a position is, then that means you can infer the importance of a position by its conservation. Identifying conserved regions of a protein can thus be tremendously useful. For instance, residues involved in an active-site or a structural core can sometimes be identified with little prior knowledge of a protein's structure or function.

4.4.1 Exercises for a conservation score

A conservation score maps a position in a multiple alignment to a number. That number represents how conserved the position is. It would be nice if there were a test set of multiple alignment positions and 'correct' numbers that you could use to judge how good a particular score is. Sadly, no such test set exists. However, you can see how a score orders a set of multiple alignment positions and compare the result with biochemical intuition.

Figure 4.2 shows some example columns from a multiple sequence alignment of functionally equivalent homologous sequences. We can order these intuitively.

First consider columns (a) to (f). Column (a) contains only D and is therefore the most obviously conserved. Column (b) also contains E, so (b) is more variable (i.e., less conserved) than (a). Column (c) contains D and E but is less dominated by any one than (b), so (c) is more variable than (b). Column (d) contains nine D and one F; it is clearly more variable than column (a), but is it more variable than column (b)? Aspartate and glutamate are both small and polar with similar properties. Tryptophan is large and non-polar, and altogether quite different. A position that tolerates only minor changes (D↔E) is likely to be more constrained by function than a position that tolerates drastic changes (D↔F). So we conclude (d) is more variable than (c). Column (e) implies both conservative substitutions (D↔E) and non-conservative ones (D↔F, E↔F). Column (e) is thus the least conserved so far. Column (f) contains the same amino acid types as (e). However, because it is less skewed towards an abundance of D and E, (f) is more variable.

Second, consider columns (g) and (h). These are equivalent in terms of the number and frequency of their amino acids. However, because (g) contains only branch-chained amino acids whereas (h) encompasses a broader mix of stereochemical characteristics, (h) is more variable. Column (i) is the most variable column encountered so far, as judged by biochemistry or amino acid frequency.

Last, consider columns (j) and (k). These illustrate the importance of gaps. Column (j) comes from an alignment of four sequences, each of which has a leucine at that position. Column (k) also contains four leucines but, because it comes from an alignment of 10 sequences, it also contains six gaps. Thus, although the conservation of (j) is unblemished,

		Alignment Positions										
		(a)	(b)	(c)	(d)	(e)	(f)	(g)	(h)	(i)	(j)	(k)
Sequences	1	D	D	D	D	D	D	I	P	D	L	L
	2	D	D	D	D	D	D	I	P	V	L	L
	3	D	D	D	D	D	D	I	P	Y	L	L
	4	D	D	D	D	D	D	I	P	A	L	L
	5	D	D	D	D	D	D	L	W	T		-
	6	D	D	E	D	E	E	L	W	K		-
	7	D	D	E	D	E	E	L	W	P		-
	8	D	D	E	D	E	E	L	W	C		-
	9	D	D	E	D	E	F	V	S	R		-
	10	D	E	E	F	F	F	V	S	H		-

Figure 4.2

Example columns from a multiple alignment

the evidence strongly suggests column (k) is not conserved. After all, six other functionally equivalent homologs seem to manage without a residue at that position. Of course, the alignment of (j) could be the same as that of (k) but with six sequences missing. This would be an example of a lack of data producing opposite conclusions about the same site, and highlights the danger of having too small an alignment.

In summary, a score that accords with biochemical intuition should reproduce the order, from most to least conserved, (a) > (b) > (c) > (d) > (e) > (f) for columns (a) to (f). Then, (g) > (h) and (j) > (k).

4.4.2   Guidelines for making a conservation score

Having looked at some example problems, we can now say what properties a good conservation score should have:

1. Amino acid frequency. The score should take account of the relative frequencies of amino acids in a column. For instance, given the columns from *Figure 4.2*, it should reproduce the ranking (a) > (b) > (c) > (e) > (f).
2. Stereochemical properties. The score should recognize conservative replacements and that some substitutions incur more chemical and physical change than others. For instance, it should score column (g) as more conserved than column (h).
3. Gaps. A lot of gaps suggests a position can be deleted without significant loss of protein function. The score should therefore penalize such positions and should rank column (j) as more conserved than column (k).
4. Sequence weighting. Sometimes a position appears conserved among a number of sequences not because of functional constraint but because those sequences have not had sufficient evolutionary time to diverge. A typical alignment often includes some sequences that are very closely related to each other. These clusters of highly similar sequences may reflect bias in the sequence databases or result from nature's irregular sampling of the space of acceptable mutations. Either way, such clusters can monopolize alignments, masking important information about allowed variability from more sparsely represented sequences. A good conservation score should find some way to normalize against redundancy and bias in the alignment without loss of evolutionary information.

4.5   Methods for scoring conservation

Some conservation scores are more easily thought of as describing the lack of conservation than its abundance. To keep the equations simple we therefore use the following notation: scores whose values increase with increasing conservation are denoted C_{name}; those that do the opposite, i.e., increase with increasing variability, are denoted V_{name}.

4.5.1   Simple scores

Fully conserved positions contain only one amino acid type whereas less conserved positions contain more. So a simple approach to measuring variability is to count the number of different amino acids present at a position and divide by the maximum number there could have been, i.e.,

$$V_{numaa} = \frac{k}{K}$$

where k is the number of amino acid types present and K is the alphabet size (e.g., $K = 20$). V_{numaa} therefore ranges from $1/K$, denoting high conservation, to 1, denoting high variability.

How does this score perform on the columns in *Figure 4.2?* It correctly identifies (a) is more conserved than (b), but fails to distinguish (b) from (c) because it incorporates no sense of how dominated a position is by one residue (criterion 1). Neither is it able to distinguish (b) from (d) owing to its ignorance of stereochemistry (criterion 2). It also has no mechanism for handling gaps. If gaps are ignored then column (k) is fully conserved. If gaps are considered another amino acid type (such that $K = 21$) then (k) scores the same as (b). So V_{numaa} is too simplistic for most applications.

A similar score was developed by Wu & Kabat in 1970. The so-called Kabat index, which became the first widely accepted method for scoring conservation, was defined as:

$$V_{Kabat} = \frac{kN}{n_1},$$

where N is the number of sequences in the alignment (i.e., the number of rows), n_1 is the frequency of the most frequently occurring amino acid and k is as before. V_{Kabat} correctly reproduces the ranks (a) > (b) > (c) > (e) but fails to distinguish (e) from (f). This is because it cares only about the frequency of the most commonly occurring amino acid and ignores the frequencies of the rest. V_{Kabat} has other problems; for one, its output is eccentric: A strictly conserved column, such as column (a), always scores 1. A column that is strictly conserved but for one aberrant amino acid always scores some number greater than 2, regardless of how many sequences are in the alignment. This discontinuity is biologically meaningless. The score also fails to consider gaps or stereochemistry and so has only an arguable advantage over V_{numaa} in that it better reflects dominance by a particular amino acid type.

The most commonly used simple score is Shannon's entropy. Like the scores above, $V_{Shannon}$ (at least in its most basic form) treats a column of residues as objects drawn from a uniformly diverse alphabet. Unlike the scores above, it provides an elegant and mathematically meaningful way of describing the diversity of these objects. Shannon's entropy is described further in *Box 4.3.* But the alphabet of amino acids is not uniformly diverse. Some pairs of amino acids are similar, whereas others pairs are quite different, and we would prefer a score that recognizes this.

4.5.2 Stereochemical property scores

In 1986, Taylor classified amino acids according to their *stereochemical properties* and their patterns of conservation in the Dayhoff mutation data matrix. He embodied them in a Venn diagram (*Figure 4.3*) in which each overlapping set represents a distinct physical or chemical property. He then devised a conservation score based on this diagram. For the amino acids present at a given position, find the smallest set or subset that contains them all. The degree of variability equals the number of amino acids contained in that minimal set. For instance, column (c) in *Figure 4.2* contains D and E. The smallest set in *Figure 4.3* containing both is the 'negative' set. This contains two amino acids, so the variability is 2. Now consider column (d), which contains D and F. A simplified version of Taylor's score, $V_{SimpleTaylor}$, might conclude 'polar' to be the minimal set such that $V_{SimpleTaylor} = 12$. In fact, Taylor's original score was more complex because it allowed a hand-edited list of set theoretic descriptions such as 'hydrophobic or negative' or 'aliphatic or large and non-polar'. We consider only the simplified version here.

The following year, Zvelebil and co-workers developed a score based on Taylor's. They reduced his Venn diagram to the less cumbersome truth-table of amino acid properties shown in *Figure 4.4*, and measured conservation as:

$$C_{Zvelebil} = \frac{n_{const}}{n_{max}},$$

Box 4.3 Shannon entropy

The *information theoretic entropy, information entropy* or *Shannon entropy*, is a popular measure of diversity that is often used to score residue conservation.

Suppose you have 10 colored balls, of which five are red, two green and three yellow. The number of distinct ways you can order them in a line is $10!/(5!2!3!) = 2520$. More generally, given N objects that fall into K types, the number of ways they can be permuted is given by the multinomial coefficient:

$$W = \frac{N!}{n_1! \times n_2! \times \ldots \times n_K!}$$

where n_i is the number of objects of type i, and $n!$ is the factorial of n (e.g., for $n = 5$, $n! = 5 \times 4 \times 3 \times 2 \times 1$). The number of permutations is smallest when all objects are of the same type ($W_{min} = 1$) and largest when every object is different ($W_{max} = N!$). If the number of types is constant, then W is highest when each type contributes an equal number of objects. For instance, with five colors and 10 balls, W_{max} is reached when there are two balls of each color. Imagine the balls are residues and the colors amino acids, and the application to scoring conservation is clear. Factorials produce big numbers and as the number of objects increases, W soon becomes awkward to calculate directly. Fortunately, we can find $\ln W$ with relative ease. Sterling's approximation states that $\ln N! \sim N \ln N - N$. So substituting this into the multinomial coefficient above gives

$$\ln W = -N \sum_{i=1}^{K} p_i \ln p_i,$$

where $p_i = n_i/N$, the fractional frequency of the ith type. Divide both sides by $N \ln 2$ and you get Shannon's entropy:

$$S = -\sum_{i=1}^{K} p_i \log_2 p_i.$$

Now if all objects are of the same type then $S = S_{min} = 0$, whereas if all types are equally represented $S = S_{max} = -\log_2 1/K = \log_2 K$. Thus, S is a convenient and intuitive measure of diversity of types among a set of symbols. When applied to amino acids (for instance, when $K = 20$) its small values indicate conservation and its high values indicate variability.

The original use of Shannon's entropy was in *information theory*, a branch of electrical engineering that examines communication and the handling of information. In that context, S is used to measure the selective information content of a communicated signal. The base 2 logarithm means that S has units of binary information digits, or 'bits'. Although published entropy-based conservation scores dwell on this fact, it should be noted that in the context of scoring conservation the units of S, and hence the base of the logarithm, are usually irrelevant. After all, the logs of one base are proportional to those of another (remember $\log_a x = \log_b x / \log_b a$) and if you normalize (e.g., by dividing through by S_{max}) the units vanish anyway.

where n_{const} is the number of properties whose state (i.e., true or false) is constant for all amino acids in the column and n_{max} is the number of properties being compared. For column (c), $C_{Zvelebil} = 0.9$ since D and E share nine out of the 10 properties. For column (d), $C_{Zvelebil} = 0.5$.

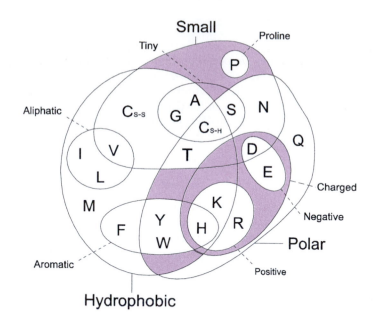

Figure 4.3

Taylor's Venn diagram of amino acid properties

ILVCAGMFYWHKREQDNSTPBZXΔ

1	●●●●●●●●●●●●●●○○○○○○●○○○●●	Hydrophobic
2	○○○○○○○○●●●●●●●●●●●●○●●●●●	Polar
3	○○●●●●○○○○○○○○○●●●●●●○○●●	Small
4	○○○○○○○○○○○○○○○○○○○●○○●●	Proline
5	○○○○●●○○○○○○○○○○○○●○○○○●●	Tiny
6	●●●○○○○○○○○○○○○○○○○○○○○●●	Aliphatic
7	○○○○○○○●●●●○○○○○○○○○○○○●●	Aromatic
8	○○○○○○○○○○○●●●○○○○○○○○○●●	Positive
9	○○○○○○○○○○○○○●○●○○○○○○○●●	Negative
10	○○○○○○○○○○○●●●●○●○○○○○○●●	Charged

Figure 4.4

Zvelebil's truth table of amino acid properties. The Δ character represents a 'gap' type.

$V_{SimpleTaylor}$ and $C_{Zvelebil}$ correctly order columns in terms of stereochemical diversity (e.g., (g) and (h)) but cannot identify when a column is dominated by one particular residue. For example, according to either score, (b) is no more conserved than (c), yet intuition would suggest otherwise. To their credit, both scores provide a mechanism for penalizing gaps. In $C_{Zvelebil}$ the gap type is set to be dissimilar to most other amino acids. Although $V_{SimpleTaylor}$ does not include gaps explicitly, a gap type could easily be incorporated as an outlier in the Venn diagram such that any position with a gap must score 21.

4.5.3   Mutation data scores

Substitution matrices provide a quantitative and reasonably objective assessment of likely amino acid substitutions. Through analogy, they can also be used to quantify amino acid conservation.

4.5.3.1   Sum of pairs scores

The most common type of mutation data score is the *sum of pairs (SP) score*. SP scores measure conservation by calculating the sum of all possible pairwise similarities between residues in an aligned column. They take the form

$$C_{SP} = \sum_{i=1}^{N-1} \sum_{j=i+1}^{N} M(a_i, a_j),$$

where N is the number of residues in the column (i.e., the number of sequences), a_i is the amino acid belonging to the ith sequence, and $M(a,b)$ is the similarity between amino acids a and b.

4.5.3.2   Performance

Sum of pairs scores measure the amount of stereochemical diversity in a column while accounting for the relative frequencies of the different amino acids. To see how they do this we can look at the calculation of C_{SP}. Taking column (b) from *Figure 4.2*,

$$C_{SP} = 36M(D,D) + 9M(D,E)$$
$$C_{SP} = 36M_{max} + 9M_{high}.$$

Out of 45 amino acid comparisons (remember: you can make $N(N–I)/2$ distinct pairs from N objects), 36 are maximally scoring and nine score highly but less than the maximum. Compare this with column (c):

$$C_{SP} = 10M(D,D) + 10M(E,E) + 25M(D,E)$$
$$C_{SP} = 20M_{max} + 25M_{high}.$$

Here a far lower proportion score the maximum. Compare these with column (e):

$$C_{SP} = 10M(D,D) + 6M(E,E) + 20M(D,E) + 5M(D,W) + 4M(E,W)$$
$$C_{SP} = 16M_{max} + 20M_{high} + 9M_{low}.$$

Now there are low scores as well as merely sub-maximal scores, which further drag down the final conservation value.

Applying the sum of pairs score to all columns in *Figure 4.2* reproduces the order suggested by biochemical intuition. In summary, although the SP score is *ad hoc* and so lacks the mathematical rigor of, say, the entropy score, it can provide more intuitive results. There are a number of other non-SP methods that use substitution matrices. For a description of these, see Further reading.

4.5.4   Sequence-weighted scores

Weighting sequences is a concern not just for conservation scoring but for any bioinformatics analysis that uses multiple sequence alignments. As a result, many weighting schemes have been devised. The simplest is described here, although see Further reading for more sophisticated methods.

4.5.4.1 A simple weighting scheme

The weight of a sequence is inversely related to its genetic distance from other sequences in the alignment. So for each sequence s_i we want to calculate a weight w_i, where $w_i = 0$ if s_i is like all the other and $w_i = 1$ if s_i is different from all the others. The simplest way is to let w_i equal the average distance of s_i from all other sequences:

$$w_i = \frac{1}{N-1} \sum_{\substack{j=1 \\ j\neq i}}^{N} d(s_i, s_j),$$

where $d(s, t)$ is the distance between sequences s and t as measured by percentage identity in the alignment or some smarter measure.

4.5.4.2 Incorporating sequence weights into conservation scores

After deciding on a weighting method, we must then find a way of incorporating our sequence weights into an existing conservation score. Most amenable to weighting are the sum of pairs scores because these include an explicit term comparing residues from two sequences, i.e.,

$$C_{SP} = \sum_{i=1}^{N-1} \sum_{j=i+1}^{N} w_i w_j M(a_i, a_j).$$

Now the effect of high scoring comparisons between very similar sequences will be diminished.

4.6 Insights and conclusions

In this chapter we have looked at a wide variety of schemes for measuring similarity and conservation between amino acids in aligned protein sequences. Measuring the degree of conservation of a residue in a protein can provide a great deal of information as to its structural or functional significance, and yet, as we have seen, there is no single best way to measure either amino acid similarity or sequence conservation. Ultimately the choice of scoring scheme or weighting scheme will come down to either the user's intuition, or simple trial and error. Hopefully, however, by being aware of the assumptions and limitations of the different schemes outlined earlier, you can make an informed choice depending on the requirements of the application in hand.

References and further reading

Altschul, S.F. (1991) Amino acid substitution matrices from an information theoretic perspective. *Journal of Molecular Biology* **219**: 555–665.

Dayhoff, M.O., Schwartz, R.M., and Orcutt, B.C. (1978) A model of evolutionary change in proteins. In: *Atlas of Protein Sequence and Structure* 5(3) M.O. Dayhoff (ed.), 345–352, National Biomedical Research Foundation, Washington.

Henikoff, S., and Henikoff, J.G. (1992) Amino acid substitution matrices from protein blocks. *Proceedings of the National Academy of Sciences USA* **89**: 10915–10919.

Valdar, W.S.J. (2002) Scoring residue conservation. *Proteins: Structure, Function, and Genetics* **48**: 227–241.

Function prediction from protein sequence

5

Sylvia B. Nagl

Concepts

- Most commonly, protein function is inferred from the known functions of homologous proteins. For homologous proteins with easily recognizable sequence similarity, this type of prediction is based on the 'similar sequence–similar structure–similar function' paradigm.
- Domains can be seen as 'units of evolution', and, therefore, both structural and functional similarity between proteins needs to be analyzed at the domain level.
- Sequence comparison is most sensitive at the protein level and the detection of distantly related sequences is easier in protein translation.
- To allow the identification of homologous domains with low sequence identity (<30%), pattern and profile methods enhance sequence information by a mutation pattern at a given position along the sequence that is derived from an alignment of homologous proteins.
- Sequence analysis techniques provide important tools for the prediction of biochemical function, but show clear limitations in predicting a protein's context-dependent functions.

5.1   Overview

Function prediction from sequence employs a combination of highly efficient sequencing techniques and sophisticated electronic searches against millions of gene and protein sequences. By providing access to biological databases and powerful tools for the prediction of the biochemical functions and cellular roles of gene products, bioinformatics can inform and focus research and enhance its efficiency. Therefore, experimental and computational methods are becoming closely integrated in biological and biomedical investigations.

This integration becomes possible when bioinformatics techniques are not only used at isolated points in the discovery process, but when theoretical predictions are directly tested in the laboratory and results are fed back for creating refinements in theoretical models. The refined model in turn becomes the starting point for further experiments, and new results then lead to further model refinement. Such an iterative approach not only supports efficiency, but also enhances the understanding of the biological system under study.

5.2   The similar sequence–similar structure–similar function paradigm

Bioinformatics seeks to elucidate the relationships between biological sequence, three-dimensional structure and its accompanying functions, and then to use this knowledge for predictive purposes. The guiding principles for the study of sequence–structure–function relationships, and the predictive techniques resulting from this analysis, are derived from the processes of molecular evolution (see Chapter 1). Most commonly, protein function is inferred from the known functions of homologous proteins; i.e., proteins predicted to share a common ancestor with the query protein based on significant sequence similarity.

For homologous proteins with easily recognizable sequence similarity, this type of prediction is based on the *'similar sequence–similar structure–similar function'* paradigm. The rule assumes a one-to-one relationship between protein sequence and unique 3-D structure, and between structure and unique biochemical function.

Whilst this usually holds true for closely related proteins, it is important to be aware that more distant relatives belonging to protein families typically exhibit a complex pattern of shared and distinct characteristics in terms of biochemical function, ligand specificity, gene regulation, protein–protein interactions, tissue specificity, cellular location, developmental phase of activity, biological role, and other features. One can expect to find an entire spectrum of function conservation in evolutionarily related genes, ranging from *paralogs* with a very high degree of functional divergence to others with strictly conserved function, the latter most likely being *orthologs* involved in core biological processes (see also Chapter 10).

Sequence alignments provide a powerful technique to compare novel sequences with previously characterized proteins. If a homologous relationship between query and target protein can be inferred based on significant sequence similarity over most of the aligned sequence, it is often possible to make quite detailed predictions regarding biochemical function and cellular location, but the biological processes the respective proteins are involved in may not be identical (see Chapter 10). In contrast, if only a significant local match to one or more protein domains is identified, function assignment for the entire protein will be less reliable. Ideally, function prediction by inferred homology ought to be confirmed by appropriate experimental analysis.

5.3   Functional annotation of biological sequences

5.3.1   Identification of sequence homologs

Functional and evolutionary information can be inferred from *sequence comparisons*. These methods are most sensitive at the protein level and the detection of distantly related sequences is easier in protein translation. The reason for this is the redundancy of the genetic code. Nucleotide sequences therefore first need to be automatically translated into their predicted primary amino acid sequence. The first step toward function identification generally involves searching protein sequence databases with a query sequence using automatic pairwise comparison tools such as *BLAST* or *FASTA* (*Figure 5.1*; see Chapter 3).

Position specific iterative BLAST (*PSI-BLAST*), developed by Altschul and co-workers, can be employed to extend the search to distantly related homologs. The PSI-BLAST server at the NCBI website is easy to use and for most protein families returns a list of putative relatives within minutes of requesting the search. The PSI-BLAST algorithm is also relatively easy to implement locally in order to run against a user's database.

PSI-BLAST accepts a query sequence as input and constructs a *profile* (or *position-specific scoring matrix, PSSM*) from a multiple alignment of the highest scoring hits in an initial gapped BLAST search. The algorithm calculates position-specific scores for every alignment position; highly conserved positions are assigned high scores and weakly conserved positions receive scores close to zero. Highly conserved positions would be expected to be present in all members of the family, and information about residue preferences at these conserved positions are therefore diagnostic for the family and can be encoded in a family profile.

PSI-BLAST uses a substitution matrix which assigns a score for aligning any possible pair of residues based on the frequency with which that substitution is known to occur among homologous proteins (see Chapter 4, for details on amino acid substitution matrices). Different substitution matrices are designed to detect similarities at certain ranges of evolutionary distance. Experimental validation showed that the BLOSUM62 matrix is very well suited to detect distant evolutionary relationships. This substitution matrix is therefore implemented as the default in PSI-BLAST, although other matrix options are also included.

The profile is used to identify more distant relatives in an iterative process which results in increased sensitivity. A conservative threshold is applied to minimize the probability of

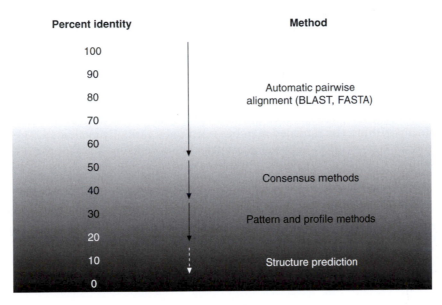

Figure 5.1

Application areas of different sequence analysis methods (adapted from Attwood and Parry-Smith, 1999).

including any unrelated sequences of false positives. In each iteration, the profile is refined by including the new members until the process converges on a unique set of proteins and no new sequences are found. For some families this may involve very many cycles, with the danger that the profile drifts beyond the actual variability for the particular family, by including false positives. Therefore, it is possible to stop the algorithm after a fixed number of cycles, typically 20, to minimize the risk of profile drift.

When performing searches with BLAST, FASTA or PSI-BLAST, a statistically significant match to a well-characterized protein over the entire length of the query sequence may be returned. However, in many cases, searches will only identify local similarities between the query sequence and a number of diverse proteins that may be otherwise unrelated. These local matches often signal the presence of shared homologous domains.

5.3.2 Identification of conserved domains and functional sites

Larger proteins are modular in nature, and their structural units, protein domains, can be covalently linked to generate multi-domain proteins. *Domains* are not only structurally, but also functionally, discrete units, i.e., specific biochemical functions are carried out by specific domains. Domain family members are structurally and functionally conserved and recombined in complex ways during evolution. Novelty in protein function often arises as a result of the gain or loss of domains, or by re-shuffling existing domains along the linear amino acid sequence. Domains can be seen as the 'units of evolution', and, therefore, both structural and functional similarity between proteins needs to be analyzed at the domain level (see Chapter 1 for evolutionary mechanisms and Chapter 7 for structural classification of domain families).

For pairs of protein domains that share the same three-dimensional fold, precise function appears to be conserved down to ~40% sequence identity, whereas broad functional class is conserved to ~25% identity (see Chapter 10).

When conducting sequence database searches with BLAST or FASTA, local matches to domains present in multi-domain proteins need to be evaluated with caution. Firstly, the domains that correspond to the query need to be correctly identified and, secondly, it is important to be aware that domain function can vary significantly depending on context. Automatic transfer of functional annotation can be misleading. The relationship between members of domain families is analogous to the relationship between genes belonging to protein families (see Chapter 1): Database search algorithms cannot distinguish between a match to an *ortholog* or a *paralog*. Since this distinction cannot be derived automatically with the available algorithms, the returned matches need to be carefully evaluated. The higher the level of sequence conservation between genomes, the more likely it is that the query and the matched sequence are orthologs (operationally defined as symmetric top-scoring protein sequences in a sequence homology search).

Current *global* and *local sequence alignment* methods, and consensus methods, are reliable for identifying sequence relatedness when the level of sequence identity is above 30% (*Figure 5.1*). *Consensus methods* occupy an intermediate place between multiple alignment and pattern/profile methods. Consensus sequences are built from multiple alignments of related sequences and the diagnostic sequence motif is represented as the majority or plurality character at each position. This short sequence alignment illustrates this method:

```
DLLFRCG
ELLAKCE
ELVFRCG
DILFKCA
DLLFRCG

DLLFRCG = consensus
```

Query sequences are then matched against a database of consensus sequences by BLAST. The *Protein Domain Database (ProDom)*, a secondary database (see below), was constructed by clustering homologous sequence segments derived from SWISS-PROT and provides multiple alignments and consensus sequences for homologous domain families. Whilst consensus sequences make some attempt at representing the pattern of mutational change at each position of the alignment to increase sensitivity, information about rarely occurring amino acids is lost and the sensitivity of such methods does not extend below 30% sequence identity.

In order to detect signals of more remote relationships in protein sequences, below a level of ~30% sequence identity, different analytical approaches have been developed which are based on representations of similarity at the level of functional protein domains. Below 20% sequence identity, evolutionary relationships are difficult to elucidate without a combination of structural and functional evidence to infer homology. Therefore, purely sequence-based comparisons fail to identify many of the evolutionary relationships that emerge once the structures of proteins are compared (see Chapter 6). The BLAST algorithm finds only 10% of distant relationships in the Protein Data Bank (PDB) of three-dimensional structures, and although much more sensitive, PSI-BLAST also misses many of these relationships.

To allow the *identification of homologous domains* with low sequence identity (<30%), *pattern and profile methods* enhance sequence information by a mutation pattern at a given position along the sequence that is derived from an alignment of homologous proteins (*Figure 5.2*). For the purpose of function assignment, the concept of a domain is defined broadly as a region of sequence homology among sets of proteins that are not all full-length homologs. Homology domains often, but not always, correspond to recognizable protein-folding domains. In multiple alignments of distantly related sequences, highly conserved regions, called *motifs*, features, signatures, or blocks, are discernible, surrounded by divergent regions

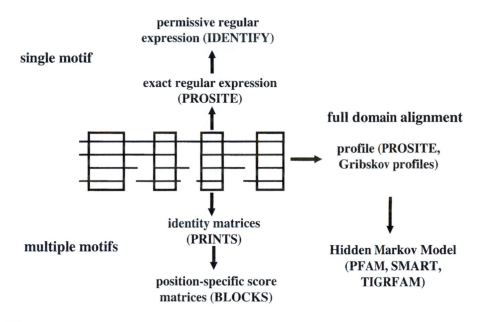

Figure 5.2

Methods for building macromolecular motif databases. The direction of the arrows indicates increased sensitivity and decreased specificity of the representation (adapted from Attwood and Parry-Smith, 1999).

exhibiting a high degree of mutational change among family members. Motifs are typically ~10–20 amino acids in length and tend to correspond to core structural and functional elements of the proteins.

Single and multiple motif methods (regular expressions, identity and position-specific score matrices) utilize motif regions in multiple alignments to build patterns diagnostic of a particular protein family. *Full domain alignment methods* (profiles and Hidden Markov Models) create complete representations of all features in the aligned sequences of a domain family (all highly conserved and divergent regions in a complete domain alignment) and thereby maximize the overall signal of the domain family.

Using pattern and profile representations, databases of diagnostic protein family motifs have been created and can be searched with a query sequence. This may reveal the presence of one or several regions corresponding to particular functions in the query sequence. Pattern and profile database searches therefore constitute an important resource for function prediction. These databases are also known as *secondary databases*, because their models of biological motifs were generated using the data contained in primary sequence databases.

5.3.3   Methods for building macromolecular motif representations

There are two general approaches to improving the sensitivity of homology searching by using family information (*Figure 5.2*). One of these restricts motif models to key conserved features to exclude the higher noise level of the more variable regions (single and multiple motif methods), the other approach includes all possible information to maximize the overall signal of an entire protein or protein domain (full alignment methods). Both approaches are valid. Within these two groupings, numerous representation methods exist and are used alone or in combination by different secondary databases. No fixed rule can be given for choosing the method to use in the analysis of a given protein sequence, as each protein family has a different conservation pattern. In practice one may decide which method is most appropriate if the protein family is known, but it is advisable to perform both types of searches.

It is also important to keep in mind that linear sequences constitute highly abstract models. Whilst the amino acid composition of proteins, their domains and functional sites, can be represented in this fashion, biochemical functionality of motif regions can only be fully understood when taking into account the three-dimensional arrangement of the involved amino acid residues within the context of a protein's structure. It is therefore instructive, and essential, to relate one-dimensional sequence models to protein structures if available. In so doing, we gain a clearer picture of what is encoded in motif patterns.

Figure 5.3 shows an example of two carboxypeptidase structures; one is a monomeric prohormone-processing carboxypeptidase from the yeast *Saccharomyces cerevisiae*, kex1(delta)p (PDB code 1ac5), the other is a dimeric human cathepsin a (PDB code 1ivy). Both proteins belong to the family of serine histidine carboxypeptidases, which can be identified from their sequences by the presence of a diagnostic pattern that maps to the active-site. The PROSITE secondary database contains an entry (PS00560) that encodes the motif in a pattern called a regular expression: [LIVF]-x(2)-[LIVSTA]-x-[IVPST]-x-[GSDNQL]-[SAGV]-[SG]-H-x-[IVAQ]-P-x(3)-[PSA]. In *Plate 2*, the sequence regions that match this pattern in each of the proteins are shown below the structural models of the active-sites with the motif residues color-coded as in the sequences. As will be explained below, the regular expression matches each of the regions exactly, whilst both sequences considerably differ from each other.

5.3.3.1 Single motif methods

Regular expressions

Sequence patterns representing a single motif can be encoded in *regular expressions* which represent features by logical combinations of characters. The secondary database **PROSITE** encodes some motifs as regular expressions. The basic rules for regular expressions are:

- Each position is separated by a hyphen '-'
- An uppercase character matches itself
- x means 'any residue'
- [ ] surround ambiguities – a string [LIV] matches any of the enclosed symbols
- A string [R]* matches any number of strings that match
- { } surround forbidden residues
- ( ) surround repeat counts.

Patterns encoded in regular expressions are restricted to key conserved features in order to reduce the 'noise' of more variable regions. Regular expressions are built by hand in a stepwise fashion from multiple alignments of protein families. Typically, the first step is to encode functionally significant residues, such as the catalytic site or a post-translational

[LIVF]-x(2)-[LIVSTA]-x-[IVPST]-x-[GSDNQL]-[SAGV]-[SG]-H-x-[IVAQ]-P-x(3)-[PSA]

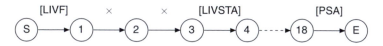

Figure 5.3

Regular expression scoring.

modification site, in a core pattern, usually 4–5 residues long. Subsequently, the length of the pattern is gradually increased to about 10–20 residues. Using regular expressions, it is possible to include gaps in a pattern. At times, a highly conserved region without known functional significance can also be diagnostic for a given protein family and may be encoded in a pattern. Such regions may be essential in maintaining the stability of the shared three-dimensional fold of protein families and thus constitute structural signature motifs.

Regular expressions, such as PROSITE patterns, are matched to primary amino acid sequences using 'finite state automata' (*Figure 5.3*). The UNIX pattern matching (word search) utilities such as *grep*, *sed*, *awk* and *vi* are examples of highly efficient implementations of finite state automata. A finite state automaton is a model composed of a number of states, and the states are interconnected by state transitions. It reads one symbol at a time from an input string. The symbol may be accepted, the automaton then enters a new state; or the symbol may not be accepted in which case the automaton stops and rejects the string. *Scoring* against a regular expression is therefore *binary* (YES/NO), the pattern either matches the sequence or it does not. If the automaton reaches a final accepting state, the input string has been successfully recognized.

Figure 5.3 depicts a finite state automaton for matching the PROSITE CARBOXYPEPT_SER_HIS (PS00560) regular expression. To check whether a sequence matches this expression, the sequence is fed to the automaton one symbol at a time (S denotes the start site). If the first symbol matches the expression [LIVF], the automaton enters state 1; otherwise it quits and rejects the sequence. If the automaton is in state 1 and it reads a symbol that matches x (any valid amino acid symbol), it moves to state 2; when in state 2, it again reads any valid amino acid symbol, it moves to state 3; when in state 3, it reads a symbol that matches the expression [LIVSTA], it moves to state 4, and so on until it successfully matches the sequence to the whole regular expression by reaching the end state E.

Regular expressions perform best when a given protein family can be characterized by a highly conserved single motif. They can also be used to encode short sequence features believed to be the result of convergent evolution, such as sugar attachment sites, phosphorylation sites, acetylation sites, etc. However, the caveat applies that the shorter the pattern (maybe as short as only 3–5 residues) the more likely purely random matches become. Such matches are inherently unreliable and can only provide a hint that such a site *may* be present.

Regular expressions possess a number of limitations. Firstly, they lose information, as only those features that are judged to be the most conserved or significant from a parent alignment are modeled. However, as more family members are identified, alignments will become more and more representative of the family and motifs earlier thought to be characteristic of the family might have to be changed considerably or completely replaced by more significant ones. Secondly, regular expressions are built from an incomplete sample of all possible members of a protein family and only observed amino acid variability is encoded. The substitution probabilities of unobserved amino acids are not modeled (in contrast to profile methods, see below). Thirdly, matches have to be exact. Even one mismatch will cause the query to be rejected, although the mismatch may be a conservative amino acid substitution or an insertion/deletion with conserved function. Regular expressions will therefore fail to detect new family members that match a partial motif (false-negative hits). Lastly, unlikely sequences may be accepted if patterns are made too inclusive in an attempt to detect distant relatives (false-positive hits).

Fuzzy regular expressions

More permissive regular expressions can be built by including information on shared biochemical properties of amino acids. Amino acids can be grouped according to hydrophobicity, charge, size, presence of aromatic rings etc., and these groupings can then be used to derive *fuzzy regular expressions* aimed at the detection of more distant relatives. The inherent disadvantage of this method is the increased likelihood of chance matches without any biological significance. An example illustrates this approach:

Alignment	A possible fuzzy regular expression
FKA...	
YPI...	[FYW]-[PKR]-[VIALG]...
FPV...	
FKV...	

Fuzzy regular expressions can be built based on different amino acid groupings, and, if constructed by hand, also contain an element of subjective judgement as to how permissive one wants the pattern to be. The same alignment can be the starting point for quite different fuzzy regular expressions. For the first column of the example alignment above, the third aromatic amino acid tryptophan (W), although unobserved, was included in the regular expression, because both phenylalanine (F) and tyrosine (Y) are aromatic residues; for the second column, arginine (R) was included because it shares a positive charge with lysine (K); for the third column, leucine (L) was included as it completes the set of aliphatic hydrophobic residue together with valine (V) and isoleucine (I), and glycine (G) shares the property of smallness with alanine (A). Fuzzy regular expressions, built from source alignments held in the *BLOCKS* and *PRINTS* databases, are employed by the *eMOTIF/IDENTIFY* sequence annotation system.

5.3.3.2 Multiple motif methods

In most protein family alignments, more than one conserved motif region will be present. *Multiple motif methods* represent several or all of these family-specific regions in order to model the mutual context provided by motif neighbors and thereby improve the diagnostic power of the pattern. The query sequence of a more distant relative may only match a subset of motifs, but the pattern of matched motifs may still allow statistically significant, and biologically meaningful, family assignments to be made.

Fingerprints

The **PRINTS** database encodes multiple motifs (called *fingerprints*) in ungapped, unweighted local alignments which, unlike regular expression or consensus methods, preserve all sequence information in aligned blocks. Usually the motifs do not overlap, but are separated along a sequence, though they may be contiguous in the three-dimensional protein structure. An advantage of providing actual alignments is that they contain all of the information needed to derive other representations (such as, regular expressions, profiles or Hidden Markov Models). A variety of different scoring methods may then be used, such as BLAST searching, providing different perspectives on the same underlying data. Fingerprints are built manually by initially starting from a small multiple alignment which provides the seed motifs. Subsequent iterative database scanning will create an increasingly informative, composite set of motifs by retrieving, at each round, all additional sequences that match all the motifs and can be used to improve the fingerprint. This iterative process is repeated until no further complete fingerprint matches can be identified.

BLOCKS

The **BLOCKS** database is derived from *PROSITE* and *PRINTS* entries. Using the most highly conserved regions in protein families contained in PROSITE, a motif-finding algorithm first generates a large number of candidate *blocks*. Initially, three conserved amino acid positions anywhere in the alignment are identified and such a spaced triplet, if found in enough sequences, anchors the local multiple alignment against which sequences lacking the triplet are aligned to maximize a block's sensitivity. In this fashion, blocks are iteratively extended and ultimately encoded as ungapped local alignments. Graph theory techniques are used to assemble a best set of blocks for a given family.

After a block is made, weights are computed for each sequence segment contained in it. These weights compensate for overrepresentation of some of the sequences; low weights are given to redundant sequences, and high weights to diverged sequences. The BLOCKS database uses position-based sequence weights, which are easy to compute and have been shown to perform well in searches. These are simple weights derived from the number of different residues and their frequency in each position in the block.

The BLOCKS search system, allowing query sequence searches against the database, converts each block to a *position-specific scoring matrix (PSSM)*, which is similar to a profile. A PSSM has as many columns as there are positions in the block, and 20 rows, one for each amino acid. It also contains additional rows for special characters: B stands for the amino acids aspartic acid (D) or asparagine (N); Z for glutamic acid (E) or glutamine (Q); X for an unknown amino acid residue; and - denotes a gap. Each PSSM entry consists of numeric scores that are based on the ratio of the observed frequency of an amino acid in a block column to its expected overall frequency in SWISS-PROT (*odds ratio*). The observed frequencies are weighted for sequence redundancy using the sequence weights in the block.

However, it is typical that most amino acids do not occur at all in a position, and so one must decide on a method to deal with these unobserved residues. One approach would be to assume that since the residue is not observed it should never occur in that position; this is clearly unrealistic, especially if the alignment contains only few sequences. Another is to use a substitution matrix to arrive at weighted average scores for all unobserved amino acids. However, this will reduce the specificity of the PSSM if indeed the variability in a motif block, that could be observed if the entire sequence family was known, is well represented by the alignment. A statistical approach to this problem of unobserved amino acids adds imaginary pseudocounts to the observed counts of each amino acid at a position, based on some belief about amino acids expected to occur there.

During a search, the query sequence is aligned with each block at all possible positions, and an alignment score is computed by adding the scores for each position. Before search scores are used to rank the blocks relative to each other, the blocks are calibrated to provide standardized scores that are comparable between blocks regardless of varying block widths and number of aligned sequences. The statistical significance of search results is reported as E-values (see Glossary).

5.3.3.3 Full domain alignment methods

Profiles

Profile analysis is used to perform whole domain alignment of a family-based scoring matrix against query sequences by dynamic programming. Profiles maximize information by including all positions in a multiple domain alignment; they allow partial matches and can detect distant relationships with only few well-conserved residues. A profile is a weight matrix comprising two components for each position in the alignment: scores for all 20 amino acids and variable gap opening and extension penalties. Specifically, the M_{kj} element of the profile is the score for the *j*th amino acid at the *k*th position of the alignment (*Figure 5.4*).

A profile can be seen as a mathematical model for a group of protein sequences. This model is more complex than a simple weight-matrix model in that it contains position-specific information on insertions and deletions in the sequence family, and is quite closely related

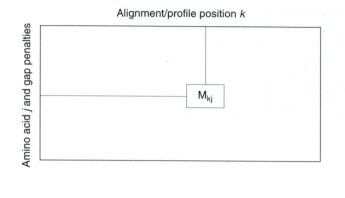

Figure 5.4

Profile matrix.

to the one employed in Hidden Markov Models for sequences (see below). However, in contrast to Hidden Markov Models, profiles allow the generation of models with good discriminatory power from a small number of sequences. As alignments often consist of many closely related sequences together with a few much more divergent ones, sequence weighting schemes which upweight divergent sequences while downweighting closely related groups can be used to improve profile sensitivity.

Earlier profile methods created the amino acid substitution scores by summing *Dayhoff exchange matrix* values according to the observed amino acids in each column of the alignment. Other residue substitution matrices, particularly the *BLOSUM* and *Gonnet series*, have subsequently been found to improve the signal-to-noise ratio in profiles. BLOSUM 45, a moderate to high divergence matrix, works well as a starting point (see Chapter 4). Ideally, several matrices should be tested during profile construction to reach an appropriate balance between specificity (ability to detect true positives and exclude false positives) and sensitivity (ability to detect divergent family members as true positives while still keeping the expected number of false positives within very low bounds). As in all sequence searches, the length of the query affects the tolerance of noise introduced by high divergence matrices. Small domains need more stringent matrices such as BLOSUM62 (or Gonnet PAM120/PAM160), whereas profiles for larger domains are often more sensitive at higher divergence (Gonnet PAM250-PAM350).

Gap penalties are reduced at positions with gaps, according to the length of the longest insertion spanning that point in the alignment. As yet there is no satisfactory way of calculating these penalties. The user should conduct several trials varying the gap parameters in order to optimize alignment of the query sequence with the profile and the detection signal-to-noise ratio.

5.3.3.4   Hidden Markov Models

Several major secondary databases, including *PFAM, SMART*, and *TIGRFAM*, represent full domain alignments as *Hidden Markov Models* (HMMs). PFAM, for example, represents each family as a 'seed alignment', a full alignment and an HMM. The 'seed' contains representative members of the family, while the full alignment contains all members of the family as detected with an HMM constructed from the seed alignment using the *HMMER2* software. Full alignments can be large with the top 20 families now each containing over 2500 sequences.

HMMs are domain family models that can be used to identify very distant relatives. In order to understand how this sensitivity is achieved, let us first consider Markov processes in a more general way. A *Markov process* is a process that can be decomposed into a succession of discrete states. The simplest Markov process is a first order process, where the choice of state is made purely on the basis of the previous state. The behavior of traffic lights is a first-order Markov process, it consists of a set sequence of signals (red–red/amber–green–amber) and the transition to the next state only depends on the present color of the signal.

However, various examples exist where the process states are not directly observable. In speech recognition, the sound we hear depends on such factors as the action of the vocal chords, the size of the throat, and the position of the tongue. The detectable sounds are generated by the internal physical changes in the person speaking; and the observed sequence of states (sounds) is somehow related to the hidden process states (physical changes). In cases like this, we can define an HMM, a statistical model of an underlying (hidden) Markov process which is probabilistically related to a set of observable state changes.

Profile HMMs are bioinformatics applications of HMM modeling that represent protein domain families. The discrete states of a profile HMM correspond to the successive columns of a protein multiple sequence alignment (*Figure 5.5*): A *match state* (M) models the distribution of residues allowed in the column. An *insert* (I) and a *delete* (D) *state* at each column allow for insertion of one or more residues between that column and the next, or for creating a gap. States have associated *symbol emission probability distributions* and are connected by state transitions with associated *transition state probabilities*. One may therefore conceptualize profile HMMs as process state models that generate (observable) sequences; they describe a probability distribution over a potentially infinite number of sequences. Each time a complete path is traversed from the begin (B) to the end (E) state, a hidden state sequence is created together with an observed (gapped) amino acid residue sequence.

A profile HMM is similar to a profile in that each column in the profile can be seen as a match state, and the values in the column as emission probabilities for each of the 20 possible amino acids. The position-specific gap weights represent transition probabilities for moving to an insert or delete state from a match state. The main difference between profile and HMM models is that the profile model requires that the transition from a match state to an insert state or a delete state have the same probability. This reflects the observation that an insertion in one sequence can be viewed as a deletion in another (see the concept of indels in Chapter 1).

In principle, HMMs can be developed from unaligned sequences by successive rounds of optimization, but in practice, protein profile HMMs are built from curated multiple sequence alignments. In the latter case, parameter estimation is simplified and observed counts of symbol emissions and state transitions can be directly converted into probabilities. These probability parameters are then usually converted to additive log odds scores before aligning and

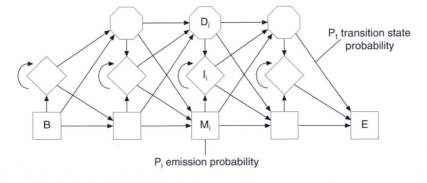

Figure 5.5

Hidden Markov Model for domain families.

scoring a sequence (p_x, probability of residue x emitted at match state; f_x, expected background frequency of x in sequence database; log p_x/f_x, score for residue x at this match state).

Once an HMM is generated, standard dynamic programming algorithms can be used for aligning and scoring sequences with the model (see also Chapter 3). HMM searches resemble later round PSI-BLAST searches, with position-specific scoring for each of the states (amino acid match state, insertion, and deletion) over the length of the sequence.

The *Viterbi algorithm* determines which sequence of hidden states most probably generated a given sequence and is used for alignment. This dynamic programming algorithm keeps a backward pointer for each state, and stores a probability δ with each state. The probability δ is the probability of having reached the state following the path indicated by the back pointers. When the algorithm reaches the end state, the δs for the final states are the probabilities of following the optimal (most probable) route to that state. Selecting the largest of these probabilities, and using the implied route, provides the best answer to the problem.

An important point to note about the Viterbi algorithm is that it does not just accept the most likely state at each instant, but takes a decision based on the whole sequence. It will therefore not matter if there is a particularly unlikely event somewhere in the sequence, provided the context is significant. Using the example of speech recognition once more, an intermediate sound may be indistinct or lost, but the overall sense of the spoken word may be detectable. The same principle applies to matching distantly related protein sequences with highly diverged positions to a family HMM.

The *forward algorithm* assigns a probability for the generation of a specific sequence by a given HMM and is used for scoring sequences. One method of calculating the probability of the observed sequence would be to find each possible sequence of hidden states that would generate it, and sum these probabilities. Calculating the probability in this manner is computationally expensive, particularly with large models or long sequences. In order to reduce the complexity of the problem, it is possible to calculate the probability of reaching an intermediate state in the HMM as the sum of all possible paths to that state. The sum of all partial probabilities is the sum of all possible paths through the model, and hence is the probability of observing the sequence given the HMM.

The forward algorithm finds this probability exploiting recursion in the calculations to avoid the necessity for exhaustive calculation of all paths. Using this algorithm, it is straightforward to determine which of a number of HMMs best describes a given sequence – the forward algorithm is evaluated for each, and that giving the highest probability is selected.

5.3.3.5   Statistical significance of profile and HMM hits

When searching profile or HMM databases with a query sequence, every match receives a numerical *raw score* generated by the search algorithm applied. Usually, only matches with scores above a certain threshold are reported, and every profile or HMM applies its own score thresholds. To facilitate the interpretation of match scores, they are mapped to a new scale, which possesses a well-defined statistical meaning. Two such scales are the *E-value* for HMMs (see Glossary) and the *normalized score* for PROSITE profiles and the Gribskov profile collection. Normalized scores have an inverse log relationship to the E-value: A normalized score of 10 corresponds to an E-value of around 1×10^{-2} and 10.5 to an E-value of around 2×10^{-3}. Any normalized score above 11 indicates a very significant match.

5.3.4   Secondary database searching: a worked example

The *InterPro* database allows efficient searching, as it is an integrated annotation resource for protein families, domains and functional sites that amalgamates the efforts of the *PROSITE, PRINTS, Pfam, ProDom, SMART* and *TIGRFAMs* secondary database projects. Release 4.0 of InterPro (November 2001) contains 4691 entries, representing 3532 families, 1068 domains,

74 repeats and 15 sites of post-translational modification. Overall there are 2,141,621 InterPro hits from 586,124 SWISS-PROT + TrEMBL protein sequences.

Figure 5.7 illustrates the results of a PFAM (version 5.6) and InterPro (version 4.0) search with the human *tissue plasminogen activator* amino acid sequence (precursor sequence; SWISS-PROT ID TPA_HUMAN). Search results are reported in PFAM as a table of significant hits showing bit scores and E-values, and also in graphical format depicting the positions of matched domains along the query sequence (*Figure 5.6a*). PFAM also allows one to retrieve a graphics table of other proteins known to share a given domain with the query, as shown by three example proteins sharing the kringle domain with human TPA (*Figure 5.6b*). The human apolipoprotein *A* provides a dramatic example of *domain duplication* – it contains 38 kringle domains. Interpro reports search results in a table that groups together hits to the same domain obtained by searching its constituent secondary databases (*Figure 5.6c*).

The graphical display allows one to clearly distinguish *short single motifs* matched by PROSITE regular expressions (PSxxxxx), *multiple motifs* matched by PRINTS fingerprints (PRxxxxx), and *whole domains* matched by HMMs from PFAM (PFxxxxx) and SMART (SMxxxxx), PROSITE profiles (PSxxxxx) and consensus sequences from PRODOM (PDxxxxx). Clicking the hyperlinked domain boxes enables the retrieval of more detailed information on matched domains.

5.4   Outlook: Context-dependence of protein function

Sequence analysis techniques provide important tools for the prediction of biochemical function, but show clear limitations in predicting a protein's context-dependent functions and role in one or more biological processes. Sequence analysis techniques are complemented and extended by function prediction from structure (see Chapters 10 and 11) and data from transcriptomics and proteomics which can enhance understanding of gene and protein function in complex biological networks (see Chapters 14, 16).

New drug targets from the human genome

The selection of molecular targets that can be validated for a disease or clinical symptom is a major problem for the development of new therapeutics. It is estimated that current therapies are based on only ~500 molecular targets. The majority of these targets are receptors such as the G-protein-coupled receptors (GPCRs), which constitute 45% of all targets. Enzymes account for a further 28%, with ion channels making up 5% and nuclear receptors 2% of the remainder. This distribution reveals that a relatively small number of target classes have so far provided most of the amenable targets for the *drug discovery* process.

For drug target discovery, reliable function assignment for novel genes and proteins is a central challenge in which bioinformatics techniques play an increasingly powerful role. *Discovery genomics* seeks to identify novel members of protein families belonging to known drug target classes by assigning genes to protein families based on shared domains in their predicted protein sequences or similarity in protein structure.

Genes identified by these techniques require careful functional validation in the laboratory or a proven relationship to a disease process. Many close relatives of known important cancer genes, for example, are not mutated in cancers so that searching for homologs of these known genes is unlikely to lead to the discovery of new candidate cancer genes. Novel complementary approaches, integrating experimental and computational methods, are needed to uncover the complex relationships between the different manifestations of disease from the molecular to the clinical level.

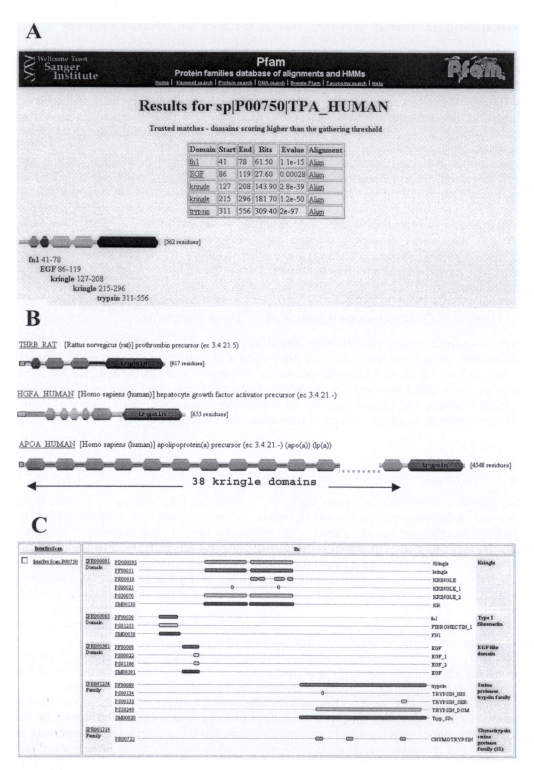

PFAM and INTERPRO search results for the human tissue plasminogen activator (TPA) amino acid sequence.

References and further reading

Attwood, T.K. and Parry-Smith, D.J. (1999) *Introduction to Bioinformatics*. Addison Weesley Longman, 218 pp.

Doolittle, R.F. (ed.) (1996) Computer methods for macromolecular sequence analysis. *Methods in Enzymology* **266**: 711.

Eddy, S.R. (1998) Profile Hidden Markov Models. *Bioinformatics* **14**: 755–763.

Weir, M., Swindells, M., and Overington, J. (2001) Insights into protein function through large-scale computational analysis of sequence and structure. *Trends in Biotechnology* **19**(10 Suppl): S61–66.

Protein structure comparison 6

Ian Sillitoe and Christine Orengo

- Mechanisms by which protein structures evolve; insertions and deletions of residues; shifts in the orientations of secondary structures.
- Use of structure comparison for recognizing distant homologs and for analyzing conformational changes associated with function, e.g. changes occurring on substrate binding.
- Approaches used for comparing protein structures; intermolecular superposition methods versus comparison of intramolecular residue distances.
- Mechanisms for dealing with extensive residue insertions/deletions occurring between distant homologs.
- Descriptions of fast methods for comparing structures, usually based on secondary structure comparison and slower, more accurate residue-based methods of structure comparison.
- Multiple structure comparison methods for comparing sets of related structures.

6.1 Concepts

Since protein structures are more highly conserved than sequences during evolution, structure comparison is very useful for establishing distant evolutionary relationships undetectable by sequence-based methods. In this chapter the concepts behind structure comparison methods will be discussed and some example methods, representative of different approaches, will be described.

6.1.1 Single domains and multidomain proteins

We will first define some of the terms used in this chapter and consider some of the challenges faced when comparing protein structures. One important fact to bear in mind is that the polypeptide chain of a protein folds into a *'tertiary'* structure, often comprising one or more compact globular regions referred to as domains. The tertiary structure associated with a domain region is also described as a protein *'fold'*. Proteins whose polypeptide chains fold into several domains, are described as *'multidomain'* structures (see also Chapters 1 and 7).

Nearly half the known globular structures are multidomain, the majority comprising two domains, though examples of 3, 4, 5, 6 and 7 domain proteins have also been determined. Generally there are fewer residue contacts between domains than there are within them, a feature which is often used to recognize them automatically (see Chapter 7). Analyses of complete genomes have suggested that a high proportion of genome sequences are multidomain proteins, particularly for eukaryotes, perhaps between 60–80% depending on the organism.

Automatic structure comparison methods were introduced in the 1970s shortly after the first crystal structures were deposited in the protein structure databank (PDB). Most commonly used methods can align or superpose both multidomain and single domain structures although it is often helpful to separate the structures into their constituent domains before comparing them as the domains can be connected differently or be oriented differently with respect to each other.

6.1.2 Reasons for comparing protein structures

6.1.2.1 Analysis of conformational changes on ligand binding

There are many reasons for wanting to develop reliable methods for comparing structures. Often the binding of ligand or a substrate to an active-site in a protein induces a structural change which facilitates the reaction being catalyzed at this site or promotes the binding of substrates at another site. Comparing the ligand bound and unbound structures can shed light on these processes and can assist rational drug design. In these cases, much of the structure may be unaffected and it is easy to superimpose equivalent positions and determine the degree of structural change that has occurred.

6.1.2.2 Detection of distant evolutionary relationships

As discussed already, protein structure is much more highly conserved than sequence during evolution. Thus protein structure comparison can be used to detect similarity between homologous proteins whose sequences have changed substantially during evolution. Proteins evolve by substitutions in their residues and by insertions and deletions (indels) of residues in their amino acid sequences (see Chapter 1). There can be large changes in protein sequences. In fact, analyses of structural families have shown that homologs frequently share fewer than 15% identical residues.

However, usually a significant portion of the structure remains similar, even between very distant homologs (see *Figure 6.1*). This often corresponds to more than half the structure, usually within the buried core of the protein. This structural fingerprint, conserved across the family, can therefore be used to recognize very distant relatives. Outside this structural core, significant variations in the structure or *embellishments* to the core, have been observed in some families (see *Figure 6.2*).

6.1.2.3 Analysis of structural variation in protein families

Comparison of structures across a protein family gives insights into the tolerance to structural change for a given family and also the impacts any changes have on the functions of the proteins. The degree to which structures diverge as their sequences change during evolution is complex (see *Figure 6.3*) and varies with the structural class and architecture of the protein (see Chapter 7) and also whether there are functional constraints. In some families (e.g. the globins, see *Figure 6.1*) structures are highly constrained and sequences can change significantly before any large structural change is observed. Recent analyses have revealed that the degree of structural variability across a protein family, also described as the *structural plasticity*, varies considerably between protein families. That is, some protein families appear to tolerate much more structural change than others.

In some highly variable families, structural change has no impact on the functions of the proteins. In other families, structural changes clearly modulate functions. For example by modifying the shape of the active-site. Multiple structure comparisons can help to identify those parts of the structure which are structurally conserved across a family or functional subgroup of a family; for example, active-sites and surface patches or protrusions involved in protein–protein interactions. Characterization of these structurally conserved regions will assist in providing more reliable functional annotations for proteins belonging to a particular structural subgroup within a family.

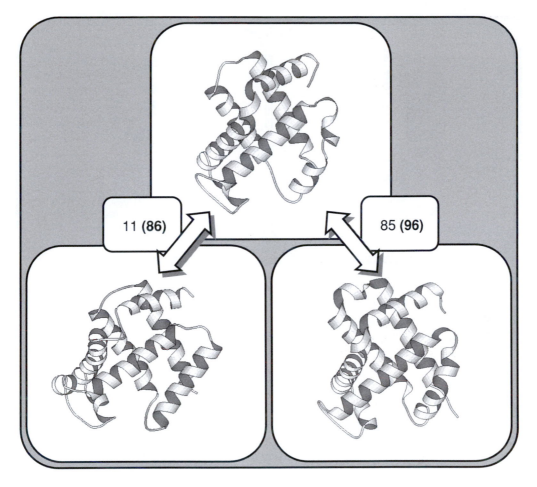

Figure 6.1

Molscript representations of diverse relatives from the globin superfamily. Pair-wise sequence identities are shown for each pair of structures and SSAP structural similarity scores are given in bold in brackets. SSAP scores range from 0 up to 100 for identical structures.

6.1.2.4   Identification of common structural motifs

Structural similarities have also been detected between proteins that do not appear to be evolutionarily related. That is, there is no other evidence for homology, i.e. no significant sequence similarity or functional similarity. Proteins adopting similar folds but having no clear evolutionary relationship are called analogs (see also Chapter 7). The structural similarities may be a consequence of constraints on secondary structure packing, such that there are a limited number of ways a protein can fold and maintain optimal packing of hydrophobic residues in the core. Therefore, analyses of these similarities will help in illuminating the physicochemical requirements on protein folding and secondary structure packing.

As well as searching for global similarity between proteins, structure comparison can also be used to look for common structural motifs. These are often described as *supersecondary* structures and generally comprise fewer than five secondary structures adjacent along the polypeptide chain (eg. β-motifs, αβ-motifs, α-hairpins, see Branden and Tooze, cited at the end of this chapter). Such motifs have been found to recur frequently in protein folds. For

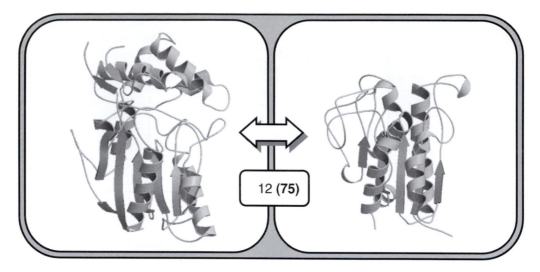

Figure 6.2

Molscript representations of diverse relatives from the αβ-hydrolase superfamily, showing superposition. Pair-wise sequence identity is shown for the pair of structures and SSAP structural similarity score is given in bold in brackets.

example, the αβ-motif recurs eight times within the αβ-barrel fold. Again, these motifs may correspond to favored packing arrangements for secondary structures, the knowledge of which can assist protein design. They may also correspond to ancient evolutionary modules which have been combined in different ways throughout evolution, e.g through DNA shuffling and gene fusion, to give rise to the variety of folds which we now observe.

6.1.3   Expansion of the Protein Structure Databank

Although the PDB is currently much smaller than the sequence databases by nearly three orders of magnitude, the international structural genomics initiatives will help to expand the database considerably and provide representative structures for many protein families (see also Chapters 7 and 11). As discussed above, structural comparisons between relatives can rationalize the mechanisms by which changes in the sequence alter the structure and thereby modulate function. Therefore, as the structural genomics initiatives proceed, and the population of known structural families increases, structure comparison will become increasingly important for analyzing the mechanisms by which function is modulated and for assigning functional properties to related proteins.

6.2   Data resources

The largest repository of protein structures is the Protein Databank (PDB) which is held in the Research Collaboratory of Structural Biology (RCSB) at Rutgers University in the States. There is a European node to this database, known as the European Macromolecular Structure Database (EMSD), which is held at the European Bioinformatics Institute (EBI) in the UK.

6.3   Algorithms

More than 30 structure comparison algorithms have been developed to date, most of which have been described in the reviews cited at the end of this chapter. Some of the more com-

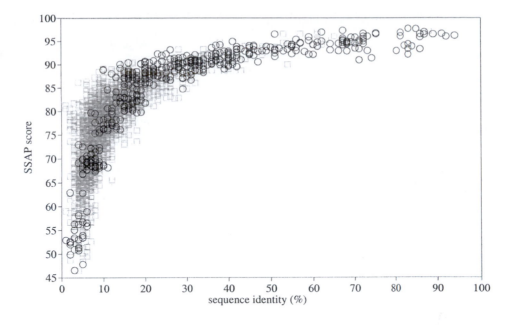

Figure 6.3

Plot of pair-wise sequence identities versus structural similarity for homologous relatives from the CATH protein structural family classification. Structural similarity was measured using the SSAP structure comparison method, which returns a score in the range of 0 up to 100 for identical protein structures. Circles denote relatives sharing the same function, whilst squares correspond to relatives whose functions have changed.

monly used methods are listed in *Table 6.1*. Many of these approaches were developed in the late 1980s to mid-1990s and employ various strategies for aligning protein structures. Methods developed since then often build on the concepts and techniques utilized in these early approaches but combine them in different ways. Therefore in this chapter, we will outline the various philosophical approaches used but give details for only one example of each.

As with sequence comparison (see Chapter 3), methods for comparing protein structures comprise two important components. The first involves techniques for scoring similarities in the structural features of proteins. The second is the use of an optimization strategy that can identify an alignment which maximizes the structural similarities measured.

The majority of methods compare the geometric properties and/or relationships of either the secondary structure elements or residues along the carbon backbone (Cα or Cβ atoms are frequently used). Geometric properties of residues or secondary structures are determined from the *3-D co-ordinates* of the structure, deposited in the PDB. Relationship information includes, for example, distances or vectors between residues or secondary structure positions.

Other non-geometric properties are sometimes included in the comparison, e.g. physico-chemical properties such as hydrophobicity. Specific bonding networks can also be compared, e.g. hydrogen-bonding patterns, disulfide bonds. However, the contributions of these characteristics need to be very carefully weighted, as they can sometimes increase noise. For example, the hydrogen-bonding properties of pairs of helices or parallel/anti-parallel β-sheets will be similar regardless of the topology of the protein structures being compared.

Table 6.1 Structure comparison algorithms

Authors	Type of method (Intermolecular/ Intramolecular)	Algorithms
Residue level		
Rao and Rossmann (1973)	INTER	Superposition
Padlan (1975)	INTRA	Distance matrices
Sutcliffe *et al.* (1987) (MNYFIT)	INTER	Superposition
Taylor and Orengo (1989) (SSAP family)	INTRA	Dynamic programming
Subbarao and Haneef (1991)	BOTH	Graph theoretic
Rose and Eisenmenger (1991) (Comp3-D)	INTER	Dynamic programming Superposition
Russell and Barton (1992) (STAMP)	INTER	Sequence alignment Superposition Dynamic programming
Nussinov group (1993)	INTRA	Geometric hashing
Fisher *et al.* (1993)	INTRA	Geometric hashing
Subbiah *et al.* (1993)	INTER	Superposition Dynamic programming
Yee and Dill (1993) (CONGENEAL)	INTRA	Distance matrices
May and Johnson (1994) (GA_FIT)	INTER	Genetic algorithm Dynamic programming Superposition
Boutonnet *et al.* (1995)	INTER	Multiple linkage clustering
Residue fragment level		
Remington and Matthews (1980)	INTER	Superposition
Rackovsky and Goldstein (1988)	INTER	Superposition
Vrend and Sander (1991)	INTER	Superposition Clustering
Holm and Sander (1993) (DALI)	INTRA	Distance matrices Combinatorics Monte Carlo optimization
Lessel and Schomburg (1994) (in BRAGI)	BOTH	Distance matrices Superposition Clustering
Feng and Sippl (1996) (ProSup)	INTER	Dynamic programming
Shindyalov and Bourne (1998) (CE)	BOTH	Combinatorial extension
Secondary structure level		
Richards and Kundrot (1988)	INTRA	Distance matrices
Abagyan and Maiorov (1988) (FAESAR)		Feature matrix Superposition
Grindley *et al.* (1993) (PROTEP)	INTRA	Graph theoretic
Rufino and Blundell (1994)	INTRA	Graph theoretic Dynamic programming
Multiple element levels		
Sali and Blundell (1990) (COMPARER)	BOTH	Dynamic programming Simulated annealing
Madej, Gibrat and Bryant (1995) (VAST)	BOTH	Graph theoretic (fast search) Monte Carlo (refinement)

6.3.1 Approaches for comparing 3-D structures

Approaches for comparing protein structures generally fall into the following categories (see also *Figure 6.4*):

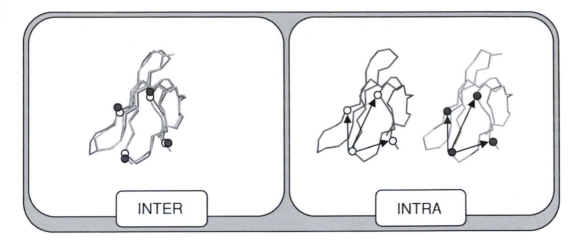

INTER   INTRA

Figure 6.4

Schematic representation of intermolecular- and intramolecular-based structure comparison: (a) intermolecular methods superpose the structures by minimizing the distance between superposed positions; (b) intramolecular methods compare the sets of internal distances between positions to identify an alignment maximizing the number of equivalent positions.

- Methods which superpose protein structures and measure *intermolecular* distances. These predominantly compare geometric 'properties', for example residue positions in 3-D co-ordinate space (see section 6.3.2).
- Methods which compare *intramolecular* distances or vectors. These align protein structures on the basis of information about internal 'relationships' within each protein. For example, the structural relationships between residues within a protein, described by the distances between them (see section 6.3.3).

Some methods combine the above two approaches in some way. The most commonly used methods based on the **intermolecular** approach employ *'rigid body superposition'* algorithms. These attempt to superimpose two protein structures on their equivalent positions by searching for an optimal superposition which minimizes the intermolecular distances between superposed positions in the two structures (see *Figure 6.4* and below).

Although these methods tend to be robust for similar structures (e.g. proteins with high sequence identity), in more distant relatives the correct alignment can often only be obtained by including information about the **intramolecular** relationships between residues or secondary structures within the proteins; for example hydrogen-bonding patterns or structural environments (see below). Example methods will be described for each approach together with the techniques they employ for coping with the extensive insertions/deletions occurring in distant relatives.

6.3.2 Intermolecular approaches which compare geometric properties (rigid body superposition methods)

Rossman and Argos pioneered these methods in the 1970s as the first crystal structures were being deposited in the Protein Databank (PDB). Their approaches employed rigid body methods to superpose equivalent Cα atoms between protein structures. The major steps of this method can be described as follows:

- Translate both proteins to a common position in the co-ordinate frame of reference.
- Rotate one protein relative to the other protein, around the three major axes.
- Measure the distances between equivalent positions in the two proteins.

Steps 2 and 3 are repeated until there is convergence on a minimum separation between the superposed structures.

Usually, the centers of mass of both structures are translated in 3-D co-ordinate space towards the origin, an operation performed by the translation vector. Subsequently, an optimization procedure is performed by which one structure is rotated around the three orthogonal axes, relative to the second structure (see *Figure 6.5*), so as to minimize the distances between superposed atoms, an operation described by the rotation matrix.

The distance between equivalent positions is generally described by a residual function, which effectively measures the distance between superposed residues (e.g. the root mean square deviation, RMSD, equation (1), see *Box 6.1*). The major difficulty lies in specifying the putative equivalent positions at the start of the optimization. If this set does not contain a sufficient number of truly equivalent positions the method can fail to converge to a solution. For proteins which are closely related (e.g. $\geq$ 35% sequence identity), the sequences can be aligned using standard dynamic programming methods (see Chapter 3) to determine an initial set of equivalent residues. This set can then be refined by the optimization procedure. For more distantly related proteins information on equivalent positions is often obtained by

---

Box 6.1   Root Mean Square Deviation (RMSD)

The residual function used to measure the similarity between two protein structures following rigid body superposition is typically the root mean square deviation between the structures (RMSD).

The RMSD between two structures is quite simply the square root of the average squared distance between equivalent atoms, defined by the equation:

$$R = \sqrt{\frac{\sum\limits_{i=1}^{N} d_i^2}{N}} \qquad (1)$$

The distances (d_i) can be visualized as:

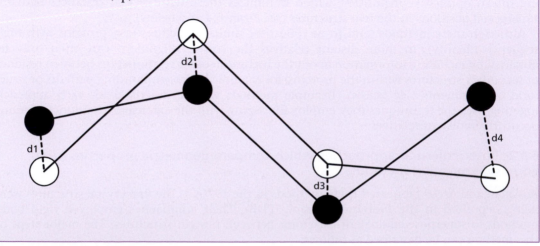

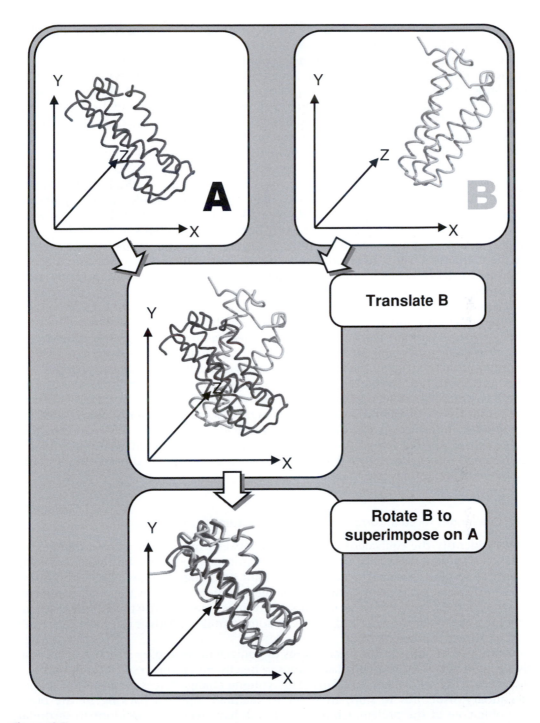

Figure 6.5

Schematic representation of the steps involved in rigid body superposition of two protein structures. (a) Translation of the centers of mass of both protein structures to the origin of a common co-ordinate frame of reference. (b) Rotation of one structure relative to the other to minimize the 'distance' between the superposed structures measured by the RMSD (see *Box 6.1*).

manual inspections of the structures or by identifying residues binding common ligands in the active-sites of the two proteins.

Once the optimal superposition has been determined the difference between the structures is commonly measured by the Root Mean Square Deviation (RMSD) shown in equation (1), *Box 6.1*. Structures having a similar fold typically give values below 3.5A although there is a size dependence to this measure. For example, distant homologs with more than 400 residues may return an RMSD above 4.0A compared to a value of <3.0A for smaller proteins with less than 100 residues, but of comparable evolutionary distance.

Sometimes RMSD is only measured over those residues which have a higher probability of being equivalent, that is residues within a certain intermolecular distance from each other after superposition, typically <3.0A. It is obviously important to consider both the value of the RMSD and the number of equivalent pairs identified, when considering the significance of a similarity. For example, an RMSD of <3A over 20 residues of a 200-residue protein would not constitute a very significant global similarity between two proteins.

RMSD is a very commonly cited measure describing structural similarity and methods which align structures using other protocols often include a final step to superpose equivalent residues by rigid body techniques, and determine the RMSD value. More recently, statistical approaches have also been developed for assessing the significance of structural similarity detected between protein pairs. These are often more reliable as they are independent of protein size and are described in more detail below.

Rigid body superposition methods are highly effective for exploring structural changes occurring on co-factor or substrate binding and also for analyzing the structural effects of point mutations in the sequence as in both these cases equivalent residues between the structures are already known. Similarly, NMR structure determination typically generates many very similar structural models, which are alternative solutions to the distance constraints obtained from the experimental measurements. Again, because the equivalent positions are already known, the structures can be superposed very easily. However, for comparing more distant protein relatives, having substantial insertions or deletions additional strategies must be incorporated to help determine the initial sets of equivalences for superposing the structures. These are discussed below.

6.3.2.1 Challenges faced in comparing distantly related structures

In order to identify the equivalent core residues between relatives in protein families and characterize structural variations, structure comparison methods must be able to cope with the extensive residue insertions occurring in distant homologs and giving rise to the *structural embellishments* to the core. Generally residue insertions occur in the *loops* connecting secondary structures. Very extensive insertions can sometimes give rise to additional secondary structures as embellishments to the core. In some families, relatives can have very extensive embellishments (see *Figure 6.2*).

There can also be shifts in the orientations of equivalent secondary structures. These are largely brought about by residue substitutions, whereby mutations result in different-sized residues occupying equivalent positions. This volume change forces neighboring residues and secondary structures to shift, in order to maintain optimal packing in the buried hydrophobic core of the protein. Chothia and Lesk have reported variations in secondary structure orientations of up to 40 degrees in the globin superfamily.

Various protocols have been developed for handling extensive residue insertions and deletions between proteins. These employ one or more of the following strategies:

- **Discard the more variable loop regions** and simply compare secondary structure elements between proteins. These methods are also much faster than residue-based methods, though they cannot give an accurate alignment.

- **Divide the proteins into fragments** and then attempt to determine the most equivalent fragments between proteins which can then be concatenated to give the global alignment.
- **Use more robust optimization methods** to identify equivalent regions and handle insertions/deletions, e.g. dynamic programming methods, simulated annealing or Monte Carlo optimization. Sometimes combinations of these optimization methods are used.

Again, some structure comparison methods combine several strategies for handling insertions/deletions. We will now consider how these techniques are used to improve the comparison of distantly related protein structures by rigid body superposition.

6.3.2.2 Superposition methods: coping with indels by comparing secondary structures

Some approaches, for example that of Rubagyan and co-workers, effectively remove all the variable loops and only attempt to superpose the secondary structure elements. There are generally an order of magnitude fewer secondary structure elements than residues, which makes it easier to consider the similarities between all secondary structures in the two proteins and converge on the correct solution. As mentioned, these methods cannot generate an accurate alignment of residue positions. However, they are very useful in fast database scans to quickly find related structures and can often identify an optimal superposition without the need to seed initial equivalences. Information on equivalent secondary structures can then be used to suggest putative equivalences for slower, more accurate residue-based superpositions.

6.3.2.3 Superposition methods: coping with indels by comparing fragments

One early approach adopted by Remington and Matthews was to divide the protein into fragments, e.g. hexapeptides and search for equivalent fragments between the structures. These can be rapidly superposed and potentially equivalent fragment pairs having low RMSD can then be concatenated to generate larger substructures whose RMSD can be re-evaluated. Obviously, fragments extracted from non-equivalent loops would not be matched, and the residues associated with them would therefore be identified as insertions. There will also be constraints on the possible combinations of these fragments dictated by the topologies of the structures. Boutonnet and co-workers revisited this approach in 1995 using a more sophisticated optimization algorithm to identify the best combinations of fragment pairs.

6.3.2.4 Superposition methods: coping with indels by using dynamic progamming

A more powerful approach for handling insertions is to apply dynamic programming techniques. These were initially developed in the early 1970s to align protein sequences (see Chapter 3). There have been various adaptations for structure comparison. For example the 2-D score or path matrix, used in dynamic programming (see *Figure 3.4*, Chapter 3), can be scored according to the proximity of the residue positions after superposition of the structures. Typically a score inversely proportional to the distance between the superposed residues, is used. In some methods, scores based on residue proximity after superposition are combined with scores obtained by comparing a host of other residue properties (e.g. phi/psi torsional angles, chirality, hydrophobicity, accessibility).

An elegant approach devised by Russell and Barton in their STAMP method, used dynamic programming to refine the set of equivalent residues given by rigid body superposition (see *Figure 6.6*). An initial residue pair set is given by a sequence alignment of the proteins and used to guide a superposition of the structures. Intermolecular distances between equivalent residues are measured and used to score a 2-D score or path matrix, analyzed by dynamic programming. The resulting path gives a new set of possible equivalent residues (see Chapter 3

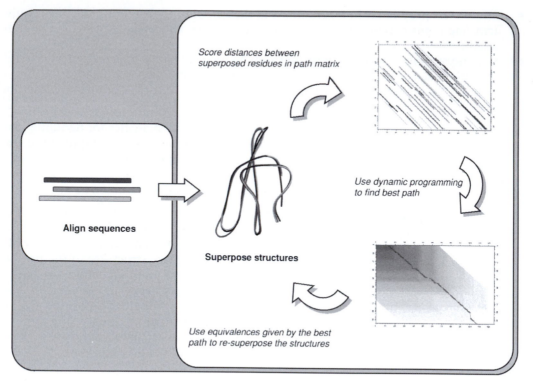

Figure 6.6

Schematic representation of the STAMP method devised by Russell and Barton. Sequence alignment is first used to determine putative equivalent positions in order to superpose the structures. Distances between the superposed positions are then used to score a 2-D score matrix within a specified window which is analyzed by dynamic programming (see Chapter 3) to obtain a better set of equivalent positions on which the structures are re-superposed.

for description of dynamic programming). Thus a new superposition can be generated using this refined residue pair set and the path matrix re-scored and re-analyzed to give a better path and so on until there is no improvement in the RMSD measured over the equivalent pairs.

The method can be summarized as follows:

- Obtain a set of putative equivalent residue positions by aligning the sequences of the two proteins.
- Employ rigid body methods to superpose the structures using this set of equivalent positions.
- Score the 2-D matrix, whose axes correspond to the residue positions in each protein, with values inversely proportional to the distances between the superposed residues.
- Apply dynamic programming to determine the optimal path through this score matrix, giving a new set of putative equivalent positions.

Steps 2 to 4 are repeated until there is no change in the set of equivalent residue positions.

6.3.3 Intramolecular methods which compare geometric relationships

We will now consider the large class of comparison methods based on comparing intramolecular relationships between protein structures. These methods do not attempt to superpose the structures. Instead internal relationships within each structure, typically distances between residues, are compared to identify the most equivalent positions between the two proteins being aligned.

6.3.3.1 Distance plots

The earliest and simplest structural comparison methods were based on visually inspecting *distance plots* between proteins. Distance plots are 2-D matrices used to visualize the distances between residues in a protein structure and can be shaded according to this separation. The related *contact maps*, are used to indicate which residues in a protein structure are in contact within an allowed distance threshold, for example <8A (see *Figure 6.7*). Therefore they readily capture information about intramolecular relationships, i.e. distances between residue positions. Contacts can be based on Cα or Cβ atoms or other atoms in the residues side-chain. By shading the cells associated with contacting residues, patterns arise in the matrix which are often very characteristic of a particular fold.

For example, patterns of diagonal lines appear in the matrix between secondary structures which pack together. Parallel secondary structures give rise to lines, which are parallel to the central diagonal, whilst anti-parallel secondary structures give rise to lines perpendicular to the diagonal. Nearly identical proteins and those with similar lengths can simply be compared by overlaying their distance matrices and searching for similarities or shifts in the patterns identified therein.

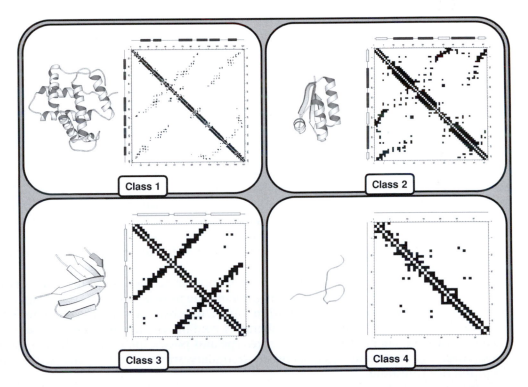

Figure 6.7

Example distant plots for each protein class, mainly-α, mainly-β and α–β and few secondary structures. Contacts between parallel secondary structures give rise to diagonal lines in the matrix, parallel to the central diagonal. Contacts between anti-parallel secondary structures give rise to lines orthogonal to the central diagonal. Different folds give rise to characteristic patterns in the matrix.

Most automated residue-based methods for comparing intramolecular geometric relationships compare distances or vectors between residues. These can be based on the Cα or Cβ atom, the latter providing more discrimination, particularly for aligning residue positions in helices and strands. For example, adjacent residues along a β-strand will have their Cβ atoms pointing in different directions relative to the β-sheet. Therefore comparing Cβ distances helps to ensure that β-strands are aligned in the correct register. For distant relatives, the strategies for handling insertions are again necessary and similar to those used for intermolecular methods, namely one or more of the following:

- Discard the variable loop regions and compare secondary structures
- Divide the proteins into fragments
- Use more robust optimization methods.

6.3.3.2 Comparing intramolecular relationships: coping with indels by comparing secondary structures: Graph theory methods

Again one of the simplest strategies for handling insertions is to discard the loop regions where insertions are more likely to occur. A number of algorithms have been developed that use this strategy to good effect. The most successful of these employ graph theoretical techniques to identify equivalent secondary structures possessing similar intramolecular relationships. Frequently these relationships are defined as distances between the midpoints of the secondary structures and orientations of the secondary structures. Such approaches were pioneered by Artymiuk, Willett and co-workers in 1993 who used graph theory to identify equivalent secondary structures. Linear representations of the secondary structures are derived; that is vectors are drawn through them using a technique known as principal components analysis to identify the central axis.

Graph theory techniques reduce the three-dimensional information embodied in a protein structure to a simplified two-dimensional map or '*graph*' of the protein (see *Figure 6.8*). In the protein graph, each secondary structure element in the structure is associated with a single node. This can be labeled according to the type of secondary structure, e.g. α-helix or β-strand.

Lines or '*edges*' connecting the nodes are labeled with information describing the relationships between the associated secondary structures. This is typically geometric information, such as the distances between the midpoints of the secondary structure vectors and angles describing the tilt and rotations of these vectors relative to each other. Some approaches also include chirality. Similarities between the graphs are then found using various protocols. These are effectively optimization procedures which search for the most equivalent nodes between the two graphs.

Graph theory is a well-established branch of computer science and there are many approaches which can be used to find matching nodes. The most widely used for comparing protein graphs is the Bron–Kerbosch algorithm which searches for common subgraphs or '*cliques*' between the two proteins. These subgraphs contain all the equivalent nodes, i.e. those labeled with the same secondary structure type and possessing edges with corresponding properties, e.g. angles, distances between secondary structure midpoints, within an allowed difference threshold.

Again, since there are few secondary structure elements, typically <20, these methods are extremely fast and have been demonstrated to be very effective in detecting fold similarities between distantly related proteins. Usually they are employed in fast database searches with newly determined structures to identify putative structural relatives. Subsequently, slower residue-based comparisons, generating more accurate alignments, can then be performed across the structural family or fold group thereby identified.

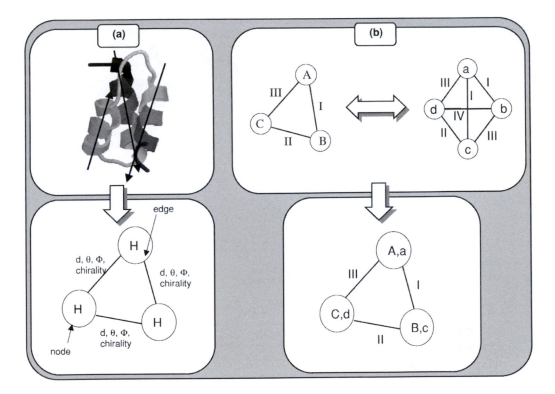

Figure 6.8

Illustration of (a) the graph used to represent a simple alpha helical protein and (b) a comparison of two graphs to reveal the common subgraph.

6.3.3.3 Comparing intramolecular relationships: coping with indels by comparing fragments

Another approach for coping with indels is to divide the protein into fragments and compare the contact maps for these fragments. For example, the DALI method of Holm and Sander, developed in 1990, fragments the proteins into hexapapetides to accomodate insertions (see *Figure 6.9*). Contact maps derived from the fragments can be rapidly compared to identify potentially equivalent fragments, which are then concatenated using a Monte Carlo optimization strategy. This explores all combinations that both satisfy the constraints on the overall topology and generate a structural unit with an acceptable RMSD. The method can be summarized as follows:

- Divide the protein into hexapeptides and derive the contact map for each hexapeptide.
- Identify hexapeptides whose contact maps match within an allowed threshold, i.e. where there is a similar pattern of distances between equivalent residues.
- Concatenate matching hexapeptide contact maps to extend the similarity between the proteins.
- Superpose the extended fragments thereby generated, and check that the difference between the fragments, as measured by RMSD, is within an allowed threshold, else reject the extension.

Steps 3 and 4 are repeated until no further fragments can be added.

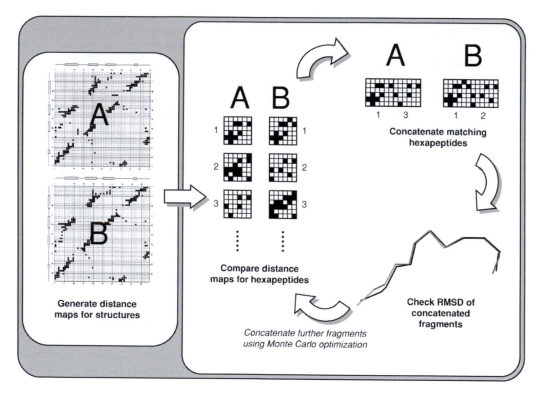

Figure 6.9

Flow chart of DALI method devised by Holm and Sander. Proteins are fragmented into hexapeptides whose contact plots are compared to identify equivalent fragments. Monte Carlo optimization is then used to identify the best order to concatenate fragments to minimize the RMSD measured between superposed fragments.

6.3.3.4 Comparing intramolecular relationships: coping with indels by applying dynamic programming techniques

The SSAP method developed in 1989 by Taylor and Orengo, employs dynamic programming at two levels; initially in the comparison of residue structural environments between proteins and finally to determine the set of equivalent residues (see *Figure 6.11*). Residue structural environments or '*views*' are defined as the sets of vectors from the Cβ atom of a given residue to the Cβ atoms of all others residues in the protein (see *Figure 6.10*). In order to compare views between residues in two proteins, vectors must be determined within a common co-ordinate frame. In SSAP, this is based on the local geometry of the Cα atom.

In the first step vector views are determined for all residue positions in each protein. These are then compared for pairs of residues, between the proteins, selected for being potentially equivalent; that is residue pairs having similar properties, e.g. accessibility, local conformation measured by phi, psi angles. Comparisons between two vector views are scored in a 2-D score matrix referred to as the *residue level score matrix*. This is analogous to the score matrix used in sequence comparison, except that the horizontal axis is labeled by the vectors associated with the view for residue (i) in structure (A). Whilst the vertical axis is labeled with vectors for the view from residue (j) in protein (B). Cells are scored depending on the similarity in the vectors. As with sequence alignment, dynamic programming is used to find the optimal path through this matrix giving the best alignment of the residue views (see Chapter 3 for a description of dynamic programming).

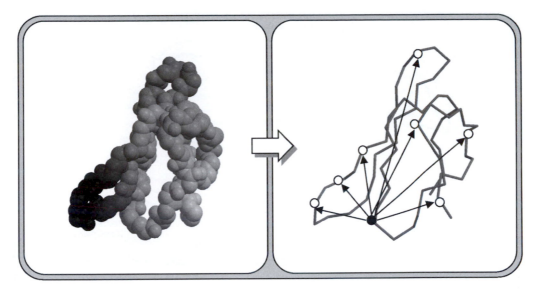

Figure 6.10

Illustration of a residue structural environment or view, employed by the SSAP method of Taylor and Orengo. The view for residue i, is given by the set of vectors from the Cβ atom in i to the Cβ atoms in all other residues in the protein. For clarity only seven of these vectors have been drawn on the figure.

For very similar residue pairs which return high scores for this optimal path, the paths are accumulated in a *summary score matrix*. Once the vector views of all residue pairs have been aligned and high scoring optimal paths added to the summary matrix, the best path through this summary matrix can be found, again using dynamic programming (see *Figure 6.11*). The algorithm is therefore often referred to as double dynamic programming and has also been used for aligning sequences against structures (see Chapter 9). A simple outline of SSAP can be summarized as follows:

- Calculate the view for each residue in the two proteins, given by the set of vectors from the residue to all other residues in the protein.
- For each potentially equivalent residue pair between the proteins (e.g. possessing similar torsional angles and accessible areas), compare vector views by using dynamic programming to find the best path through a *residue level score matrix*, scored according to the similarity of vectors.
- For residue pairs scoring highly, add the scores along the optimal path obtained in step 2, to a 2-D *summary score matrix*.
- Repeat steps 2 and 3 until all potentially equivalent pairs have been compared.
- Use dynamic programming again to determine the optimal path through the 2-D *summary score matrix*, giving the equivalent residues between the proteins.

Various modifications of this approach have been developed over the years but the basic concepts remain the same.

6.3.4 Combining intermolecular superposition and comparison of intramolecular relationships

In the powerful COMPARER algorithm developed by Sali and Blundell in 1989, intermolecular superposition is performed and intramolecular relationships are compared (see *Figure*

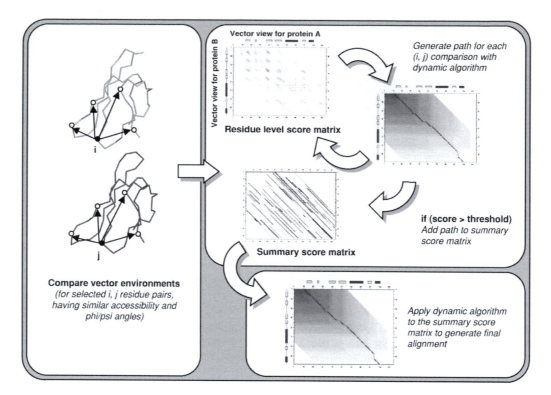

Vector view for protein A

Vector view for protein B

Residue level score matrix

Generate path for each
(i, j) comparison with
dynamic algorithm

if (score > threshold)
Add path to summary
score matrix

Summary score matrix

Compare vector environments
(for selected i, j residue pairs,
having similar accessibility and
phi/psi angles)

Apply dynamic algorithm
to the summary score
matrix to generate final
alignment

Figure 6.11

Flowchart of SSAP algorithm, illustrating the use of dynamic programming at two levels, firstly to obtain the best alignment of vector views between potentially equivalent residue pairs. All high-scoring paths from vector view alignments are added to the summary matrix within a specified window. Finally, dynamic programming is used to find the best path through the summary score matrix within a specified window.

6.12). Similarities are used to score a 2-D score matrix which is subsequently analyzed by dynamic programming. Property information includes residue secondary structure type, torsional angles, accessibility, side-chain orientations and local conformations whilst relationships include intramolecular residue distances, disulfide and hydrogen-bonding patterns. Distances from each residue to the center of mass of the protein are included.

Rigid body superposition is used to compare geometric properties of residues, i.e. intermolecular distances between equivalent positions after superposition. Similarities in the relationships of potentially equivalent residues are identified using an optimization technique known as simulated annealing and also used to score the 2-D score matrix analyzed by dynamic programming. Secondary structure properties and relationships can also be included and compared between proteins.

6.3.5   Searching for common structural domains and motifs

Most structure comparison algorithms search for global similarities in topology between proteins, thus both 3-D conformation and connectivity are important. Removing this constraint at the residue level makes little biological sense and would result in computationally demanding algorithms. However, it is clear from analysis of structural families that secondary structure elements are sometimes shuffled between relatives and proteins from different families can share large common supersecondary motifs which are combined in different

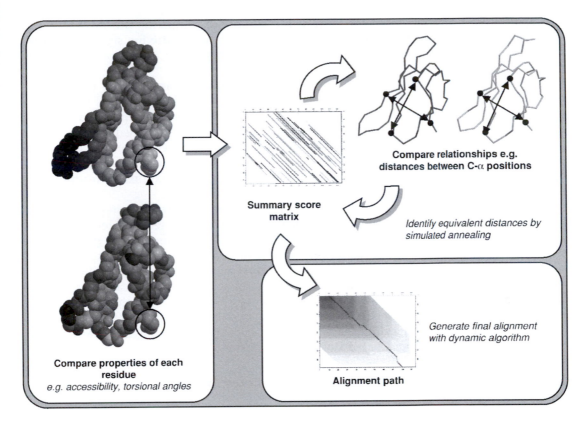

Compare relationships e.g.
distances between C-α positions

Summary score
matrix

Identify equivalent distances by
simulated annealing

Generate final alignment
with dynamic algorithm

Alignment path

Compare properties of each
residue
e.g. accessibility, torsional angles

Figure 6.12

Flowchart of the COMPARER algorithm devised by Sali and Blundell. A 2-D score matrix is
scored by comparing various structural features (e.g. accessibility, torsional angles,
distances between superposed positions) and intramolecular relationships between the
two proteins. Relationships are compared using simulated annealing. Finally, dynamic
programming is used to analyze the summary score matrix and find the best alignment of
the structures.

ways. It is interesting to study these alternative combinations both from the perspective of
possible evolutionary mechanisms and also to facilitate protein design. However, because of
the combinatorial explosion in alternative alignments resulting from the removal of constraints on connectivity, protocols have only been implemented for comparing secondary
structure elements.

6.4 Statistical methods for assessing structural similarity

Although RMSD is one of the most commonly cited measures for describing structural similarity, it is not so useful for comparing distant structures where extensive embellishments or
secondary structure shifts can often mask the underlying similarity. The length dependence
also means that smaller structures often return much lower RMSD than larger proteins, for
similar levels of sequence identity. Another serious consideration is the redundancy of protein structural motifs. Within each protein class there are common favored motifs and super-
secondary structures (e.g. β-hairpins, α-hairpins, αβ motifs) sometimes associating to form
larger motifs e.g. β-meander and extensive αβ motifs $(\alpha\beta)_3$, which may recur in proteins having different overall topologies. These repetitive units can sometimes increase the scores

between proteins adopting different folds. If the purpose of comparing the structures is to identify such motifs, this will not be a problem. However, for identifying significant global structural similarities perhaps to infer homology between proteins, these local similarities can be regarded as noise.

The problem can be addressed to some extent by using statistical approaches. These rely on comparing the pair-wise alignment score for putative relatives with the scores returned by scanning the same structures against a database of unrelated structures. Typically a Z-score is calculated to assess the statistical significance of a match. This is determined by dividing the pair-wise comparison score between putative relatives by the number of standard deviations by which that score differs from the mean score for comparisons involving unrelated structures (see also Chapter 3 and Glossary). The mean and standard deviations are readily calculated from the distributions of scores from the database scans.

Although the Z-score is often more reliable than the RMSD and some protein family classifications use it for assigning relatives (e.g. Dali Domain Database, see Chapter 7) score distributions returned by database scans are best described as extreme value distributions, similar to those commonly observed for scanning sequence databases. This is not surprising as both algorithms are attempting to find the optimal alignment, i.e. the maximum score between two proteins. There are established numerical solutions which can be used to fit these extreme value distributions and thereby calculate the probability that a given score could have been given by unrelated proteins, i.e. by chance. These probability (P) scores or expectation (E) values for random matches can be calculated in a straightforward manner and used to assess the significance of a match (see Glossary).

6.5   Multiple structure comparison and 3-D templates for structural families

It is often useful to compare relatives across a family to reveal any conserved structural features associated with the function or stability of the protein. Alternatively, some characteristics, e.g. specific turns in the polypeptide chain may ensure that the protein folds rapidly. Multiple comparisons can identify a structural fingerprint which is present in all relatives and is therefore a powerful diagnostic for the family. Use of this structural fingerprint, or rather a structural template for the family, therefore gives greater sensitivity in recognizing distant relatives than pair-wise comparison to any of the individual members. Generally the 'core' of the protein structure is very highly conserved across a protein family (see also Chapter 7).

There have been fewer developments in this area than for pair-wise structure comparison and in contrast to developments in multiple sequence alignment algorithms. This is largely because the data are so sparse. In many structural families a high proportion of relatives are nearly identical structures (see Chapter 7). Though undoubtedly many families will be more highly populated with diverse relatives, as the structural genomics initiatives progress.

Both the STAMP and COMPARER algorithms described above can be used to align sets of structures. The most common strategies for performing multiple structure alignments resemble protocols developed for multiple sequence alignment and use a progressive alignment approach (see Chapter 3). That is, all the structures in the set are compared pair-wise against each other and a phylogenetic tree is derived based on the pair-wise similarity scores. Most methods use single linkage hierarchical clustering (see Chapter 15) to generate the tree. In other words the most similar proteins are clustered first and subsequent proteins added to these clusters in an order dictated by their similarity to the best-matched structure in the cluster. The tree therefore gives the order in which individual structures should be added to a growing multiple alignment.

Some multiple structure comparison methods, like the COMPARER algorithm of Sali and Blundell, effectively chain together successive pair-wise alignments. Whilst others, like MNY-

FIT developed by Sutcliffe and co-workers or the CORA algorithm developed by Orengo, compare each structure added to an average or consensus structure derived from those already aligned.

6.5.1 3-D templates

Multiple structural alignments can also be used to derive 3-D profiles or templates for particular families, in a manner analogous to that used in sequence analysis (see Chapter 5). In this case residue positions having highly conserved structural properties, e.g. torsional angles, or relationships, e.g. structural environments or views, can be selected to make the 3-D-template. In the CORA method of Orengo, average properties and environments are calculated for these positions together with information on the variability of these features. This variability information can be used to weight the importance of matching these positions in any new structure aligned against the template. Because 3-D templates reduce the number of comparisons performed in a database scan, and increase sensitivity, it is likely that template-based methods will become more important as the PDB grows in size.

6.6 Conclusions

Although the PDB is currently much smaller than the sequence databases by nearly three orders of magnitude, the international structural genomics initiatives will help to populate the database and provide representative structures for many protein families.

Structure comparison allows us to recognize distant relatives belonging to the same protein family. Since function is thought to be conserved across protein families, classification of structures in this way will help in suggesting functional properties for experimentally uncharacterized relatives, with certain caveats in some families (see Chapter 10). Several groups have suggested that there may be only a few thousand domain families. Therefore clustering proteins into families will undoubtedly help in determining to what extent function is conserved across a protein family and can be predicted on the basis of local or global structural similarity.

References

Branden, C., and Tooze, J. (2000) *Introduction to Protein Structure*. Garland Publishing.

Brown, N., and Orengo, C.A. (1996) A protein structure comparison methodology. *Computers Chem* **20**: 359–380.

Gibrat, J., Madej, T., and Bryant, S. (1996) Surprising similarities in structure comparison. *Curr Op Struct Biol* **6**: 377–385.

Johnson, M. (2000) In: Higgins D., and Taylor W. (ed) *Bioinformatics, Sequence, Structure and Databanks*. Oxford University Press.

Koel, P. (2001) Protein structure similarities. *Curr Op Struct Biol* **11**: 348–353.

Orengo, C. (1994) Classification of protein folds. *Curr Op Struct Biol* **4**: 429–440.

Wood, T.C., and Pearson, W.R. (1999) Evolution of protein sequences and structures. *J Mol Biol* **291**: 977–995.

Protein structure classifications

Frances Pearl and Christine Orengo

- Methods for recognizing domain folds within multidomain structures are described.
- Methods for clustering proteins structures into fold groups and evolutionary related families are considered.
- Hierarchical classifications of protein structures are reviewed together with the statistics of structural classifications, i.e. population of homologous superfamilies, fold groups and protein architectures.
- The use of protein structure classifications to benchmark sequence comparison methods is described.

7.1   Concepts

We have already seen, in Chapter 1, how changes in the sequences of proteins during evolution; residue mutations, insertions and deletions, can give rise to families of homologous proteins related to a common ancestral gene. Because structure is much more highly conserved during evolution than sequence, very distant evolutionary relationships are more easily detected by comparing structures and a number of protein family classifications have arisen, providing information on the evolutionary relationships of proteins, based on comparisons of their known structures.

However, because sequence data are more extensive than structural data by nearly two orders of magnitude, protein family classifications were first established in the 1970s by Dayhoff and co-workers, based on sequence similarities (see Chapter 5 for detailed descriptions of sequence-based protein family resources). Structure classifications developed later in the 1990s as the protein databank (PDB) increased to more than a thousand structures. At present, in 2002, there are about 16,000 entries in the PDB, corresponding to 30,000 protein chains. However, some of these entries are synthetic or model proteins, peptides and DNA fragments.

Clustering proteins into structural families enables more profound analysis of evolutionary mechanisms, e.g. tolerance to structural change, recurrence of favored structural motifs, evolution of protein functions in families. Analysis of structural families and groups also reveals constraints on secondary structure packing and favored topological arrangements. Although the structural core appears to be highly conserved across most protein families, considerable embellishments to this core have been detected in some families sometimes leading to changes in the functions of the relatives (see Chapter 10).

Unlike sequence-based family resources, most of the structure classifications have been established at the domain level, as this is thought to be an important evolutionary unit and it is easier to determine domain boundaries from structural data than from sequence data. Chapter 11 describes the extent to which domain shuffling and recombination of domains appears to be occurring from a recent analysis of genomic data. Structural classifications are therefore an important resource for studying domain organization in the genomes as many of the protein families in sequence-based resources comprise more than one domain.

International structural genomics initiatives, which aim to solve representative structures for all the protein families identified in the genomes, have recently received funding and will yield thousands more structures over the next decade. More importantly, these initiatives will particularly target sequence families for which there are currently no known structures. Inevitably these will reveal distant relationships between sequence families. Clustering of the GenBank sequence database currently containing approximately half a million non-redundant sequences yields between 15,000 and 20,000 protein families, depending on the method used. However, the structural data currently enable relationships to be revealed between many of these families. It is possible to assign structures to nearly 10,000 of these sequence families, allowing them to be merged to approximately 1200 structural superfamilies, on the basis of structural similarities.

In fact, several groups have speculated that there may only be a few thousand protein superfamiles in nature and thus it is reasonable to hope that with the progress of the structure genomics initiatives, we shall soon have structural representatives for the majority of these superfamilies. Membrane-based structures are still underrepresented in the PDB as are other proteins which are difficult to crystallize and not amenable to solution by NMR. However, the structural genomics initiatives will fund research designed to improve basic technologies associated with structure determination, and may therefore also facilitate determination of these more difficult targets.

Other initiatives which will inevitably improve our understanding of evolutionary mechanisms are the recently established collaborations between some of the major structure-based and sequence-based protein family classifications. The InterPro database is, in fact, an integrated resource linking major sequence-based family databases (PRINTS, PROSITE, TIGRFAM, Pfam, ProDOM, SMART, SWISS-PROT) (see Chapter 5). These initiatives aim to generate an integrated and consensus view of protein family data, whilst maintaining complementary resources associated with search facilities and analysis of the data. InterPro is now being extended to integrate the protein structural classifications, SCOP and CATH.

7.2 Data resources

There are now several protein structure classifications accessible over the web (see *Table 7.1*). These have been constructed using a variety of different algorithms and protocols for recognizing similarities between the proteins and for clustering relatives into structural groups. Some classifications build partly on the classifications developed by other groups, adding their own specialized data. For example the HOMSTRAD database of Sowdhamini *et al.* uses the superfamilies identified by the SCOP classifications but provides multiple structural alignments for relatives in these families using an in-house comparison algorithm, COM-PARER, developed by Sali and Blundell.

Some of the larger classifications are described in more detail in the text to illustrate the concepts used and the challenges faced in generating protein family resources. Further details of some of the classifications (e.g. SCOP, CATH) can be obtained from the textbook by Bourne, cited at the end of this chapter.

7.3 Protocols used in classifying structures

7.3.1 Removing the redundancy and identifying close relatives by sequence-based methods

A large proportion of structures in the Protein Databank (PDB) are nearly identical, corresponding to single-residue mutants, which have been determined to establish the effect of a mutation on the structure and function of the protein. There are also many identical proteins crystallized with different ligands or co-factors. Relatives with highly similar sequences can be easily detected using the sequence comparison techniques described in Chapter 3. Currently this reduces the size of the dataset by an order of magnitude. Sequence-based alignment methods are one or two orders of magnitude faster than structure-based methods, depending on the algorithms, so this is an important step. However, if the complete chains are being compared it is important that the alignment shows significant overlap between the two protein sequences, especially as many known structures are multidomain (see below) and may possess unusual combinations of domains not present in any other known structure. Failure to impose this condition may result in proteins being clustered on the basis of similarity between individual domains despite differences in domain compositions and/or organization.

Standard pairwise sequence comparison methods are generally used (e.g. based on dynamic programming or hashing to compare larger datasets, see Chapter 3). These algorithms are reliable provided there is significant sequence identity over the complete sequence (e.g. ≥35% identity). However, it is important to bear in mind that smaller proteins can yield higher sequence identities, simply by chance (see Chapter 3, *Figure 3.1*). To ensure a low error rate the equation describing the relationship between the sequence identity exhibited by homologous proteins as a function of size, established by Rost and Sander, can be used. This was revised recently using a much larger dataset.

More recently, powerful profile-based methods such as PSI-BLAST and Hidden Markov models such as SAM-T99 (reviewed in Chapter 5) have been used to identify potential homologs having lower levels of sequence identity in the twilight zone (<30%), which can then be verified structurally. Structural classifications have been used to benchmark these methods, since distant homologs are more reliably detected by structure and have often been manually validated, for example in the SCOP and CATH resources.

Both these databases have been used to determine optimum parameters for some of the more widely used profile-based methods and Hidden Markov models (see below). About 70% of very distant homologs, can currently be found using these profile-based methods, tolerating a very low error rate (<0.1%). Again, this is important in reducing the number of structural comparisons which must be performed to classify newly determined structures.

These approaches will become increasingly sensitive as the international genome initiatives proceed and the structure and sequence databases expand, in turn populating the protein family databases with a broader range of sequence relatives. Currently, nearly three-quarters of all known structural relatives can be identified using sequence-based methods, both pairwise (nearly 60%) and profile-based (a further 15%).

7.3.2 Identifying domain boundaries

At present (2002), 40% of known protein structures are multidomain. However, recent genome analysis suggests that a much higher proportion of genes are multidomain, estimates range from ~60% in the unicellular organisms to nearly 80% in metazoa and over 90% in eukaryotes (see Chapter 11 for more details). In the early days of crystallography, many large polypeptide chains were fragmented at putative domain boundaries to assist structure determination. However, as the technologies associated with X-ray crystallography improved, a higher proportion of multidomain proteins have been solved, and as the technologies improve further this trend is likely to increase.

Detecting domain boundaries by sequence-based methods is unreliable for distant homologs. Once the 3D structure is known, putative boundary definitions can be refined by manual inspection using graphical representations. The program RASMOL by Sayle, which is freely available, provides excellent 3D representations of protein structures which can be easily rotated. However, in some large multidomain proteins it can be difficult to detect boundaries by eye, particularly for discontiguous domains in which one domain may be disrupted by insertion of another domain (see *Figure 7.1*).

Table 7.1 Protein structure classification resources

Database	Location and author	Coverage (in January 2002)	Structure comparison method	Type	Description	URL
3Dee	EBI, Cambridge, UK Barton	7231 PDB entries	STAMP (Russell and Barton, 1993)	Fully automatic – utilizes the SCOP classification	Multi-hierarchical classification of protein domains	http://jura.ebi.ac.uk:8080/3Dee/help/help_intro.html
CAMPASS	Cambridge University, UK Sowdhamini	4612 protein domains 1067 superfamilies	COMPARER (Sali and Blundell, 1990)	Mixture of SCOP superfamilies and those derived from the literature	**CAM**bridge database of **P**rotein **A**lignments organized as **S**tructural **S**uperfamilies. Contains a compilation of sequence alignments of proteins that belong to a superfamily	http://www-cryst.bioc.cam.ac.uk/~campass/
CATH	UCL, London, UK Orengo	13938 PDB entries. 34287 fully classified domains. 1386 superfamilies. 3285 sequence families.	SSAP (Taylor and Orengo 1989)	Semi-automatic. Some manual validation of homologous superfamilies required	CATH is a hierarchical classification of protein domain structures, which clusters proteins at four major levels, **C**lass, **A**rchitecture, **T**opology and **H**omologous superfamily	http://www.biochem.ucl.ac.uk/bsm/cath_new/
CE	SDSC, La Jolla, CA, USA Bourne	14878 PDB entries. 28687 chains.	CE, Shindyalov and Bourne (1998)	Fully automatic. (Nearest neighbor)	Combinatorial **E**xtension of the optimal path Databases of pairwise alignments for all polypeptide chains, kept current within the PDB	http://cl.sdsc.edu/ce.html
DDD	EBI, Cambridge, UK Holm	11886 PDB entries. 21493 chains. 35492 domains.	Dali (Holm and Sander, 1993)	Fully automatic classification	**D**ali **D**omain **D**ictionary A structural classification of recurring protein domains. Domain boundaries are identified automatically by searching for topological recurrence in large, compact units from known protein structures	http://www2.ebi.ac.uk/dali/domain/

	Location/Author	Statistics	Method/Algorithm	Classification	Description	URL
DHS	UCL, London, UK Bray	903 Homologous superfamilies. 22295 domains.	SSAP (Taylor and Orengo, 1989) CORA (Orengo 1999)	Fully automatic, relies on CATH classification	Dictionary of Homologous Superfamilies Structural alignment of members of homologous superfamilies defined in the CATH database. Data are augmented with SWISS-PROT keywords and EC numbers	http://www.biochem.ucl.ac.uk/bsm/dhs
ENTREZ/ MMDB	NCBI, Bethesda, MD, USA Bryant	~15,000 PDB entries. ~35,000 chains. ~50000 domains.	VAST (Madej et al., 1995; Gibrat et al, 1996)	Fully automatic (Nearest neighbor)	MMDB contains pre-calculated pairwise comparison for each PDB structure. Integrated into ENTREZ	http://www.ncbi.nlm.nih.gov/Structure
FSSP	EBI, Cambridge UK Holm	2977 sequence families. 27946 protein structures (chains).	Dali (Holm and Sander, 1993, 1996)	Fully automatic (Nearest neighbor)	Fold classification based on Structure-Structure alignment of Proteins. Integrated sequence alignments	http://www.embl-ebi.ac.uk/dali/fssp/
HOMSTRAD	Cambridge University Blundell	2898 structures. 864 families.	COMPARER (Sali and Blundell, 1990)	Manual classification. Relying on SCOP, Pfam, PROSITE and SMART	(HOMologous STRucture Alignment Database) Database of annotated structural alignments for homologous families	http://www-cryst.bioc.cam.ac.uk/~homstrad/
SCOP	LMB, MRC, Cambridge, UK Murzin	14729 PDB entries. 1007 superfamilies 35685 domains. 1699 families.	None	Manual classification	A Structural Classification Of Proteins. Hierarchical classification of protein structure manually curated. The major levels in the classification are family, superfamily, fold and class	http://scop.mrc-lmb.cam.ac.uk/scop/

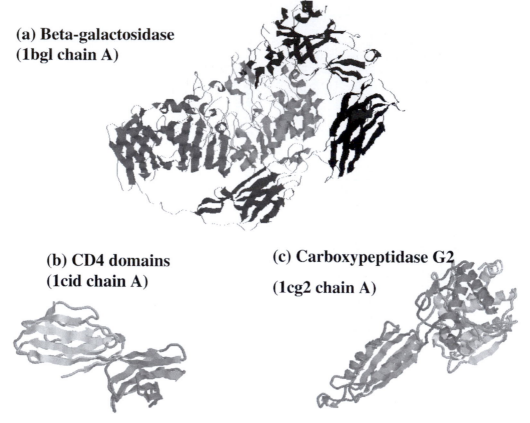

(a) Beta-galactosidase (1bgl chain A)

(b) CD4 domains (1cid chain A)

(c) Carboxypeptidase G2 (1cg2 chain A)

Figure 7.1

Example of some multidomain structures: (a) large complex multidomain; (b) simple two-domain structure; (c) discontiguous multidomain.

However, automatic methods for discerning domain boundaries are also hindered by the lack of any clear quantitative definition of a domain. Domains were first suggested to be independent folding units by Richardson. Various qualitative definitions exist, for example the following criteria are often used in assessing domain regions within a structure:

- the domain possesses a compact globular structure
- residues within a domain make more internal contacts than to residues in the rest of the polypeptide
- secondary structure elements are usually not shared with other regions of the polypeptide
- there is evidence for existence of this region as an evolutionary unit. For example in other structural contexts (i.e. different multidomain structures) or as a single domain.

This latter concept of recurrence is used strictly in generating the SCOP classification and domains are identified manually by visually inspecting the structures. In other resources such as CATH, automated procedures have also been implemented to identify putative domains which may not yet have been observed in other contexts. In the DALI Domain Database (DDD), an automatic algorithm is used to identify putative domains which are then validated by searching for recurrences in other structures.

Numerous automated methods have been developed for identifying domain boundaries

Table 7.2 List of commonly used domain boundary assignment programs

Acronym	Description	URL or reference
DBS	**D**omain **B**oundary **S**uite Comparison of the outputs from three different domain boundary programs, DETECTIVE, DOMAK and PUU. The consensus (if present) is taken as the domain definition	Jones *et al.* (1998)
DETECTIVE	A procedure for **DETECT**ing structural domains in proteins. Identifies hydrophobic core of each domain	Swindells (1995)
DOMAK	**DOM**ain **MAK**er Algorithm based on the premise that a domain will make more internal than external contacts. A 'split value' is calculated when the protein is arbitrarily divided into two parts. The value is large when the two parts are distinct	Siddiqui & Barton (1995)
STRDL	**STR**uctural **D**omain **L**imits Uses the Kernighan–Lin graph heuristic to partition the protein into residue sets that display minimum interactions between them. The interactions are deduced from a weighted Voronoi diagram, and the generated partitions are accepted or rejected on the basis of optimized criteria, representing the basic expected physical properties of structural domains	Wernisch, Hunting & Wodak (1999)
DOMS	The basic method is similar to an Isling model in which the structural elements of the model changes state according to a function of the state of their neighbors. Each residue in the protein chain is assigned a numerical label. If a residue is surrounded by neighbors with (on average) a higher label its label increases, otherwise its label decreases	Taylor (1999)
PUU	Parser for **P**rotein **U**nfolding **U**nits Algorithm based on the premise that a domain will make more internal than external contacts. Uses a harmonic model to approximate inter-domain dynamics	Holm & Sander (1994)
DAD	Algorithm based on the premise that a domain will make more internal than external contacts. Chain is divided to minimize the density of inter-domain contacts	Islam, Luo & Sternberg (1995)

and with varying degrees of success (see *Table 7.2*). Some approaches such as 'Detective' by Swindells, attempt to identify clusters of inaccessible hydrophobic residues which could potentially form the core of a structural domain. These clusters can then be extended by including residues in contact with them and also involved in multiple contacts with other residues. Other methods, such as Holm and Sanders' PUU and Siddiqui and Barton's DOMAK, attempt to divide the polypeptide chain at positions which would maximize the ratio of internal residue contacts within the putative domains to external contacts between them.

Most of the algorithms perform well for simple multidomain structures in which the separate domains have minimal contact (see *Figure 7.1b*). However, for more complex multidomains in which the domains are more intimately connected and which may also contain discontiguous domains (*Figure 7.1c*), the performance is more disappointing. A recent comparison of several available algorithms using a validated domain dataset suggested an average accuracy of between 70–80%. Consensus approaches can be used which seek agreement between the results obtained from different algorithms, preferably using complementary philosophies. Such protocols typically result in agreement, within about 10 residues, for less than 15% of the multidomains analyzed. However, since many of the algorithms identify reasonable boundaries for a large proportion of multidomains, and since they are generally very fast (<1 s), these methods can be used to suggest boundaries which can then be validated manually.

As discussed above, the concept of domain recurrence can also be used to validate domain boundaries within multidomain proteins. Currently about 70% of domain structures within multidomain proteins recur in different domain contexts or as single-domain proteins. As the structure genomics initiatives proceed and there are increasing representatives for each domain family, these approaches will become more successful.

7.3.3 Detecting structural similarity between distant homologs

7.3.3.1 Pairwise structure alignment methods

Chapter 6 reviews many of the concepts employed in structure comparison and describes some of the most commonly used methods. *Table 7.1* lists the structure comparison algorithms adopted by the protein structure classification resources accessible over the web. Approaches range from the largely manual protocol adopted by the SCOP database through to the completely automated protocol used for constructing the DALI Domain Dictionary. Many groups have developed fast approximate methods for rapidly scanning new structures against a library of structural representatives (for example SEA (Rufino and Blundell), VAST (Bryant *et al.*), GRATH (Harrison *et al.*), TOPSCAN (Martin)). Usually these compare secondary structures between proteins to identify putative relatives which can then be realigned using more accurate methods. A large proportion of structural relatives, at least 90%, can be recognized in this way and since these methods are two to three orders of magnitude faster than residue-based comparison, this considerably reduces the time required for classifying new structures.

Dali Domain Dictionary, CATH (GRATH algorithm) and CAMPASS (SEA algorithm) use fast database-searching methods based on secondary structure to identify potential relatives which are then subjected to more reliable, but computationally demanding, residue-based methods. Many of these algorithms employ graph-theoretic approaches to recognize common motifs between the structures, comparing simplified graphs of the proteins derived from the relationships between their secondary structures (see Chapter 6 for a description). Several of the slower but more accurate residue-based methods used in comparing structures (COMPARER - HOMSTRAD, SSAP - CATH, STAMP - 3Dee) employ dynamic programming algorithms to handle the extensive insertions/deletions expected between distant homologs.

Differences between the classifications can arise due to variation in the sensitivity and accuracy of these methods, but recent initiatives to integrate the sequence- and structure-based family resources within InterPro will have the benefit of combining many of the resources thereby maximizing the number of distant relationships detected.

7.3.3.2 Multiple structure alignment and profile-based methods

Several of the classifications (HOMSTRAD, DaliDD, CATH, 3DEE) provide multiple structure alignments within families and superfamilies. Multiple alignments enable detection of conserved structural features across a family which can be encoded in a template for that family. In the HOMSTRAD database, multiple alignments have been generated for each superfamily and used to derive substitution matrices describing preferences for residue exchange in different structural environments. Although, to date, such matrices have not performed as well as those derived using sequence-based families (e.g. BLOCKS database for the BLOSUM matrices), recent updates performed well and as the structure databank expands with the structure genomics initiatives and more diverse representatives are determined, these approaches will become increasingly powerful.

The comprehensive FSSP resource generated using the DALI algorithm, provides multiple alignments for all related structures identified as having either local or global structure similarity. In CATH, multiple structure alignments are provided for well-populated superfamilies, currently ~900, and are accessible through the web-based dictionary of homologous superfamilies (DHS). Sowdhamini's CAMPASS database contains multiple alignments for ~1000 superfamilies extracted from the SCOP database.

7.4   Descriptions of the structural classification hierarchy

Most of the larger classifications (e.g. SCOP, CATH, DaliDD, 3Dee) describe similar hierarchical levels in their structural classifications, corresponding largely to phylogenetic and phonetic, that is purely structural, similarities between the data. These are summarized in *Table 7.3*.

At the highest level in the hierarchy, proteins are grouped according to their structural class, where class refers to the composition of residues in α-helical or β-strand conformations. There are some differences between the resources. Many of the classifications recognize four major classes; mainly-α, mainly-β, alternating α-β in which α-helices and β-strands alternate along the polypeptide chain and α plus β in which mainly-α and mainly-β regions are more segregated. The CATH classification, which uses an automatic protocol for assigning class (see *Table 7.1*) merges the last two categories into a single α-β class. Other groupings at this top level in the hierarchy of these classifications are mostly bins containing proteins having

Table 7.3 Description of each level in hierarchical structural classifications

Level in hierarchy	Description
Protein class	The class of a protein structure reflects the proportion of α-helices or β-strands within the three-dimensional structure. The major classes are mainly-α, mainly-β, alternating α/β and α+β. In CATH the α/β and α+β classes are merged
Protein architecture	This is the description of the gross arrangement of secondary structures (α-helices and β-strands) in three-dimensional space, independent of their connectivity
Protein fold/topology	This is the description of the gross arrangement of secondary structures in three-dimensional space, and is dependent on the orientation of secondary structures and the connectivities between them
Homologous superfamily	A homologous superfamily is a group of proteins whose structures and functions suggest a common evolutionary origin
Homologous family	Proteins clustered together into families are clearly evolutionarily related. Generally, this means that pairwise residue identities between the proteins are 30% and greater. However, in some cases similar functions and structures provide definitive evidence of common descent in the absence of high sequence identity

few secondary structures or model or synthetic proteins. Currently, 98% of known structures can be classified into the four major protein classes.

The CATH classification contains a hierarchical level beneath class, which describes the architecture of the protein. This refers to the way in which the secondary structure elements are packed together, that is the arrangement of secondary structures in 3D space regardless of the connectivity between them. Other resources, such as SCOP, frequently assign an architectural description to a particular structural family, e.g. β sandwich, without formally recognizing a separate hierarchical level within the classification. There are currently 28 major architectural groups in the CATH classifications. *Figure 7.2* shows examples of representative structures from each, some of which are more highly populated than others (see below). Structures outside these categories are generally highly irregular, complex packings of secondary structures or contain few secondary structures.

At the next level in the classification, the fold group or topology level classifies structures according to both the orientation of secondary structures and the connectivity between them. *Figure 7.3* shows an example of two structures possessing a similar 3-layer α-β sandwich architecture but which have different secondary structure connectivities and which are therefore classified into separate fold groups.

In many fold groups, structural relatives are all clearly related by divergent evolution (see below). However, over the last 10 years there have been increasing examples of proteins adopting similar folds but possessing no other characteristics indicative of homology, that is no similarities in their sequences or in functional attributes. Therefore within each fold group proteins are further classified according to their homologous superfamily. Protocols for recognizing evolutionary relatives differ between the various classifications and are discussed in more detail below.

Within each superfamily, most classification resources group together more closely related proteins into families. These may be identified by sequence similarity as in the DDD or CATH databases, where close homologs are clustered at more than 35% identity. Alternatively, families may be recognized by manually considering similarity in functional properties as in the SCOP database. Many resources also cluster at several further levels according to sequence identity, for example at 60%, 95% and 100% identity to facilitate the selection of representatives.

Several groups have suggested that protein folds may be constrained by the requirement to maintain optimal packing of secondary structures and preserve the hydrophobic core of the structure and this may restrict the number of possible folds. Estimates based on the number of sequence families identified in the genomes and the frequency with which novel folds are detected for apparently unique sequence families currently range from less than a thousand to a few thousand. Proteins sharing structural similarity but no other evolutionary characteristics are described as analogs. Analysis of both the SCOP and CATH databases has demonstrated that some fold groups are particularly highly populated with apparently diverse superfamilies. These have been referred to as superfolds by Orengo and co-workers or 'frequently occurring domains' – FODs, by Brenner and co-workers. It is clear from *Figure 7.5* below that some protein folds have considerably more analogs than others. In fact, approximately 20% of the homologous superfamilies currently known belong to fewer than 10 different fold groups.

7.4.1 Neighborhood lists

Some of the data resources do not attempt to cluster structural relatives explicitly into fold groups or homologous superfamilies but rather provide the user with lists of putative structural neighbors (see *Table 7.1*). The DALI resource was the first to provide this type of information through a web-based server which accepted 3D co-ordinate data and searched a non-redundant subset of the PDB, returning lists of structurally similar proteins. The

α Bundle (2ccy)

α Non-Bundle (1eca)

α Horseshoe (1lrv)

α Solenoid (1pprM)

αα Barrel (1cem)

β Roll (1pht)

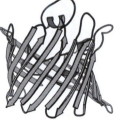
β Barrel (2por)

β Clam (3bcl)

β Sandwich (2hlaB)

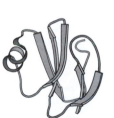
β Distorted Sandwich (1cdq)

β Trefoil (1afcA)

β Orthoganol Prism (1msaA)

β Aligned Prism (1vmoA)

β 4-Propellor (1hxn)

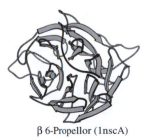
β 6-Propellor (1nscA)

Figure 7.2

Molscript representations of representatives from the 28 major regular architectures in the CATH domain structure classification.

β 7 Propellor (2bbkH)

β 8 Propellor (3aahA)

β 2 Solenoid (1tsp)

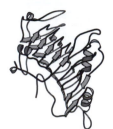
β 3 Solenoid (2pec)

β Complex(1ppkE2)

αβ Roll(1std)

αβ Barrel (4timA)

αβ 2-Layer Sandwich (1brsD)

αβ 3-Layer Sandwich(aba) (1ı

αβ 3-Layer Sandwich(bba) (1pya

αβ 4-Layer Sandwich (2dnjA)

αβ Box (1plq)

αβ Horseshoe (1bnh)

αβ Complex (1pyp)

αβ Propellor (1h70A)

Figure 7.2 – *continued*

Molscript representations of representatives from the 28 major regular architectures in the CATH domain structure classification.

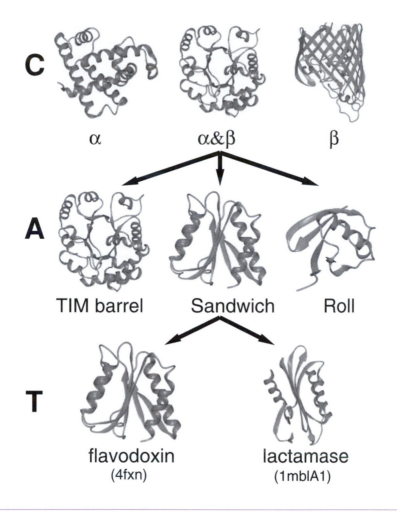

C

α   α&β   β

A

TIM barrel   Sandwich   Roll

T

flavodoxin   lactamase
(4fxn)   (1mblA1)

Figure 7.3

Illustration of top levels class, architecture and topology, in the hierarchy of the CATH domain classification; class, architecture, topology or fold group.

significance of the matches was described by a Z-score (see Glossary) calculated from the database scan. More recently, the Entrez database has established a similar resource based on the VAST structure comparison algorithm and the PDB also has its own search facility using the CE algorithm. In the CATH server, a fast search algorithm (GRATH) is used to provide lists of structural neighbors whose similarity can be assessed by the expectation E-values returned by the method.

7.4.2 Comparisons between the different classification resources

A recent analysis of three of the most comprehensive classifications (SCOP, DALIDD, CATH) undertaken by Hadley and Jones, revealed that there was considerable agreement between the different resources, even at the fold group (>75%) and superfamily (>80%) levels. Differences identified were due to the alternative protocols employed and particularly to differences in the automated thresholds used in recognizing structural relatives and clustering them into fold groups and homologous superfamilies.

In the SCOP database, structural similarity is assessed manually and there are no quantitative criteria for determining whether proteins should be assigned to the same structural group. This allows more flexibility in recognizing some of the very distant relatives, where other resources using numeric criteria are constrained by the need to maintain a low error rate. Considerable human expertise is brought to bear in maintaining the SCOP resource in contrast to other classifications which rely on more automated approaches.

Most other classifications (e.g DALI, CATH, 3Dee) calculate a structural similarity score (e.g. RMSD see Chapter 6) and thresholds are set for recognizing related proteins. Generally, these have been established by empirical trials using validated data on known structural relatives. However statistical approaches for assessing significance are more reliable than raw scores like RMSD (see Chapter 6 and Glossary). Both the DALI Domain Dictionary and the Entrez resources calculate a Z-score from the distribution of scores returned from scanning a new structure against non-redundant representatives from the database. However, several analyses have suggested that the distributions returned from database scans are best represented as extreme value distributions. Therefore more recent approaches for assessing significance have employed numerical approaches to fitting these distributions in order to calculate the probability (p-value) of unrelated structures returning a particular score. These can be corrected for database size to give an expectation value (E-value) of random match, for assessing significance (see Glossary).

As well as differences in methods used for assessing the significance of the structural match, variations in clustering procedures also result in some differences in fold groups between the resources. Single linkage clustering is the most widely used approach since the structural data are still relatively sparse and newly determined structures may represent outliers to existing structural families and groups. However, in the SCOP database some of the larger fold groups are maintained as smaller separate groups by clustering according to characteristics which are shared by many of the relatives.

The problem of identifying distinct fold groups is compounded by the recurrence of favored structural motifs in the different protein classes. These are often described as super-secondary motifs as these comprise two or more secondary structures (e.g. β-hairpins and β-meanders in the mainly-β class, α-hairpins in the mainly-α class and α–β and split α–β motifs in the α-β class). This phenomenon, which has been described as the structural continuum, can lead to many fold groups being linked by virtue of their common motifs depending on the degree of structural overlap required for clustering two proteins. In this context, the notion of a fold group is rather subjective and may be somewhat artificial, though it is often an extremely useful means of recognizing structural similarities which may reflect constraints imposed on the folding or stability of protein structures.

It is also clear from analysis of some superfamilies, that although much more conserved than sequences, structures can vary considerably in very diverse relatives. Although most analyses have shown that generally at least 50% of the residues, usually in the buried core of the protein, will have very similar conformations often these 'core fold motifs' are structurally embellished in different ways (see, for example, *Figure 7.4*). This can complicate the assignment of homologous relatives to the same fold group and Grishin and others have shown examples of clearly homologous proteins adopting slightly different folds.

Neighborhood lists provided by some of the resources can be used to view alternative clusterings generated by selecting different thresholds on structural similarity. Harrison and co-workers have shown that the continuum effect is more pronounced for certain architectural types e.g. mainly-β and αβ sandwiches. Folds adopting these architectures often share β-hairpin and α/β motifs, whilst some 15% of fold groups (e.g. the trefoil) were clearly very distinct, sharing no super-secondary motifs with other fold groups.

7.4.3 Distinguishing between homologs and analogs

Other significant differences between the resources can be attributed to methods used for distinguishing homologs and analogs. Again, the outstanding expertise of Murzin has led to the

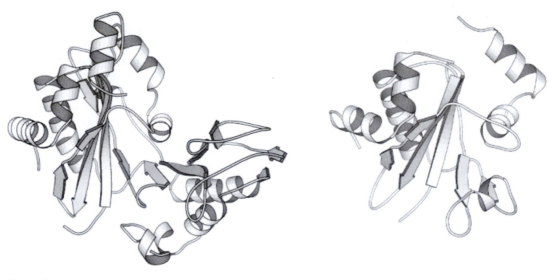

Figure 7.4

Structurally diverse relatives from the ATP Grasp superfamily in the CATH database.

recognition of some extremely diverse homologous relatives being recognized in the SCOP database. Several other classification resources have opted to benefit from the quality of these assignments by extracting superfamily relatives from the SCOP database to seed their own structural families (e.g. 3Dee, HOMSTRAD, see *Table 7.1*).

In the DALI domain database, completely automated protocols are employed. These seek significant structural similarity between proteins as measured by a Z-score and also require similarity in functional annotations using various data-mining methods based on keyword matching and text analysis. In the CATH database a combination of automated and manual procedures are used. Significant structural similarity as measured by E-values returned from a structure comparison algorithm, together with matches to specific 3D templates for homologous superfamilies and matches to Hidden Markov models associated with homologs sharing common functional attributes. Manual validation is based on expert analysis of associated functional annotations from various public resources (e.g. SWISS-PROT, Enzyme database, Entrez).

Templates encoding conserved sequence or structural attributes within a superfamily, only rarely occurring in non-relatives are amongst the most powerful approaches for recognizing homologs. Sequence-based approaches using 1D profiles have already been described above. Murzin has suggested that unusual structural characteristics such as β-bulges or rare β-turns can often be used to identify homologs and has also proposed methods for assessing homology based on the statistical significance of common sequence motifs. Grishin has recently proposed a similar approach.

7.4.4 Recognizing homologs to structural families in the genomes: Intermediate sequence libraries

Several structural classification resources have recently recruited sequence relatives from the genome initiatives and sequence databanks (e.g. SWISS-PROT, TREMBL, GenBank) (see also Chapter 11). These were identified using various search algorithms (e.g. BLAST, PSI-BLAST) and Hidden Markov models. Generally the classifications have been expanded 10-fold or more by including these relatives which are either explicitly recruited into superfamilies or provide lists of putative sequence relatives for each structural superfamily. In addition to increasing the functional annotations available within each superfamily, this expansion increases the probability of obtaining a match when searching the database with the sequence of a putative relative. These expanded resources have been described as 'intermediate sequence libraries' because they can improve recognition of distant relatives through a chain of pairwise sequence relationships between 'intermediate' sequences.

Several groups have demonstrated the increased sensitivity which can be obtained by scanning intermediate sequence libraries for each structural superfamily rather than simply scanning a dataset of representative structural relatives (see Chapter 3 for more discussion). For example, for both the CATH (PFDB) and SCOP (PDB-ISL) ISS resources the performance in recognizing very distant homologs using sequence-based methods such as PSI-BLAST or SAM-T99 increased from between 30 and 50% depending on the search method.

7.5   Overview of the populations in the different structural classifications and insights provided by the classifications

As well as providing insights into the mechanisms of structural evolution in different protein families, the structural classifications, have proved valuable resources for validating sequence-based search algorithms. Since structure is much more highly conserved than sequence in protein families, distant relatives can be much more readily identified by structural means and where these relationships have also been validated by functional similarity assessed by human expertise, these relatives provide useful datasets for validating relationships suggested by sequence-based methods such as 1D profiles, Hidden Markov models and threading algorithms. Both SCOP and CATH resources have been used for benchmarking a variety of different sequence search algorithms.

7.5.1   Populations of different levels in the classification hierarchy

Table 7.1 summarizes the number of structural entries in the publicly accessible classifications as of December 2001. The DALI domain database is currently the most comprehensive containing all current entries from the PDB which meet the acceptance criteria for the database. The SCOP and CATH databases also contain a large proportion of the known structures. The CATH resource is the only classification to explicitly assign architectures to all domain entries and recent analyses of this resource revealed that nearly half the superfamilies in the database belonged to six simple, regular architectures, comprising α-bundles, β-barrels, β-sandwiches, αβ-two layer sandwiches, αβ-three layer sandwiches and αβ-barrels. However, at least 28 distinct architectures are currently described in this resource (see *Figure 7.2*).

The numbers of fold groups currently identified vary between 500–750 for the different resources, depending on the number of entries classified and protocols for comparing and clustering them. Similarly, the number of homologous superfamilies currently identified is in the range of 1000–1200 for the more comprehensive SCOP, DDD and CATH classifications. These superfamilies are not evenly populated amongst fold groups. *Plate 3* shows that some fold groups are particularly highly populated and at present contain nearly one-fifth of all known structural superfamilies. It is not clear whether their recurrence reflects a particularly stable arrangement of secondary structures or an ability to fold rapidly. Alternatively, it may become clear as new structures are determined that many of the superfamilies within these fold groups are in fact evolutionary related but possess the ability to diverge considerably acquiring new functions.

There appears to be considerable variation in the degree to which the structure is conserved across relatives in each family, although the data are currently limited in that some families have very few representatives. In the CATH resource, there are at present only 500 superfamilies with three or more diverse representatives, 100 superfamilies with five or more.

Analysis of these more highly populated families, reveals that some families exhibit much greater structural plasticity than others. That is there is considerable structural change accompanying the same degree of sequence variation. Although it is likely there will be more constraints on the structural embellishments between orthologous relatives, in order to maintain the function of the protein, than between paralogs. In some cases variation within homologous superfamilies is greater than differences between some of the fold groups. However, for at least three-quarters of these families 50% or more of the secondary structures are highly conserved across the family, generally within the protein core.

Structural change can be restricted to inserted residues in loops or in some cases to elaborate embellishments decorating the conserved core found in all the relatives (see *Figure 7.4*). Consequently, it may be better to use different criteria or thresholds for assessing structural similarity depending on the family or class of structure being considered. In this context multiple structural alignments of representatives from the family can be used to identify particular characteristics specific to the superfamily.

In some superfamilies, structural embellishments are clearly associated with modification to the function of the protein, sometimes resulting in changes in the shape and accessibility of the active site and also in providing additional surface contours associated with regulation or novel protein–protein interactions or oligomerization states, acquired by these relatives. Chapter 10 discusses mechanisms by which functions evolve in protein superfamilies. In some superfamilies, embellishments appear to have no effect on the function but are presumably tolerated because they do not impact on protein stability or function. In the future, the projected expansion of the structure databank with the structural genomics initiatives, should allow much more detailed analysis of the means by which structural changes modulate function.

References

Hadley, C., and Jones, D.T. (1999) A systematic comparison of protein structure classifications: SCOP, CATH and FSSP. *Structure Folding and Design* **7**: 1099–1112.

Holm, L., and Sander, C. (1996) Mapping the protein universe. *Science* **273**: 595–603.

Holm, L., and Sander, C. (1998) Dictionary of recurrent domains in protein structures. *Proteins* **33**: 88–96.

Murzin, A.G., Brenner, S.E., Hubbard, T., and Chothia, C. (2000) SCOP: a structural classification of proteins for the investigation of sequences and structures. *J Mol Biol* **247**: 536–540.

Orengo, C.A. (1994) Classification of protein folds. *Current Opinion in Structural Biology*. **4**: 429–440.

Orengo, C.A., Michie, A.D., Jones, S., Jones, D.T., Swindells, M.B., and Thornton, J.M. (1997) CATH – a hierarchical classification of protein domain structures. *Structure* **5**: 1093–1108.

Sowdhamini, R., Burke, D.F., Deane, C., Huang, J.F., Mizuguchi, K., Nagarajaram, H.A., Overington, J.P., Srinivasan, N., Steward, R.E., and Blundell, T.L. (1998) Protein three-dimensional databases: domains, structurally aligned homologues and superfamilies. *Acta Crystallogr* D. **54**: 1158–1177.

Comparative modeling

Andrew C.R. Martin

- Techniques are introduced for building a three-dimensional model of a protein of known sequence, but unknown structure, using known related structures as templates – a technique known as 'comparative modeling'.
- The powers and limitations of the method are discussed. We aim to give a feel for the level of accuracy that may be achieved and factors that influence the quality of a model are described.
- Pointers on how to evaluate the quality of a model are given.

8.1 Concepts

Comparative modeling allows us to build a three-dimensional (3-D) model for a protein of known amino acid sequence, but unknown structure, using another protein of known sequence and structure as a template.

Comparative modeling is also known as *'homology modeling'*, implying that models are always generated from homologous proteins. Generally this is indeed the case. However, techniques such as fold recognition and threading (Chapter 9) allow us to recognize that a sequence may adopt the same fold as a non-homologous protein and, providing a satisfactory alignment can be achieved, a 3-D model can be built. However, in these cases, the quality of the model will be much lower.

We present two major approaches to comparative modeling: (i) 'traditional' methods based on splitting a structure into conserved and variable regions and modeling these separately; and (ii) a restrained molecular dynamics 'simulated annealing' approach in which the whole structure is modeled simultaneously using constraints derived from known parent structures. Other more complex methods are mentioned briefly. We then go on to discuss the major factors which contribute to the accuracy of a model and evaluate which regions are likely to be accurately modeled and which regions will be poor.

Every 2 years, an international experiment, the *'Critical Assessment of Structure Prediction'* (CASP) is conducted to assess the ability of groups around the world to perform structure prediction and modeling. Some of the discussion in this chapter is based on analysis of the comparative modeling section of the CASP experiments.

8.2 Why do comparative modeling?

The alternative to comparative modeling is to solve the structure of a protein using techniques such as *X-ray crystallography* or *nuclear magnetic resonance* (NMR). In both cases, it is important to note that the 'structure' obtained is still a model – one that best fits the data available from the X-ray or NMR experiment. As such, these 'structures' can, and do, contain errors. Proteins are, of course, dynamic. The process of 'freezing' a protein into a crystal limits the dynamics to some extent, but the set of coordinates one obtains from an X-ray crystal structure is still a time average of a dynamic structure. In the case of NMR structures, the constraints imposed by crystallization are not present and one is seeing a picture of the protein in solution. In addition to this true reflection of increased dynamics, the NMR method does not give data as precise as those available from X-ray crystallography and, as a result, structures solved by NMR are generally presented as an ensemble of models which all satisfy the available data.

Thus it is important to remember that models resulting from comparative modeling are themselves based on other models. What one normally considers to be the 'true structure' is itself a model – albeit (in the case of a high resolution X-ray crystal structure) one derived from large amounts of high-quality data and which satisfies the constraints implied by those data.

So, given that one is performing comparative modeling based on a gold standard which is itself a model, why bother? Are the results in any sense valid or useful? Before beginning any modeling exercise, one should ask why the model is being created and for what it will be used. 3-D models do not give answers. All they can do is offer hypotheses that can then be tested by experiment. That said, where the sequence identity between target and parent is high and there are few insertions and deletions required in forming the alignment, the differences between a model and the crystal structure are similar to the differences between two structures of the same protein solved in different space groups.

The major reason for performing comparative modeling is that X-ray crystallography and NMR are both time-consuming processes requiring expensive equipment. They also require relatively large amounts of pure protein and, in the case of crystallography, one must also be able to generate high-quality crystals capable of diffracting to high resolution. Despite recent advances and the availability of crystallization kits, this is still something of a 'black art' with the conditions under which any one type of protein will crystallize being apparently arbitrary and not predictable. In some (relatively rare) cases, soluble proteins simply cannot be purified in sufficient quantities and crystallized, while integral membrane proteins (making up some 20% of eukaryotic genomes) are immensely difficult to crystallize. Although methods for high-throughput crystallography are currently in development, the whole process of extracting and purifying protein, growing crystals, collecting data and solving the structure typically takes a minimum of 6–12 months and can take much more.

In contrast, given a protein sequence, it is possible to generate some kind of model in half an hour using services available over the Web. Assessing the quality of that model, refining and improving it, can take rather more time (in the order of days or weeks), but this still compares very favorably with the alternative experimental methods.

The volume of sequence data (which is now fast and easy to obtain) far outstrips structural data, so modeling is an important technique to span the sequence to structure gap. *The Protein Databank* (the repository of solved protein structures) currently contains:

- around 15,000 structures
- fewer than 10,000 distinct sequences (many structures are of the same protein)
- fewer than 5000 distinct structures (as defined in CATH or SCOP)

whereas Genbank currently contains more than 600,000 distinct protein sequences. As the genome projects continue apace, this gap will continue to widen. However, the advent of the structural genomics projects (see Chapter 10) will start to fill in the gaps in our structural database. Once we have three-dimensional structures for representatives from all fold classes and homologous families (Chapter 7), we will have parents available to make the modeling process more straightforward and reliable throughout the repertoire of globular proteins.

Integral membrane proteins are an immensely important class of proteins, which include the *G-protein coupled receptors* (GPCRs) which are targets for many drugs. Unfortunately, these remain an immense problem for experimental techniques as they are generally not amenable to X-ray crystallography or to NMR. Unfortunately, because few known structures are available, comparative modeling is also difficult.

8.3   Experimental methods

We will describe the traditional method for building a protein model in some detail. Much of this technique may be implemented manually, but is also automated in software such as *COMPOSER* (from Mike Sutcliffe and co-workers) and *SwissModel* (from Manuel Peitsch and co-workers) – the latter is freely available for use over the Web. We will then go on to give a brief overview of *MODELLER*, a comparative modeling package from Andrej Săli which uses a simulated annealing approach and has been shown to do consistently well in the CASP experiments. MODELLER is freely available for non-commercial use. The assessments of the biennial CASP experiments are presented in special issues of the journal *PROTEINS: Structure, Function, Genetics*, and the reader is encouraged to consult these for further information.

8.3.1   Building a model: Traditional method

Central to the traditional approach to comparative modeling, is the observation by Jonathan Greer in 1981 that insertions and deletions are generally accommodated by small local changes in *'Structurally Variable Regions'*, while the rest of the structure (the *'Structurally Conserved Regions'*) remains largely unchanged. Thus one identifies the conserved and variable regions and models these separately.

The procedure may be split into eight stages:

1. Identify one or more (homologous) sequences of known structure to use as *'templates'* or *'parents'* (these terms are used interchangeably). The template structures need not be homologous if they have come from fold recognition or threading, but in most cases they are. They will almost certainly have to be homologs of high sequence identity if a good-quality model is to be built.
2. Align the target sequence to be modeled with the parent(s).
3. Determine the boundaries of the 'framework' or 'structurally conserved regions' (SCRs) and the 'structurally variable regions' (SVRs). The latter are generally loops.
4. Inherit the SCRs from the parent(s).
5. Build the SVRs.
6. Build the sidechains.
7. Refine the model.
8. Evaluate errors in the model.

This type of approach may be implemented manually or automated with systems such as *SwissModel* (over the web) or *COMPOSER*. We will now go through each of these stages in turn.

8.3.1.1   Identifying parent structures

Normally this is done by searching the target sequence against the sequences in the *Protein Databank* (PDB, http://www.rcsb.org/) using standard sequence search tools such as *FASTA* and *BLAST* to identify homologs (see Chapter 3).

More distant homologs may be identified using *PSI-BLAST* or *Intermediate Sequence Searching* (see Chapter 3). Since PSI-BLAST builds *'profiles'* (position-specific mutation matrices) based on what sequence hits are found, in order to be effective, it requires a large non-redundant sequence database. One therefore tends to use protein sequences from a non-redundant version of *Genbank* or TrEMBL together with PDB sequence data. This enables links to sequences in the PDB to be identified which might not be found if the additional sequences were not included. The sequences in Genbank or TrEMBL allow the scoring matrix to evolve more effectively.

Parent structures may also be identified by threading or by fold recognition. However, the quality of resulting models is likely to be low since the subsequent alignment step will be difficult to achieve accurately.

8.3.1.2 Align the target sequence with the parent(s)

When the sequence identity between the target and the parent is high (>70%) the alignment is generally trivial. However, as the identity gets lower (in particular as it drops below 40%) and the number of insertions and deletions ('*indels*') gets higher, obtaining the correct alignment becomes very difficult. As described below, obtaining the correct alignment is the major factor in obtaining a good model.

The 'correct' alignment is the one which would be achieved if one had structures of the two proteins and performed a structural alignment, then derived a sequence alignment from that structural alignment.

Optimal sequence alignment is almost always achieved using a variation of the '*dynamic programming*' algorithm for finding the optimum path through a matrix, first applied to protein sequence alignment by Needleman and Wunsch in 1970 (see Chapter 3). However, a sequence alignment obtained from dynamic programming is unable to account for factors which come into play when the sequence folds into three dimensions.

In practice it is often necessary to hand-correct the resulting automatic alignment. Given that one knows the structure of one of the proteins in the alignment, one can make changes such that indels occur in loop regions rather than in regions of secondary structure. If one has multiple parent structures, it can be an advantage to align these structurally first, calculate a multiple sequence alignment based on this structural alignment and then align the target sequence to that multiple alignment.

Figure 8.1 shows an example where an apparently obvious automated alignment would require large structural rearrangements in order to make a four-residue deletion while a manually corrected alignment will lead to a deletion resulting in a very simple chain closure. If alignment Aln1 (the automatic alignment) were used then it would be necessary to close the large 3-D gap between the proline (P) and methionine (M) in the parent structure. If however, the manual alignment, Aln2, were used then the 3-D gap between the glutamine (Q) and leucine (L) is much smaller and easier to close. The latter manual alignment will almost certainly give a better-quality model than that obtained from the automatic alignment.

8.3.1.3 Identify the structurally conserved and structurally variable regions

A reliable definition of Structurally Conserved Regions (SCRs) is only possible if one has multiple parent structures, i.e. one is able to define which regions are conserved in structure between all the parents and have the same lengths in the target. Collectively, the structurally conserved regions are often termed the '*core*' region. Those regions which differ in structure between the parents are termed the Structurally Variable Regions (SVRs).

Where only one parent is available, then initially we assume that all regions are structurally conserved except where indels occur (almost exclusively in loop regions). Where an indel occurs in a loop, the whole loop is defined as an SVR. An alternative approach is to define all loop regions as SVRs.

8.3.1.4 Inherit the SCRs from the parent(s)

When building from a single parent, the SCR regions are simply copied from this parent and used for the model.

Where multiple parents are used, different approaches may be made. In any case, the first step is to fit the multiple parents to one another in three dimensions. Each SCR region may then be selected based on a number of factors. The relative importance of these factors in making the choice varies between the modeling software used and the preferences of the individual modeler. Examples are:

- the local sequence similarity to the target sequence
- local RMSD to an averaged structure of all parents

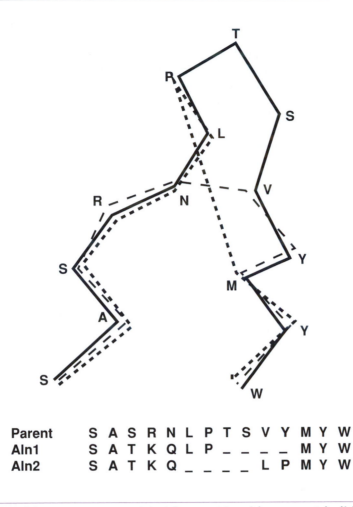

Parent	S	A	S	R	N	L	P	T	S	V	Y	M	Y	W
Aln1	S	A	T	K	Q	L	P	_	_	_	_	M	Y	W
Aln2	S	A	T	K	Q	_	_	_	_	_	L	P	M	Y	W

Figure 8.1

Example of a 10-residue sequence modeled from a 14-residue parent (solid lines). The automatic alignment (Aln1, dotted lines) would result in a very long distance between the 7th and 8th residues requiring substantial changes to the structure. However, the manually corrected alignment (Aln2, dashed lines) requires only minimal structural changes.

- low temperature factors in the crystal structure
- the length and/or sequence of the adjacent SVR(s).

This is one of the areas in which it is the skill and intuition of the modeler that plays an important factor in determining the quality of the resulting model.

The results of the CASP2 (1996) experiment suggested that using multiple models does not, in general, improve the quality of the resulting model and simply using the parent with the highest sequence identity with the target (the '*primary parent*') gives the best results. Other analysis has suggested that multiple parents do improve model quality.

8.3.1.5 Build the SVRs

SVRs are almost always *loop regions*. When these vary in length from the loops present in the parent structure(s), they will inevitably be built with lower accuracy than the rest of the structure. Even where loop regions are conserved in length they can adopt quite different conformations.

Three major approaches have been used:

- **by hand:** modified loops are built by hand using molecular graphics. Gaps are closed or residues introduced by modifying backbone torsion angles. Frequently this is followed by cycles of *energy minimization* (see 'Refining the Model', below). This type of approach was popular in the 1980s.
- **knowledge-based methods:** candidate loops are identified through database searching. Criteria for selecting loops must be defined and most frequently the separation of the ends of the framework where the loop is attached is used as a primary screen. Other factors such as sequence similarity within the loop and geometry of the takeoff region of the framework may also be used. The loops are then fitted to the SCRs using some set of geometric criteria.
- *ab initio* **methods:** candidate loops are built through conformational search and/or molecular dynamics. Such techniques, exemplified by the work of Bob Bruccoleri and Martin Karplus in the mid-1980s, attempt to saturate conformational space in an intelligent manner. The correct conformation is then selected using energy and solvent accessibility filters.
- **combined approaches:** these combine elements of knowledge-based and *ab initio* methods. Part of the loop may be built by database searching while the region around an indel, a region known to be functionally important, or simply the tip of the loop, is built by *conformational search*. Screening the generated conformations may then use a combination of energetic and knowledge-based filters. Such techniques were introduced by the author in the late 1980s.

The effectiveness of knowledge-based methods depends on the level of saturation of conformational space in the Protein Databank. In 1994, John Moult's group analyzed this and showed that conformational space for loops of up to seven amino acids is well saturated, making it possible to build loops of up to five amino acids if two residues are used as anchor points in the framework. Since then the Protein Databank has grown hugely in size, but more recent partial analyses have suggested that the lengths of loop for which conformational space are saturated has not grown very much.

A variation on the knowledge-based procedure is to predict the general conformational class to which a loop will belong. The *SLOOP database* from Tom Blundell's group classifies loops according to their length and the bounding secondary structures. For example a loop of 10 residues with a strand at either end would be placed in the class E10E ('E' is the standard label used for 'extended' secondary structure – i.e. β-strands). Loops in each class are then superimposed and clustered to define conformational classes for the backbone based on seven regions of the *Ramachandran plot*. For each resulting class, a multiple sequence alignment is calculated and information on the separation of the endpoints and the difference in the vectors described by the adjoining secondary structures together with accessibility class is also stored. A novel sequence can then be scored against each SLOOP template family to suggest a conformational class into which the loop will fit. A Web-based server for accessing the SLOOP database and making predictions of loop conformation is available.

Manual methods of loop modeling are always subject to error while conformational search is limited by available CPU time (a full conformational search of a fragment of 17 amino acids would take many hundreds or even thousands of years to complete). In 1988, Bob Bruccoleri and co-workers applied a *'real space renormalization'* procedure where the bases of long loops were constructed first. Low-energy conformations for these were selected and used as starting points for constructing the next part of the loop. This technique inspired the first combined approach in which the base of the loop was constructed by database searching and the mid-section of the loop by conformational search. Both procedures suffer from relatively arbitrary decisions being made as to which combinations of residues should be built at each stage and, while almost invariably capable of generating conformations similar to the 'correct' conformation, in both cases it is very difficult to select the best conformation from the many thousands generated. Thus one's ability to model insertions or deletions in loops longer than about 10 amino acids is very limited.

When loop regions are not too long and are one or maybe up to three residues shorter than observed in the parents, they may often be built quite well. Deletions can often be accommodated by relatively small structural changes providing care is taken in the alignment stage as shown above. Such deletions may be made by any of the above methods.

When target loops are longer than the parent(s), however, it is very difficult to achieve a good model. Making insertions involves introducing new amino acid residues with no clear guide as to where they should be placed. Knowledge-based methods are the methods of choice, providing loops are not too long.

8.3.1.6 Build the sidechains

Various protocols are available for building sidechains. The simplest is known as the *'Maximum Overlap Protocol'* (MOP). Here the torsion angles of the sidechain are inherited from the parent sidechain where possible and additional atoms are built from a single standard conformation.

In the *'Minimum Perturbation Protocol'* (MPP) proposed by Martin Karplus's group in 1985, each substitution is followed by a rotation about the sidechain's torsion angles to relieve clashes. The *'Coupled Perturbation Protocol'* (CPP) introduced by Mark Snow and Mario Amzel in 1986 is similar, but the sidechain torsion angles of structurally adjacent residues are also rotated.

In principle, CPP can be extended such that more distant shells of residues are also spun around their sidechain torsion angles. However, one rapidly faces a problem of combinatorial explosion with every additional torsion angle multiplying the number of possible conformations by $360°/x$ (where x is the sample step size in degrees). For example, if one samples the torsion angle every 30°, then each torsion angle added will cause there to be 12× more conformations.

Sidechains are most effectively built my making use of *'rotamer libraries'*. Owing to steric effects within the sidechain, only certain sidechain torsion angles are favored. *Figure 8.2* illustrates a simple case, looking along a bond (here from Cβ to Cα) where atoms attached to the Cβ are staggered with respect to the atoms attached to the Cα.

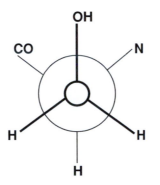

Figure 8.2

A staggered atomic conformation looking from a Cβ to Cα atom for a serine sidechain.

Of the three favored staggered positions, some are more favored than others and this can be dependent on the secondary structure environment of the amino acid. *Figure 8.3* illustrates the conformations of 50 phenylalanine residues taken from a random selection of structures.

Further information on rotamers is available in papers by Roland Dunbrack and on his web site. Dunbrack's *SCWRL* software makes use of backbone-dependent rotamer libraries for automated *sidechain modeling* and is one of the most effective sidechain modeling algorithms. Each of the potential rotamers is built and the rotamer set is searched for the minimum steric clash to create the final structure. The protocol proceeds as follows:

1. Any possible disulfide bonds are identified and frozen. The most favored rotamer positions are then set for all other sidechains.
2. The sidechains are searched for steric clashes with mainchain; when a clash is found the rotamer is changed through progressively less favorable rotamers until one is found which makes no clash with mainchain.
3. Some sidechain–sidechain clashes will now remain. Clashing sidechains are placed in a group and each is rotated through other allowed rotamers which do not clash with mainchain. Any additional residues which clash during this procedure are added to the group and are also allowed to explore their rotamers. In this way the groups or 'clusters' of residues grow and can merge if members of two such clusters make a clash.
4. When each cluster stops growing the optimum set of rotamers is selected through a combinatorial search to find the minimum steric clash energy.

8.3.1.7 Refining the model

Models are generally refined through a process called *'Energy Minimization'* (EM). This is related to the process of *'Molecular Dynamics'* (MD) which may also be used in refinement. There are many good books describing these topics in depth including McCammon and Harvey's 1987 textbook. The descriptions below only give a brief outline of the concepts involved.

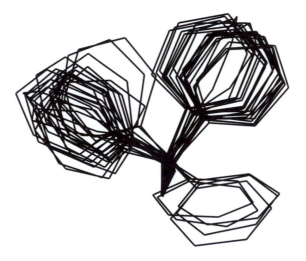

Figure 8.3

Fifty phenylalanine sidechains selected at random from the Protein Databank and overlapped on the N, Cα and C backbone atoms.

Both EM and MD require some form of potential able to describe the energy of the system (see *Box 8.1*). More complex quantum effects are ignored and the molecules are treated as simple sets of springs which obey the laws of classical mechanics defined by Newton. Collectively, the methods are thus known as *molecular mechanics*.

In EM, the aim is to move the atoms such that a minimum energy conformation is found. Standard mathematical minimization algorithms such as *steepest descents, conjugate gradients* or *Newton–Raphson* are used in this procedure. However, EM is only able to move atoms such that a local minimum on the energy surface is found – it is not able to make large movements that effectively explore conformational space (see *Figure 8.4*). In addition, in EM, movements are generally made in Cartesian (XYZ) space, so groups of atoms do not move effectively

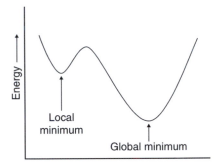

Figure 8.4

A simple representation of a one-dimensional energy surface with a local and a global energy minimum.

Box 8.1   A simple potential

The first term is the sum of covalent bond energies, each of which is treated as a simple spring obeying Hooke's law with a force constant k_b and equilibrium bond length b_0. The second term accounts for deformation of valence bond angles. The third term represents rotation about dihedral angles between covalently bonded atoms and also includes 'improper torsions', similar terms which are introduced to keep groups such as rings and peptide bonds planar. The fourth term accounts for van der Waals (vdW) repulsion, dispersion and Coulombic interactions between non-bonded atoms separated by a distance r. This potential does not show hydrogen bonds which are treated implicitly within the vdW and electrostatic terms, but may also be treated explicitly.

$$V = \frac{1}{2}\sum k_b(b - b_0)^2$$

$$+ \frac{1}{2}\sum k_\theta(\theta - \theta_0)^2$$

$$+ \frac{1}{2}\sum k_\phi[1 + \cos(m\phi - \delta)]$$

$$+ \sum_{i<j}\left[\frac{A_{i,j}}{r^{12}} - \frac{B_{i,j}}{r^6} + \frac{q_i q_j}{4\pi\varepsilon_0\varepsilon_r r}\right]$$

together. For example, if a tyrosine ring is making a bad contact at its end, one would expect it to move as a group and swing about a torsion angle in order to relieve the stress. However, since standard EM has no concept of atom groups, it simply has the effect of bending the ring. The increase in the 'improper torsion' energy (a term used to try to keep groups of atoms that should be co-planar in a planar configuration) is more than compensated by the reduction in van der Waals repulsive energy. In such cases, manual intervention is necessary to restore the ring planarity, possibly followed by an iterative cycle of more EM and manual refinement until one reaches convergence.

EM is only useful to identify major clashes (which generally have to be fixed by hand) and resolve minor clashes. It rarely improves the model in terms of accuracy to any significant degree.

As in EM, MD treats the protein as a set of Newtonian springs. However, rather than trying to reach an energy minimum, the aim is to simulate the dynamics of the protein either *in vacuo* or in the environment of a solvent. Each atom is assigned an initial velocity from a temperature-dependent Boltzmann distribution and Newton's laws of motion are then solved iteratively to keep updating the coordinates, forces, acceleration and velocities, thus integrating the laws of motion. Various refinements are necessary to keep either the temperature, or the pressure, constant and to minimize cumulative errors in the integration.

Because MD provides a dynamic simulation, it is able to move over energy barriers that are inaccessible to EM. Thus a more effective way to attempt to find the minimum energy of a protein is to take snapshots from an MD run and to minimize each snapshot. This provides a much more effective sampling of conformational space.

8.3.1.8   Evaluating the model

Assessment of the quality of the model is described in detail in the later section on model quality.

8.3.2   Building a model: Using MODELLER

One of the most successful programs for generating protein models is *MODELLER* written by Andrej Săli. It works on a rather different principle from that described above.

The first two stages are the same as before: identifying one or more parents and generating an alignment between the parents and the target sequence. However, rather than the conventional 'spare parts' approach described above, a molecular dynamics approach is used to assemble the SCRs and SVRs in a single step.

A conventional molecular dynamics (MD) force-field is used and Newton's laws of motion are integrated in the conventional manner. However, additional restraints are imposed in the force-field, known as a *'probability density function'* (PDF). This is essentially a set of preferred inter-Cα distances calculated from the parent structure(s). Where more than one parent structure is used, the PDF is weighted such that regions which are highly conserved in structure have stronger restraints while regions which vary more in structure have weaker restraints. Thus the approach taken can be viewed as an *ab initio* one which is heavily guided by information from knowledge-based analysis of known structures.

MODELLER also allows the user to explore local areas of the structure (generally SVRs) in more detail if desired. Options are available to control the level of refinement of generated models and one can generate a single model or a range of possible solutions.

MODELLER is available free for non-commercial use from Andrej Săli's website where more information may be obtained. For commercial use, MODELLER may be purchased as an add-on for the Quanta and Insight II molecular graphics packages. It is a complex and powerful program, but comes with a very extensive manual and is supported by an active mailing list.

The MODELLER approach neatly side-steps the problems of identifying SCRs and SVRs as well as providing a feasible route to modeling longer loop regions. There are two problems in using MODELLER: (i) the primary factors influencing the quality of the final model are unchanged – the level of similarity between the parent structure(s) and the structure of the target and the quality of the alignment between the target sequence and the sequences of the parent structure(s); (ii) on a more simplistic level, the control language is complex and, given the size of the package, it can be hard to know exactly what the program is capable of and how to achieve the desired results.

8.3.3   Building a model: Other methods

Other methods for comparative protein modeling have, thus far, shown less success. Like MODELLER, these methods rely on attempting to satisfy sets of distance restraints inherited from parent structures as well as energetic criteria. However, rather than solving the problem by using simulated annealing methods in normal three-dimensional space, projection into higher dimensional space is used, applying *distance geometry*-based methods. Such techniques are inspired by the methods used by NMR spectroscopists to derive three-dimensional coordinates from distance information resulting from 2-D or 3-D NMR techniques. Examples of such methods are *DRAGON* from Andras Aszódi and Willie Taylor and *AMMP* from Irene Weber.

8.4   Evaluation of model quality

Root Mean Square Deviation (RMSD) is the conventional measure used to say how similar one structure is to another (see Chapter 6). It is thus useful as a measure of the accuracy of a model if one has a crystal structure of the protein with which to compare the model. Since atomic coordinates in proteins are generally expressed in Ångströms (where $1 \text{ Å} = 10^{-10} \text{ m} = 0.1 \text{ nm}$), RMSDs are also expressed in Å units. Two crystal structures of the same protein (solved by different groups or in slightly different crystallization conditions) typically have a Cα RMSD of ~0.6 Å and values in this range are thus the goal of comparative modeling.

RMSD as a measure of protein modeling accuracy does have a number of problems. A few very poorly placed atoms can mask the fact that all other atoms are placed extremely well. Similarly if a structure is locally correct in two separate regions, but a single bond between the two regions is rotated, the local accuracy will be masked. Even though RMSD is calculated as an average over all equivalent atom pairs and thus accounts for the number of atoms in the comparison, an RMSD of, say, 4 Å over a stretch of four amino acids is regarded as much worse than the same RMSD over 100 amino acids. It is clearly a difficult problem to represent differences in two sets of coordinates in three dimensions by a single number! In assessing the accuracy of comparative modeling, alternative measures have been suggested, for example, both by Tim Hubbard and by Alwyn Jones and Gerard Kleywegt in assessing the CASP3 (1998) modeling experiment published in 1999. These give a better feeling for the true accuracy and spread of errors, but by necessity, they are considerably more complex than citing a single number and have yet to gain wide acceptance.

In comparing descriptions of the accuracy of comparative modeling, care must be taken since RMSDs are often quoted over 'all atoms', 'backbone atoms' or 'Cα atoms'. Different authors interpret 'backbone' differently; it may mean N,Cα,C or N,Cα,C,O or N,Cα,Cβ,C,O. In addition when modeling loop regions, both local and global RMSD figures may be used. The local RMSD is calculated by fitting just the loop region in question, whereas the global RMSD is obtained by fitting the whole structure, or just the structurally conserved regions, and then calculating the RMSD over the loop region. All these variations can make comparison between results in different publications very difficult.

8.5   Factors influencing model quality

The following discussion comes from the author's assessment of the comparative modeling section of the CASP2 (1996) experiment published in 1997.

The overall CASP1 (1994) and CASP2 (1996) results are summarized in *Figure 8.5*. As can be seen from the graph as the sequence identity gets higher than 80%, models of <1.0 Å RMSD can be generated, equivalent to the differences between two crystal structures of the same protein. As the sequence identity drops below 40%, the model quality falls rapidly. Results from subsequent CASP meetings show marginal improvements.

As the sequence identity drops, the accuracy of the models falls and the RMSD increases. This is largely because it becomes more difficult to obtain the structurally correct alignment as shown in *Figure 8.6*. This graph compares sequence alignments derived from structural alignments performed with *SSAP* (see Chapter 6) with Needleman and Wunsch sequence alignments. The percentage of Needleman and Wunsch aligned residue pairs that are identical to the structurally aligned pairs is plotted against the sequence identity calculated from

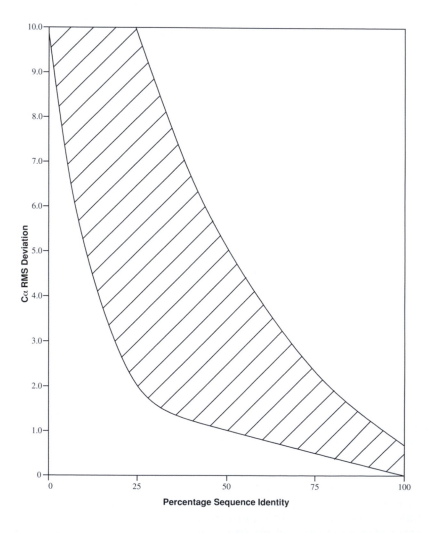

Figure 8.5

Schematic summary of results from the CASP1 and CASP2 modeling experiments. The shaded band represents the expected range of RMS deviations.

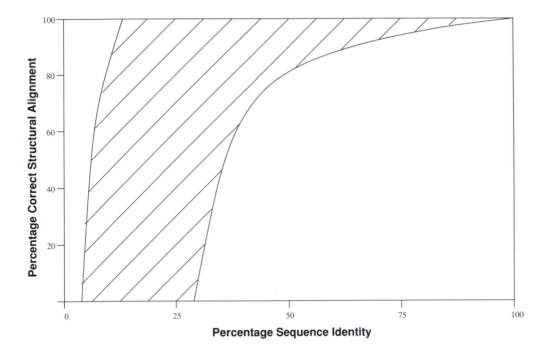

Figure 8.6

Accuracy of automated sequence alignment plotted against percentage sequence identity. This schematic graph summarizes results from comparison of approximately 56,000 pairs of protein sequences and structures. The shaded band represents the range of commonly observed results.

the Needleman and Wunsch sequence alignment. The schematic graph is based on an analysis of 56,000 pairs of domains defined as homologs in the CATH classification of protein structures.

As shown in *Figure 8.7*, when one makes mistakes in the alignment, the RMSD increases rapidly. Given that the distance between adjacent Cα atoms in a protein is approximately 3.8 Å, then an average shift of one residue in the alignment will lead to a minimum possible RMSD of 3.8 Å.

As the graph shows, in practice an average shift in alignment of one residue leads to an RMSD of ~5.0 Å. An average shift of <0.1 residues is required to get a high-quality model.

8.6   Insights and conclusions

The analysis of the CASP2 (1996) experiment included a comparison of the results that modelers obtained with how well they could have done if they got the sequence alignment 100% correct. For each target being modeled, a 'primary parent' showing the highest sequence identity with the target to be modeled, was defined. For each model generated, an RMSD with the correct structure was calculated. In addition, the primary parent was structurally aligned to the target structure and the RMSD was calculated representing the best model one could hope to obtain without making further refinements to the parent structures (Best-RMSD). Finally, the target sequence was aligned with the primary parent sequence using a fully automated Needleman and Wunsch sequence alignment and the RMSD based on this alignment was calculated (Worst-RMSD) – a completely automated procedure should therefore do at least this well and manual modifications should improve the situation.

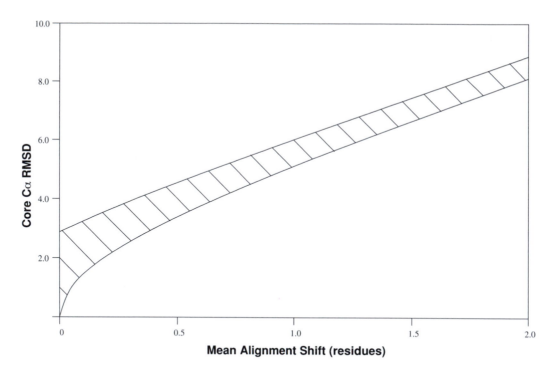

Figure 8.7

Schematic graph of the dependence of model accuracy (as assessed by the Cα RMS deviation of the core region; i.e. the structurally conserved regions) on alignment error. The shaded band represents the range of observed results.

One would thus hope that the models all have RMSDs falling between Worst-RMSD and Best-RMSD. In practice, this was generally the case, and, in a couple of cases, manual modifications and refinements even obtained an RMSD slightly better that Best-RMSD. However, in a number of cases, the manual modification of the alignment actually made the model worse than Worst-RMSD. Attempts to combine SCRs from multiple parents and additional errors introduced into the alignment during manual modification made the resulting model worse than a fully automated procedure could have achieved.

In summary, five features are critical in determining the quality of a protein model by comparative modeling:

- The sequence identity with the primary parent
- The number and size of insertions and deletions ('indels')
- The quality of the alignment between the target and the parent(s) (which depends on the above factors)
- How structurally similar is the parent or parents used to build the model
- The amount of change which has been necessary to the parent(s) to create the model.

References

Marti-Renom, M.A., Yerkovich, B., and Săli, A. (2002) Comparative protein structure prediction. *Current Protocols in Protein Science*, 2.9.1–2.9.22. This and other reviews may be downloaded from Andrej Săli's MODELLER Web site.

McCammon, J.A., and Harvey, S.C. (1987) *Dynamics of Proteins and Nucleic Acids*. Cambridge University Press, Cambridge.

Protein structure prediction

David T. Jones

- There are three main categories of protein structure prediction methods: comparative modeling methods, fold recognition methods and *ab initio* methods. Methods can also be characterized according to whether they are based on physical principles or are based on statistics derived from known structures (knowledge-based approaches).
- Secondary structure prediction methods attempt to decide which type of secondary structure (helix, strand or coil) each amino acid in a protein sequence is likely to adopt. The best methods are currently able to achieve success rates of over 75% based on sequence profiles.
- Ultimately we would hope that accurate folding simulations will allow us to predict the structure of any protein. However, not only is this approach impractical due to limitations of computing power, but our understanding of the principles of protein folding are far short of the level needed to achieve this.
- Fold recognition is currently the most effective way of predicting the tertiary structure of proteins with no clear homology to any proteins of known 3-D structure. The idea is to recognize the correct fold of a protein from a set of alternatives (known as a fold library). However, fold recognition is a difficult problem and results can sometimes be misleading. The CASP (Critical Assessment of Structure Prediction) experiments offer an unbiased way of assessing the current state-of-the-art in protein structure prediction.

9.1 Concepts

The 'Protein Folding Problem' is probably the greatest remaining challenge in structural molecular biology (if not the whole of biology). How do proteins go from primary structure (sequence) to tertiary structure? How is the information encoded? Basically, how do proteins fold? Often the protein-folding problem is seen as a computational problem – do we know enough about the rules of protein structure to program a computer to read in a protein sequence and output a correct tertiary structure? Aside from the academic interest in understanding the physics and chemistry of protein folding, why are so many people interested in finding an algorithm (i.e. a method) for predicting the native structure of a protein given just its sequence? The answer is that experimental methods for determining protein structure are difficult and time consuming, and knowing the structure of a protein can give an immense amount of information about the function of a protein.

9.2 Strategies for protein structure prediction

9.2.1 Comparative modeling

Comparative modeling, sometimes called homology modeling, is by far the most reliable technique for predicting protein structure, though it is often not thought of as being a structure prediction method. In comparative modeling, the structure of the new protein is predicted by comparing its sequence with the sequences of proteins of known structure. If a strong similarity is found then it can be assumed that the proteins have similar overall structures. If no strong similarities can be found then comparative modeling cannot be used. The assumption is that proteins with similar sequences have almost identical structures, and this has already been discussed. Even between very similar proteins, however, there are differences, and some of these differences might be functionally important (different binding loop conformations for example). Predicting what the effects of these small structural changes might be is the real challenge in modeling. Comparative modeling is discussed thoroughly in Chapter 8, and so it will not be discussed further here.

9.2.2   *Ab initio* prediction

Ab initio (i.e. 'from first principles') methods are so-called because they use only the information in the target sequence itself. There are two branches of *ab initio* prediction.

9.2.2.1   Knowledge-based methods

Knowledge-based prediction methods attempt to predict structure by applying rules. These rules are based on observations made on known protein structures. A trivial example of such a rule might be that proline residues are uncommon in α-helices. Whilst nowhere near as reliable as comparative modeling, knowledge-based methods are rapidly becoming more useful as the number of known protein structures increases.

9.2.2.2   Simulation methods

The most ambitious approach to predicting protein structure is to simulate the protein-folding process itself using basic physics. Much research effort is expended in developing such simulation techniques, but as yet simulation is only useful for short peptides and small molecules. Simulation techniques are, however, very useful for predicting unknown loop conformations as part of comparative modeling. The basic principle of simulation-based protein structure prediction is that the native fold of a protein can be found by finding the conformation of the protein which has the lowest energy as defined by a suitable potential energy function.

Unfortunately, the exact form of this energy function is as yet unknown, but it is reasonable to assume that it would incorporate terms pertaining to the types of interactions observed in protein structures, such as hydrogen bonding and van der Waals effects (see Chapter 8). The conceptual simplicity of this model for protein folding stimulated much early research into *ab initio* tertiary structure prediction. A successful *ab initio* approach necessitates the solution of two problems. The first problem to solve is to find a potential function for which the conformation of the protein is the conformation of lowest energy. The second problem is to construct an algorithm capable of finding the global minimum of this function.

The case against proteins searching conformational space for the global minimum of free energy was argued by Cyrus Levinthal in 1968. The Levinthal paradox, as it is now known, can be demonstrated fairly easily. If we consider a protein chain of N residues, we can estimate the size of its conformational space as roughly 10^N states. This assumes that the main chain conformation of a protein may be adequately represented by a suitable choice from just 10 main chain torsion angle triplets (phi, psi and omega angles) for each residue. This of course neglects the additional conformational space provided by the side chain torsion angles, but is a reasonable rough estimate, albeit an underestimate. The so-called paradox comes from estimating the time required for a protein chain to search its conformational space for the global energy minimum. Taking a typical protein chain of length 100 residues, it is clear that no physically achievable search rate would enable this chain to complete its folding process. Even if the atoms in the chain were able to move at the speed of light, it would take the chain around 10^{82} seconds to search the entire conformational space, which compares rather unfavorably to the estimated age of the Universe (10^{17} seconds).

Clearly proteins do not fold by searching their entire conformational space. There are many ways of explaining away Levinthal's paradox. Perhaps the most obvious explanation is that proteins fold by means of a folding pathway encoded in the protein sequence. Despite the fact that chains of significant length cannot find their global energy minimum, short-chain segments (5–7 residues) could quite easily locate their global energy minimum within the average lifetime of a protein, and it is therefore plausible that the location of the native fold is driven by the folding of such short fragments. Thus, Levinthal's paradox is only a paradox if the free energy function forms a highly convoluted energy surface, with no obvious downhill paths leading to the global minimum. The folding of short fragments can be envisaged as the traversal of a small downhill segment of the free energy surface, and if these paths eventually converge on the global energy minimum, then the protein is provided with a simple means of rapidly locating its native fold.

One subtle point to make about the relationship between the minimization of a protein's free energy and protein folding is that the native conformation need not necessarily correspond to the global minimum of free energy. One possibility is that the folding pathway initially locates a local minimum, but a local minimum which provides stability for the average lifetime of the protein. In this case, the protein in question would always be observed with a free energy slightly higher than the global minimum *in vivo*, until it is degraded by cellular proteases, but would eventually locate its global minimum if extracted from the cell, purified, and left long enough *in vitro* – though the location of the global minimum could take many years. Thus, a biologically active protein could in fact be in a metastable state, rather than a stable one.

9.2.3   Fold recognition or threading

As we have seen, one of the difficulties in *ab initio* protein structure prediction is the vast number of possible conformations that a polypeptide chain can adopt. Searching for the correct structure from so many alternatives is a very challenging problem, and one that, as yet, has not been solved.

In fact, it is now evident that there is a limited number of naturally occurring protein folds. As we have seen, if we consider a chain of length 50 residues we might naively calculate the number of possible main chain conformations as $\sim 10^{50}$. This is a high estimate, however, as clearly most of these conformations will not be stable folds, and many will not be even physically possible (e.g. parts of the chain will overlap with other parts). We know from looking at protein structures that in order to form a compact globular structure, a protein chain forms regular secondary structures, and it is this constraint, along with the constraints imposed from a requirement to effectively pack the secondary structures formed, that limits the number of stable conformations for a protein chain. Beyond the physical constraints there also exist evolutionary constraints on the number of naturally occurring folds (i.e. many proteins around today have evolved from distant common ancestors and therefore will share a common fold with other relatives).

From these observations we can expect the number of possible folds to be limited, and indeed this is exactly what we see when we compare the structures of proteins (see Chapters 6 and 7). At the time of writing, from comparisons of all known protein 3-D structures, it is estimated that there is an $\sim 70\%$ chance that a newly characterized protein with no obvious common ancestry to proteins with a known structure will in fact turn out to share a common fold with at least one protein of known structure.

These limits to the number of protein folds offer a 'short-cut' to protein structure prediction. As already described, attempting tertiary structure prediction by searching a protein's entire conformational space for a minimum energy structure is not a practical approach, but if we know that there are only a few thousand possible protein folds, then an intelligent way to search a protein's conformational space would be to first look at folds which have already been observed. This is analogous to the difference between an exam requiring the writing of an essay and an exam requiring multiple-choice questions to be answered. Clearly a person with little or no knowledge of the subject at hand has a much greater chance of achieving success with the multiple-choice paper than with an essay paper.

9.3   Secondary structure prediction

Before we continue the discussion on methods for predicting protein tertiary structure, we will take some time out to discuss the prediction of protein secondary structure. Although predicting just the secondary structure of a protein is a long way from predicting its tertiary structure, information on the locations of helices and strands in a protein can provide useful insights as to its possible overall fold. Also, it is worth noting that the origins of the protein structure prediction field (and to some extent the whole bioinformatics field) lie in this area.

9.3.1   Principles of secondary structure prediction

Secondary structure prediction methods have been around for over a quarter of a century, but the earliest methods suffered from a severe lack of structural data. Predictions were performed on single sequences, and there were relatively few known 3-D structures from which to derive parameters. Probably the most famous early methods are those of Chou and Fasman in 1974 and Garnier, Osguthorpe and Robson (GOR) in 1978. Although the authors originally claimed quite high accuracies (70–80%), under careful examination, the methods were shown to be only between 56 and 60% accurate. An early problem in secondary structure prediction had been the inclusion of structures used to derive parameters in the set of structures used to assess the accuracy of the method.

There are two aspects of protein structure that are exploited in predicting secondary structure from just a single protein sequence: the intrinsic secondary structure propensities of amino acids, and the hydropathic nature of amino acids.

9.3.2   Intrinsic propensities for secondary structure formation of the amino acids

Are some residues more likely to form α-helices or β-strands than others? Clearly, yes. We have already mentioned that proline residues are not often found in α-helices, but the other amino acids also show striking differences in propensity to form α-helices or β-strands. To determine what these propensities were, Chou and Fasman in 1974 performed a statistical analysis of 15 proteins (all that were available in 1974!) with known 3-D structures. For each of the 20 amino acids they calculated the probability of finding each amino acid in α-helices and in β-strands. They also calculated the probability of finding any residue in α-helices and in β-strands. For example, suppose there was a total of 2000 residues in their 15 protein data set. Analyzing these residues, suppose the following counts were made:

Total number of residues	2000
Number of alanines	100
Number of helical residues	500
Number of alanines in helices	50

We would calculate the propensity of alanine for helix formation as follows:

P(Ala in Helix) = 50/500 = 0.1
P(Ala) = 100/2000 = 0.05
Helix Propensity (PA) of Ala = P(Ala in Helix)/P(Ala) = 0.1/0.05 = 2.0

A propensity of 1.0 indicates that an amino acid has no particular preference for the specified secondary structure. A propensity <1.0 indicates that an amino acid disfavors the specified secondary structure, and >1.0 indicates a preference for that secondary structure.

The values calculated by Chou and Fasman are shown in the *Table 9.1*. PA is the propensity to form α-helix, PB the propensity to form β-strands and PT the propensity to form turns or loops.

From *Table 9.1* we can see that as expected, proline has the lowest propensity to form α-helices, but more interestingly, we can see that glycine also disfavors the helical conformation. Along these lines, Chou and Fasman divided amino acids into different classes based on their analysis – for example, alanine was classed as a 'strong alpha former' and proline as a 'strong helix breaker'. *Table 9.1* shows how Chou and Fasman classified the 20 amino acids.

Using these parameters, Chou and Fasman proposed an algorithm for deciding which residues in a sequence are most likely to be in helices and strands. Essentially, the propensities of the residues were summed up across segments of the sequence. By comparing the sums for α, β and coil propensities for the same segment a decision could be made as to whether the segment was more likely to be part of an α helix, a strand or a coil region.

A much better method for predicting secondary structure by amino acid propensities was proposed by Garnier, Osguthorpe and Robson. The details of this method, which used principles of information theory, are beyond the scope of this chapter, but the key difference between the GOR method and the Chou and Fasman method is that whilst the Chou and Fasman method just takes into account the propensities of single amino acids in making a prediction, GOR takes into account a window of 17 amino acids – i.e. a value is calculated that involves eight residues either side of the residue you are trying to assign to helix, strand or coil states.

Table 9.1 Chou and Fasman Conformational Parameters

PA			PB			PT	
Glu	1.51	(HA)	Val	1.70	(HB)	Asn	1.56
Met	1.45	(HA)	Ile	1.60	(HB)	Gly	1.56
Ala	1.42	(HA)	Tyr	1.47	(HB)	Pro	1.52
Leu	1.21	(HA)	Phe	1.38	(hB)	Asp	1.46
Lys	1.16	(hA)	Trp	1.37	(hB)	Ser	1.43
Phe	1.13	(hA)	Leu	1.30	(hB)	Cys	1.19
Gln	1.11	(hA)	Cys	1.19	(hB)	Tyr	1.14
Trp	1.08	(hA)	Thr	1.19	(hB)	Lys	1.01
Ile	1.08	(hA)	Gln	1.10	(hB)	Gln	0.98
Val	1.06	(hA)	Met	1.05	(hB)	Thr	0.98
Asp	1.01	(IA)	Arg	0.93	(iB)	Trp	0.96
His	1.00	(IA)	Asn	0.89	(iB)	Arg	0.96
Arg	0.98	(iA)	His	0.87	(iB)	His	0.95
Thr	0.83	(iA)	Ala	0.83	(iB)	Glu	0.74
Ser	0.77	(iA)	Ser	0.75	(bB)	Ala	0.66
Cys	0.70	(iA)	Gly	0.75	(bB)	Met	0.60
Tyr	0.69	(bA)	Lys	0.74	(bB)	Phe	0.60
Asn	0.67	(bA)	Pro	0.55	(BB)	leu	0.59
Pro	0.57	(BA)	Asp	0.54	(BB)	Val	0.50
Gly	0.57	(BA)	Glu	0.37	(BB)	Ile	0.47

Key to Helical assignments: HA, strong α former; hA, α former; IA, weak α former; iA, α indifferent; bA, α breaker; BA, strong α breaker.

Key to β-sheet assignments: HB, strong β former; hB, β former; IB, weak β former; iB, β indifferent; bB, β breaker; BB, β breaker.

9.3.3   Hydropathy methods and transmembrane helix prediction

In fact, there is an even easier method than even the Chou & Fasman method to predict α-helices in proteins, and that is to draw the amino acid sequence on a 'helical wheel' *Figure 9.1*. A helical wheel is a diagrammatic representation of the positions of amino acids where each amino acid is plotted in a circular fashion corresponding to the pitch of an ideal α-helix. α-Helical regions generally exhibit clustering of hydrophobic residues along a single sector of the wheel. Though of some vintage, this technique still remains a powerful means for identifying and comparing α-helical regions in proteins.

A specific use of hydropathy information in protein structure prediction, which is worth special mention, is the prediction of membrane spanning helices in integral membrane proteins. Because of the apolar environment within the lipid bilayer, helices which span the membrane are almost entirely composed of hydrophobic amino acids. Therefore, a good way to predict transmembrane helices, is simply to calculate the average hydrophobicity over a sliding window (usually 18–22 amino acids long) moved along the protein's sequence. Regions of 18–22 amino acids within a protein which have a high average hydrophobicity are very likely to be transmembrane helices as such hydrophobic regions are rare in water-soluble globular proteins.

9.3.4   Predicting secondary structure from multiple sequence alignments

As we have seen, it has long been recognized that patterns of hydrophobic residues are indicative of particular secondary structure types. α helices have a periodicity of 3.6, which

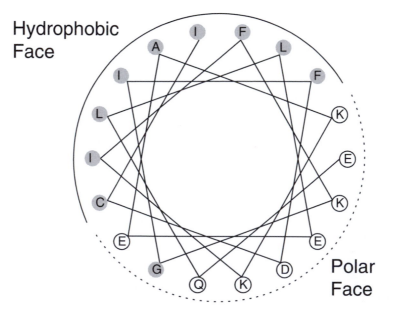

EQLLEEAKKIFKGIFDCI

Figure 9.1

This diagram shows a short amino acid sequence plotted in the form of a helical wheel. The hydrophobic amino acids are underlined in the sequence and shown shaded on the helical wheel. Glycine can be thought of as either hydrophobic or polar as it has no side chain.

means that in helices with one face buried in the protein core, and the other exposed to solvent, residues at positions i, i + 3, i + 4 and i + 7 (where i is a residue in an α-helix) will lie on one face of the helix. Thus hydrophobic residues in such an i, i + 3, i + 4, i + 7 pattern, are highly indicative of an α helix. It is this pattern that shows up clearly when the sequence is plotted on a helical wheel.

By applying this kind of analysis to an aligned family of proteins we see a marked improvement by the averaging out of evolutionary noise. Pairs of conserved hydrophobic amino acids separated by pairs of unconserved, or polar residues (HHPPHHPP) are very strong indicators of an α-helix. Other structural features can also sometimes be deduced from looking closely at multiple alignments of protein families. For example:

- The positions of insertions and deletions suggest regions where surface loops exist in the protein.
- Conserved glycine or proline residues suggest the presence of a β-turn.
- Alternating polar/hydrophobic residues suggest a surface strand.
- A short run of hydrophobic amino acids (say four) suggests a buried β-strand.

These patterns cannot be readily seen in a single sequence, but given a suitably divergent multiple sequence alignment, they often stand out and allow secondary structure to be assigned with a high degree of confidence.

9.3.5 Predicting secondary structure with neural networks

Although a sufficiently expert human being can study a multiple sequence alignment and identify regions which are likely to form helices and strands, this is clearly not a practical approach if we wish to analyze many sequences. Also, the results of human analysis may not be easily replicated by others, and so it is important that automated prediction techniques be developed.

Many methods for secondary structure prediction have been proposed over the years, but the most effective methods that have been developed have all made use of neural networks to analyze substitution patterns in multiple sequence alignments. Neural networks (more properly called Artificial Neural Networks or ANNs) are based on models of the human brain. Neural networks are collections of mathematical models that attempt to mimic some of the observed properties of biological nervous systems, and were originally aimed at modeling adaptive biological learning. A simple neural network is illustrated in *Figure 9.2*. It is composed of a number of highly interconnected processing elements that are intended to be analogous to neurons and which are linked with weighted connections that are analogous to the synapses in a brain. This is the simplest type of ANN, and is known as a 'feed-forward network' as all the connections are in one direction, i.e. the information feeds forward from the inputs through to the outputs. The essential idea of training such a neural network is that the connection weights are adjusted so that the network produces a particular output pattern when it is presented with a particular input pattern. With many processing elements and connections, a neural network can learn many such pattern relationships, and thus neural networks are some of the most powerful pattern recognition methods that are currently available in the field of machine learning.

9.3.6 Secondary structure prediction using ANNs

We will now look at a typical methodology used to train a feed-forward network for secondary structure prediction, based on the approach of Qian & Sejnowski (1988). To train the network we need many different input/output pattern pairs. In other words we need to give the network many different input patterns (based on the amino acid sequence) and with each pattern we need to give it the required output pattern (based on the observed secondary structure).

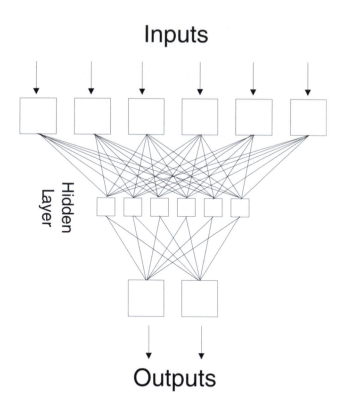

Inputs

Hidden Layer

Outputs

Figure 9.2

A simple feed-forward neural network is shown with a single hidden layer. Patterns presented on the inputs are processed by the connections or weights (shown as lines in the diagram) between the activation units (squares), and result in a certain output pattern.

For example, suppose we are looking at the following segment of sequence from sperm whale myoglobin:

VLSEG **GEWQLVLHVWAKV** EADV ...

From the experimentally determined structure of this protein we know that this sequence forms an α-helix, so ideally we should like to train the neural network to produce a helix output whenever it sees a sequence similar to this. In practice we specify a window of amino acids (typically 13–17 consecutive amino acids) and ask the network to predict the secondary structure of the middle residue. So in the above example, we would be interested in whether the central Leu residue is part of an α-helix, a β-strand or coil.

Before we can feed this information into the neural network we need to *encode* the information in some way. Most neural networks are designed such that their inputs and outputs can represent values from 0 to 1 (actually they can be made to represent any real value – although for now we will gloss over that fact for simplicity), and so, in the above case we need to convert amino acids into a pattern of 0s and 1s and also we need to encode 'helix' as a similar pattern.

Although there are many ways of encoding amino acids into strings of 0s and 1s, the best way turns out to be to use 20 inputs to represent the 20 amino acid types. So, if we assign the amino acids alphabetically, input 1 represents Ala and input 20 represents Val. The above amino acid sequence is therefore encoded thus:

```
GEWQLVLHVWAKV
(G) -> 0000000100000000000
(E) -> 0000001000000000000
(W) -> 0000000000000000100
```

. . . and so on.

To encode the three possible secondary structure types (helix/strand/coil) we can use three outputs. So in the above case, the total number of inputs would be $13 \times 20 = 260$, and there would be three outputs, and in this particular instance the output pattern would be 100 to represent helix (010 would represent strand and 001 would represent coil). It is usual to define a fake 21st type of amino acid to represent positions which are beyond the N or C terminus of the protein when the window is positioned near either end of the protein, and so in a practical secondary structure prediction method, $13 \times 21 = 273$ inputs would be used (assuming again a window of 13 amino acids).

9.3.7 Training the network

In order for the neural network to learn *general* rules about secondary structure, we must feed it as many examples as possible. The set of examples used to train a neural network is called the *training set*. Typically this training set would comprise examples from several hundred different proteins of known structure. After presenting the training set to the network many times, it eventually 'learns' how to relate the inputs (encoded amino acid sequences) to the outputs (the predicted secondary structure). More accurately, the training algorithm (e.g. steepest decent backpropagation which attempts to modify the weights in the network by essentially running the network in the opposite direction – outputs back through to the inputs) is trying to adjust the weights in the network so that it can reproduce the input/output pattern pairs as well as possible. In practice, of course, it is not possible for the network to learn how to reproduce the patterns with 100% reliability. Partly this is because of limitations in the machine learning method, but mostly this is because protein secondary structure cannot be predicted without taking into account the overall structure of the protein and so a window of say 13 residues just does not contain all the necessary information. In theory it might be possible to feed the entire sequence into the network (i.e. use a very large window size), but this is not practical as the resulting network would not only be far too large and complicated to train properly, but there would not be enough available protein structures to train it.

9.3.8 Cross-validation

One serious problem with any pattern recognition method or indeed any knowledge-based protein structure prediction method is how to go about estimating its accuracy. The first idea that comes to mind is simply to feed the training set back into the network and see how well it reproduces the expected outputs. This is a seriously flawed approach, because it is possible that the method has learned some unique features of the training set and thus the obtained results will not be as good when completely new examples are tried (this effect is known as *overfitting*). To avoid this, it is essential that cross-validation or 'jack-knifing' be carried out. To do this, the total set of examples is divided into a training set and a testing set. The testing set is kept aside and is not used at all for training. It is essential that the testing set be entirely unrelated to the training set, i.e. there should be no proteins in the testing set which are homologous to any protein in the training set. After training is complete, then the method can be tried out on the testing set and its accuracy estimated based on examples it has not been trained on. It cannot be emphasized enough that this is the only reliable way to evaluate a knowledge-based prediction method. Methods which have not been tested using rigorous cross-validation have not been tested at all.

9.3.9   Using sequence profiles to predict secondary structure

We have seen how neural networks can be used to predict secondary structure based on a single amino acid sequence. We have also seen that information from multiple sequence alignments can be used to predict secondary structure based on a set of simple rules. In fact, neural networks can also make use of information from multiple sequence alignments in order to make highly accurate predictions. The first example of this approach to secondary structure prediction was proposed by Rost and Sander in 1993. Rather than training a neural network using binary-encoded single amino acid sequences, their method, known as PHD, used sequence profiles based on aligned families of proteins to train a network. Although neural networks are often presented with strictly binary inputs (i.e. 0 or 1), in fact they can be trained with arbitrary real values. In the case of profile-based secondary structure prediction, the training inputs to the network are typically probabilities of occurrence for the 20 amino acids. Based on a single sequence, the training inputs for a certain position might look like this:

$$0\ 0\ 0\ 1\ 0\ 0\ 0\ 0\ 0\ 0\ 0\ 0\ 0\ 0\ 0\ 0\ 0\ 0\ 0\ 0$$

This would indicate that this position in the sequence contains an Asp residue. However, a similar training input based on a sequence profile might look like this:

$$0.0\ 0.0\ 0.0\ 0.2\ 0.0\ 0.0\ 0.4\ 0.0\ 0.0\ 0.0\ 0.0\ 0.0\ 0.0\ 0.0\ 0.0\ 0.2\ 0.2\ 0.0\ 0.0\ 0.0$$

This would indicate that in this position in the sequence alignment, Asp occurs with a probability of 0.2, Glu occurs with a probability of 0.4 and Ser and Thr both occur with probability 0.2.

By using sequence profiles rather than single sequences, Rost and Sander were able to improve the average accuracy of protein secondary structure prediction from around 62% to at least 70% (see next section). More recent methods, such as PSIPRED by Jones (1999) have raised this accuracy to at least 77% by generating better sequence profiles with which to train the networks.

9.3.10   Measures of accuracy in secondary structure prediction

Papers describing methods for secondary structure prediction will always quote estimates of their accuracy, hopefully based on cross-validated testing. The most common measure of accuracy is known as a Q_3 score, and is stated as a simple percentage. This score is the percentage of a protein that is expected to be correctly predicted based on a three-state classification, i.e. helix, strand or coil. Typical methods which work on a single protein sequence will have an accuracy of about 60%. What does this mean? It means than *on average* 60% of the residues you try to assign to helix, strand or other will be correctly assigned and 40% will be wrong (a residue predicted to be in a helix when it is really in a β-strand for example). However, this is just an average over many test cases. For an individual case the accuracy might be as low as 40% or as high as 80%. The Q_3 score can be misleading. For example, if you predicted every residue in myoglobin to be in a helix you would get a Q_3 score of 80% – which is a good score. However, predicting every residue to be in a helix is obviously nonsense and will not give you any useful information. Thus, the main problem with Q_3 as a measure is that it fails to penalize the network for over-predictions (e.g. non-helix residues predicted to be helix) or under-predictions (e.g. helix residues predicted to be non-helix). Hence, given the relative frequencies of helix, strand and coil residues in a typical set of proteins, it is possible to achieve a Q_3 of around 50% merely by predicting everything to be coil.

A more rigorous measure than Q_3, introduced by Matthews in 1975, involves calculating the *correlation coefficient* for each target class, e.g. the correlation coefficient for helices can be calculated as follows:

$$C_h = \frac{ab - cd}{\sqrt{(a + c)(a + d)(b + c)(b + d)}}$$

where:

a   is the number of residues correctly assigned to helix,
b   is the number of residues correctly assigned to non-helix,
c   is the number of residues incorrectly assigned to helix,
d   is the number of residues incorrectly assigned to non-helix.

The correlation coefficients for helix (C_h), strand (C_e) and coil (C_c) are in the range +1 (totally correlated) to −1 (totally anti-correlated). C_h, C_e and C_c can be combined in a single figure (C_3) by calculating the geometric mean of the individual coefficients.

Although the correlation coefficient is a useful measure of prediction accuracy, it does not assess how protein-like the prediction is. How realistic a prediction is depends (to some extent) on the lengths of the predicted secondary structural segments; a prediction of a single residue helix, for example, is clearly not desirable. However, just as important is that the correct order of secondary structure elements is predicted for a given protein structure. Taking myoglobin again as an example, in this case we would expect a good prediction method to predict six helices and no strands. Predicting eight helices or even one long helix might give a good Q_3 score, but clearly these predictions would not be considered good. A measure which does take the location and lengths of predicted secondary structure segments into account is the so-called segment overlap score (Sov score) proposed by Rost and, like the Q_3 score, this is expressed as a percentage. The definition of the Sov score is rather complicated, however, and so has not been as widely used at the more intuitive Q_3 score. Furthermore, at high accuracy levels (e.g. >70%) the Q_3 scores and Sov scores are highly correlated, and so to compare current prediction methods, Q_3 scores are usually sufficient.

9.4   Fold recognition methods

As we have seen, *ab initio* prediction of protein 3-D structures is not possible at present, and thus a general solution to the protein-folding problem is not likely to be found in the near future. However, it has long been recognized that proteins often adopt similar folds despite no significant sequence or functional similarity and that there appears to be a limited number of protein folds in nature. It has been estimated that for ~70% of new proteins with no obvious common ancestry to proteins with a known fold there will be a suitable structure in the database from which to build a 3-D model. Unfortunately, the lack of sequence similarity will mean that many of these will go undetected until after 3-D structure of the new protein is solved by X-ray crystallography or NMR spectroscopy. Until recently this situation appeared hopeless, but in the early 1990s, several new methods were proposed for finding these similar folds. Essentially this meant that results similar to those from comparative modeling techniques could be achieved without requiring there to be homology. These methods are known as fold recognition methods or sometimes as threading methods.

9.4.1   The goal of fold recognition

Methods for protein fold recognition attempt to detect similarities between protein 3-D structure that are not necessarily accompanied by any significant sequence similarity. There are many approaches, but the unifying theme is to try and find folds that are compatible with a particular sequence (see *Figure 9.3*). Unlike methods which compare proteins based on

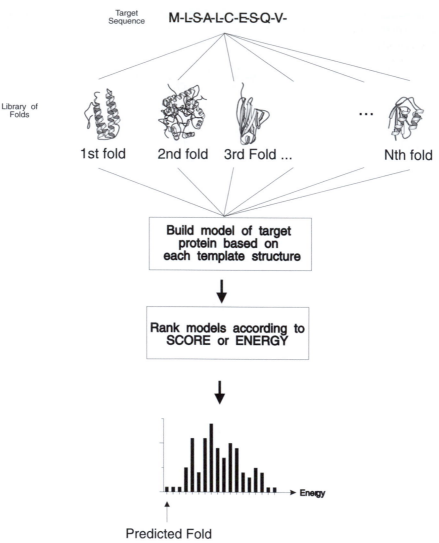

Figure 9.3

The basic components of a fold recognition method are shown. A target amino acid sequence is fitted onto each template fold in a fold library and the goodness of fit evaluated by some kind of scoring or energy function. The lowest energy or highest scoring fold forms the basis of the resulting prediction.

sequences only, these methods take advantage of the extra information made available by 3-D structure. This now provides us with three ways of comparing two proteins:

Method	Application
Sequence–Sequence Comparison	Sequence alignment, homology modeling
Structure–Structure Comparison	Flexible structural alignment, rigid body methods (superposition)
Sequence–Structure Comparison	Fold recognition, threading

In effect, fold recognition methods turn the protein-folding problem on its head: rather than predicting how a sequence will fold, they predict how well a fold will fit a given sequence. There are three main components of a fold recognition method (as illustrated in *Figure 9.3*): a library of folds (i.e. a set of proteins with known folds), a scoring function which measures how likely it is that the target sequence adopts a given fold and some method for aligning the target sequence with the fold so as to best fit as judged by the scoring function. The best scoring alignment on the best scoring fold is used to generate a 3-D model for the target protein in the same way as simple sequence alignments are used to guide comparative modeling techniques.

9.4.2   Profile methods

One of the earliest methods for protein fold recognition was proposed by Bowie, Lüthy and Eisenberg in 1991. Here sequences are matched to folds by describing each fold in terms of the environment of each residue in the structure. The environment of a residue can be described in many different ways, though Bowie *et al.* opted to describe it in terms of local secondary structure (three states: strand, helix and coil), solvent accessibility (three states: buried, partially buried and exposed), and the degree of burial by polar rather than apolar atoms. Propensities for the 20 amino acids to be found in each of the environment classes were calculated, in a somewhat analogous way to the calculation of secondary structure propensities described earlier, though with 20 classes rather than just three. With this table of propensities, a profile could be defined based not on protein sequence but on the 3-D structure of the protein.

Unfortunately, although they are conceptually simple, these environment-based methods were not very effective for detecting structural similarities between extremely divergent proteins, and between proteins sharing a common fold through convergent evolution (analogous folds). The reason for these limitations is down to the loss of structural information. A given residue in a protein typically interacts with many other residues. Classifying these interactions into a simple environment class such as 'buried and in a helix' will inevitably result in the loss of useful information, reducing the *specificity* of sequence–structure matches evaluated in this way.

9.4.3   Threading methods

A more effective method for detecting compatible folds for a target protein, called *threading*, was proposed by Jones, Taylor and Thornton in 1992. In threading, a sequence is matched to a structure by considering the plethora of detailed pairwise interactions in which each residue participates, rather than averaging them into an environmental class.

To represent these pairwise interactions, an energy function is used, similar to those used for *ab initio* protein-folding simulations. Thus, rather than using simplified descriptions of the environments of each residue in a structure, the network of pairwise interactions between each residue and its close neighbors is modeled explicitly.

9.4.4   Potentials of mean force

Although any kind of pairwise potentials, even potentials based on physical principles, could be used for threading, it was found that the best results were obtained by using potentials based on statistics or more accurately *potentials of mean force*. One very elementary principle in statistical physics is that the difference in energy between two states of a system can be related to the transition probability, i.e. the probability of moving from one state to the other. In the case of protein folding the states we are interested in relate to the folded and unfolded protein chain. Suppose we wish to estimate an energy term for a close contact between two alanine residues in a folded protein. Let us suppose that we calculate the probability of observing this interaction in many different folded proteins (i.e. all the proteins taken from the Protein Data Bank). Let us call this value $P_{folded}(Ala-Ala)$. Let us also suppose that we have access to the structures of many different unfolded proteins and we calculate a similar value $P_{unfolded}(Ala-Ala)$. We can estimate a value for the free energy change upon folding for an Ala-Ala interaction using the inverse Boltzmann equation as follows:

$$\Delta E(Ala - Ala) = -kT \ln \left(\frac{P_{folded}(Ala - Ala)}{P_{unfolded}(Ala - Ala)} \right)$$

where k is the Boltzmann constant and T is the temperature (typically room temperature). It should be noted that for threading we can ignore kT as it is simply a multiplicative constant that gives the energy in particular units (kcal per mole usually).

Of course, in practice we have no knowledge of the structures of unfolded proteins, and so we have to estimate this probability based on the assumption that unfolded proteins adopt completely random conformations.

For most practical threading potentials, the interactions between residue pairs are usually subclassified according to their sequence separation (i.e. how far apart they are in the amino acid sequence) and their spatial separation (i.e. how far apart they are in 3-D space). The reason that the sequence separation is important is that the folding properties of a short chain are different from those of a long chain because of the differing number of bonds (the ends of a long chain can obviously get further apart than the ends of a short chain) – therefore the interactions between two amino acids close together in the sequence will be different from those which are far apart.

9.4.5 Threading algorithms

Having defined a set of potentials which will allow us to determine the 'energy' of a particular protein model, or rather the compatibility of a sequence with a particular fold, the next problem to solve is how to find the best way of fitting a sequence onto a particular fold template. This is obviously an alignment problem, where rather than aligning two sequences, we are now trying to align a sequence with a particular protein structure. We have seen that for sequence alignment, dynamic programming (see Chapter 3) can be used to find an optimal solution given a particular scoring matrix and gap penalty terms. In the case of threading, however, the score function is based on pairwise interactions between amino acids. Because of this pairwise dependency in the threading score function, it is not possible to use simple dynamic programming to find the best scoring alignment. In fact it can be shown that it is impossible to find the global optimum scoring threading alignment without considering all possible alignments between the sequence and the structure (in computer science this kind of problem is called an NP-complete problem). As the number of possible threading alignments is extremely large (for a structure with N secondary structure elements where we restrict gaps to loop regions only, the number of possible threadings is of the order 10^N), searching through all possible threading alignments is not practical for all but the very smallest of structures. In practice, a number of heuristic algorithms have been applied to the threading problem, that promise to find good, but not necessarily optimal alignments.

9.4.6 How well do threading methods work?

In tests, threading methods have been found to be capable of detecting quite remote sequence–structure matches. One of the earliest successes for threading methods was to be able to detect the structural similarity between the globins and the phycocyanins, which had proven beyond the ability of sequence profile methods. Other interesting early successes included detecting similarities between some $(\alpha\beta)_8$ (TIM) barrel enzymes and also the β-trefoil folds: trypsin inhibitor DE-3 and interleukin 1β for example. The degree of sequence similarity between different $(\alpha\beta)_8$ barrel proteins and between trypsin inhibitor DE-3 and interleukin 1β is extremely low (5–10% sequence identity), and again, sequence profile methods had not proven able to detect these folds. It is therefore clear that threading methods are able to detect structural similarities which are entirely undetectable by purely sequence-based methods.

Despite the success of threading methods, they do have a number of problems. The sensitivity of threading does depend on the size of the protein structure and also its secondary structure content. For example, threading methods are better at detecting similarities between proteins in the all-α class of protein structure than in the $\alpha\beta$ class, with the all-β class being the most difficult of all. Also, proteins with very large relative insertions and deletions prove difficult to match. Another problem which is frequently observed with threading methods is that despite being able to recognize a particular structural similarity, the accuracy of the resulting alignments can often be quite poor. Therefore, although threading methods offer a great deal of scope for making useful structure predictions, some caution is required in their application.

9.5   *Ab initio* prediction methods

Fold recognition or threading methods are now widely used tools for building 3-D models of newly characterized proteins. However, no matter how much these methods are improved, they suffer from a fundamental limitation: namely, that they can only predict folds which have already been observed.

Of course, as we have seen, the problem of predicting the structure of a protein from first principles, or at least without reference to a template structure is one of the great unsolved problems in modern science. So, what are the prospects of making predictions for proteins with novel folds.

A typical *ab initio* prediction method can be described along the following lines:

1. Define a (possibly simplified) representation of a polypeptide chain and the surrounding solvent.
2. Define some kind of energy function which attempts to model the physicochemical forces found in real proteins.
3. Search for the simplified chain conformation which has the lowest energy – generally by some kind of Monte Carlo search, or by perhaps assembling fragments of existing protein structures in order to build plausible new structures.

This approach to *ab initio* protein structure prediction has a long history, starting with the pioneering work by Scheraga and co-workers in the 1970s, Warshel and Levitt's work in 1975, Kolinski and Skolnick's work with simplified lattice models in the 1990s, and many others. The reason for there remaining such an interest in these methods for folding proteins is that there is a general feeling that this kind of approach may help to answer some of the questions pertaining to how proteins fold *in vivo*. The basic idea is that by keeping as close to real physics as possible, even simplified models of protein folding will provide useful insights into protein-folding processes. However, despite a few recent successes on small proteins, *ab initio* methods are still not a practical proposition for predicting protein structure.

9.6   Critically assessing protein structure prediction

If protein structure prediction methods are to be accepted widely by the general biology community, the reliability of these methods must be evaluated in such a way as to convince a hostile jury. Of course it is usual for authors of prediction methods to attempt to devise realistic 'benchmarks' for their methods, and many such methods do indeed provide useful measures of confidence alongside the actual predictions. However, a sceptic might argue that as the developers of the methods know the expected answers (i.e. the structures of the proteins being predicted are already known), there remains the possibility that an unscrupulous or incompetent developer might bias the results in some way to demonstrate better than expected success rates. Published benchmarking results might not therefore be representative of the results that might be expected on proteins of entirely unknown structure. Alongside these concerns, there is also the fact that, historically, a number of groups have erroneously claimed to have solved or almost solved the protein-folding problem. How then can we evaluate the reliability of prediction methods and spot baseless claims of success?

To tackle these difficulties, John Moult and colleagues initiated an international experiment to validate the reliability of protein structure prediction methods by *blind testing*. The 1st Critical Assessment in Structure Prediction (CASP) Experiment was carried out in 1994, and has since been run every 2 years (at the time of writing CASP5 is currently underway in 2002). In CASP, crystallographers and NMR spectroscopists make available to the prediction community the sequences of as yet unpublished protein structures. Each research group attempts to predict the structures of these proteins and deposits their predictions *before* the structures are revealed to the public. After a few months of this work, all the predictions are collated and evaluated by a number of independent assessors. In this way, the current state-of-the-art in protein structure prediction can be evaluated without any possibility of bias or cheating. CASP has proven to be an extremely important development in the field of protein structure prediction, and has not only allowed progress to be accurately measured, but has also greatly stimulated a massive expansion of interest in the field itself. More information on CASP can be found at the Livermore Laboratory Prediction Center web site (URL http://predictioncenter.llnl.gov).

9.7   Conclusions

In this chapter we have seen that predicting protein structure from amino acid sequence is possible in certain cases. Although a general solution to the protein-folding problem is not yet in hand, knowledge-based techniques, which exploit information from known 3-D structures to make predictions on proteins of unknown structure, are becoming more and more effective. Elsewhere in this book we have seen that protein structure and protein function have a close relationship, and so it is hoped that further developments of knowledge-based protein structure prediction techniques may have a great deal of impact on the functional characterization of new gene sequences. Eventually, perhaps, we will gain some understanding of how proteins actually fold, and this may well have a direct impact on a number of important biological and medical problems. For example, it is now know that a large number of diseases (e.g. Creutzfeld–Jacob disease) are caused by protein *mis*folding, and future protein structure prediction methods may be able to model these misfolding events and even predict how particular drugs might influence them. Time will tell.

References

Bishop, C.M. (1996) *Neural Networks for Pattern Recognition*. Oxford University Press.

Bowie, J.U., **Lüthy**, R., and **Eisenberg**, D. (1991) A method to identify protein sequences that fold into a known three-dimensional structure. *Science* **253**: 164–170.

Jones, D.T. (1999) Protein secondary structure prediction based on position-specific scoring matrices. *J Mol Biol* **292**: 195–202.

Jones, D.T. (2000) Protein structure prediction in the postgenomic era. *Curr Opin Struct Biol* **10**: 371–379.

Jones, D.T., **Taylor**, W.R., and **Thornton**, J.M. (1992) A new approach to protein fold recognition. *Nature* **358**: 86–89.

Qian, N., and **Sejnowski**, T.J. (1988) Predicting the secondary structure of globular proteins using neural network models. *J Mol Biol* **202**: 865–884.

Rost, B., and **Sander**, C. (1993) Prediction of protein secondary structure at better than 70% accuracy. *J Mol Biol* **232**: 584–599.

From protein structure to function

Annabel E. Todd

- Structural genomics projects aim to determine three-dimensional protein structures on a genome-wide scale. The structures of many proteins will be determined prior to having any knowledge of their function, so there is a pressing need to accurately determine function from structure.
- What is function? The role of a protein may be described in terms of its biochemical, cellular or phenotypic function. Only biochemical features can be deduced directly from structure.
- The complexities of protein structure/function relationships present a major challenge in the inference of function from structure. One protein-folding topology may support a variety of functions and, conversely, one function may be associated with several different folds. Ancestral genes have been combined, mixed and modulated to generate the many functions necessary for life. In addition, some proteins are multifunctional.
- There are many functional classification schemes now available. Most schemes that exist fall into one of two categories: (i) genome-related schemes; and (ii) schemes related to the interaction networks of gene products.
- To infer function from structure, the knowledge-based approach that involves the comparison of a protein fold to other structures has the most potential. One can derive functional features by identifying evolutionary homologs, general fold similarities and three-dimensional functional motifs that exist in related and non-related proteins. The functional features that can be derived from the basic structure in isolation are more limited.
- Thus far, attempts at structure-based functional assignment have illustrated the importance of experimental work to complement the structural data if detailed functional information is to be obtained.

10.1 Introduction

The molecular activity of a protein is critically defined by its fold. Indeed, structure determination typically follows the functional characterization of a protein in order to uncover the details of its molecular mechanism at the atomic level. The fold reveals binding sites, interaction surfaces and the precise spatial relationships of catalytic residues.

An increasingly pressing question is to what extent can function be derived from structure? Genome sequencing has seen the advent of structural genomics initiatives, which aim to derive a structural representative for all homologous protein families. As such, biologists are presented with a complete reversal of the classical approach to protein characterization. That is, the three-dimensional structure of a protein is determined prior to having any knowledge of its function.

If structural genomics projects are to achieve their full potential, accurate methods to predict function from structure are needed. In particular, a complete understanding of the complex relationships between protein sequence, structure and function is critical.

10.2   What is function?

Protein function is a somewhat vague term in that it may be described at a variety of levels from biochemical through to phenotypic function. For example, thrombin is referred to as both a serine endopeptidase and as an extracellular protein in plasma involved in blood coagulation.

Biochemical, or molecular, function refers to the particular catalytic activity, binding properties or conformational changes of a protein. At a higher level, the complex, or metabolic or signal transduction pathway, in which a protein participates describes the cellular function. This is always context-dependent with respect to tissue, organ and taxon. Lastly, the phenotypic functions determine the physiological and behavioral properties of an organism. A particular phenotypic property may be attributed to a single gene, or more usually to a collection of genes which operate together within the organism.

Whilst the native structure may hint at little more than biochemical function, this functional assignment provides a valuable first step towards the experimental elucidation of function at all levels. In this chapter, the word 'function' largely refers to biochemical function.

10.3   Challenges of inferring function from structure

When an uncharacterized protein is sequenced, it is conventional to compare its sequence to those of known function using bioinformatics sequence search tools (see Chapter 3). Evolutionary relationships are exploited to infer function from sequence homologs, on the basis that family members commonly exhibit some similarity in function. Having three-dimensional structural data, however, is more advantageous than sequence data alone. Not only do the structures reveal spatial motifs and binding sites of functional significance, they uncover evolutionary relationships between hitherto apparently unrelated proteins, since protein structure is conserved even after all trace of sequence similarity disappears. Such distant relationships can provide functional insights that are impossible to glean at the sequence level. These two benefits that structural data bring are together the principal driving force behind the structural genomics initiatives.

However, there is no straightforward relationship between structure and function. A particular fold may perform a variety of functions and, conversely, one function may be associated with several different folds. Homologs sharing the α/β hydrolase fold have a rich variety of enzyme and non-enzyme functions, whilst the glycosyl hydrolase enzyme activity is associated with at least seven different protein scaffolds (see *Table 10.1*). This 'one structure, many functions and one function, many structures' paradox is not just true for proteins but for things we may use or see in everyday life. Four-wheeled vans serve as ambulances and ice-cream stalls. Meanwhile a story may be described on CD or in a book. These protein structure complexities present a major challenge to biologists attempting to infer function from structure.

10.4   Methods of functional evolution

In the discussion of protein structure and function, it is useful to consider the possible routes to new functions (see *Figure 10.1*). Conceptually, the simplest route to a new function is to create a new gene *ab initio*, that is, from scratch. Instead, old genes have been adapted for new functions, as exemplified by the α/β hydrolase fold superfamily in the above section.

10.4.1 Gene duplication

Homologous superfamilies provide evidence for multiple gene duplication events, and these may be the predominant source of new gene functions. Even in the very small genome of *Mycoplasma genitalium* 60% of genes arose via gene duplication, with one ancestral domain

Table 10.1 One structure, many functions and one function, many structures paradox.

One structure: many functions	One function: many structures
Four-wheeled van	**Telling a story**
Ambulance	Book
Ice-cream stall	CD
α/β hydrolase fold[a]	**Glycosyl hydrolase**
Triacylglycerol lipase	α/α toroid
Cholesterol esterase	Concanavalin A-like 2-layer β sandwich
Dienelactone hydrolase	Double psi β-barrel
Haloalkane dehalogenase	6-bladed β-propeller
Serine carboxypeptidase	$(\beta\alpha)_8$ or TIM barrel
Non-heme chloroperoxidase	Cellulase-like β/α-barrel
Neurotactin (cell-cell adhesion)	Orthogonal α-bundle

[a]Despite their differences in function, enzymes of the α/β hydrolase fold nevertheless have similar catalytic triads in their active-sites.

having more than 50 copies. In the more complex eukaryotic organisms the proportion of domains produced by duplications increases significantly, to 88% for *Saccharomyces cerevisiae* and 95% for *Caenorhabditis elegans*. With two identical copies of the same gene, one of them can retain its function whilst the other, free of functional constraints, can assume a new biological role through incremental mutations.

10.4.2 Gene fusion

Individual domains are fundamental 'building blocks' and they often function in combination, existing in oligomeric structures or in combination with one or more additional domains on the same polypeptide chain. This 'mix and match' method is probably a fast route to new functions, and appears to be more prevalent in the more complex eukaryotes in which average gene lengths are greater.

10.4.3 One gene, two or more functions

Gene recruitment, or gene sharing, refers to the acquisition of a new function by an existing gene product, rendering the protein multifunctional. This evolutionary strategy is exemplified by the recruitment of enzymes as crystallins, structural proteins in the eye lens. The new role in the eye has been acquired by modifications in gene expression. There are several methods by which proteins 'moonlight', that is, perform more than one function. In addition to differential expression, changes in oligomeric state, cellular localization and substrate concentration can all lead to functional variations.

Other genes owe their multifunctionality to post-translational modifications, alternate splicing and alternate translation initiation. The gene products do not moonlight; the functions are carried out by *non-identical* proteins which are nevertheless derived from the same gene.

The multifunctionality of genes results in constraints on adaptability. In identifying such genes we may be observing an early 'snapshot' of an evolutionary process; subject to adaptive pressures their functions may specialize in time, following gene duplication. The use of one gene for two or more functions clearly simplifies the genome, but complicates the process of genome annotation.

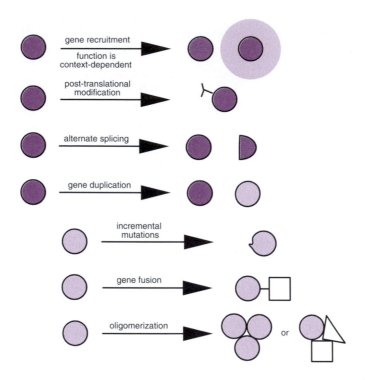

Figure 10.1

The mechanisms by which new functions are created. In practice, new functions often evolve by a combination of mechanisms. In particular, gene duplication provides two identical copies of the same gene and one, free of functional constraints, can assume a new biological role through incremental mutations, gene fusion or oligomerization. Note, however, that a change in oligomerization state without prior duplication or mutation events can provide a route to an alternative gene function and this represents a method by which proteins 'moonlight'.

10.5 Functional classifications

Genome sequencing projects have driven the development of functional classification schemes. Given this chapter's interest in function, it would be timely to digress a little to discuss these schemes. Classification of proteins by function is essential for consistency in database annotations and to facilitate the functional comparison of different proteins and organisms. It is a more problematic task than classifying proteins by sequence or structure owing to:

- the multifunctionality of many genes
- the need to describe complex biological processes with a controlled vocabulary
- the requirement for a classification applicable to widely differing organisms
- the 'apples and oranges' of function; it is useful to distinguish the different aspects of protein function, such as biochemical and cellular function, and to classify them separately and independently.

Only recently have all of these different issues been properly addressed with the development of the Gene Ontology scheme. This scheme, and other functional classifications currently available, are outlined in *Box 10.1*.

Box 10.1 Functional classifications and related schemes

- Organism-specific databases

A number of organism-specific gene function databases exist, such as FlyBase (*Drosophila*), Mouse Genome Informatics (MGI), *Saccharomyces* Genome Database (SGD) and MIPS (yeast), and GenProtEC (*Escherichia coli*).

- Gene product interactions

Some databases classify the function of a gene product according to the regulatory or metabolic pathway, or complex, in which the protein participates. Three such databases are the literature-derived EcoCyc (*Escherichia coli*) and the cross-species classifications, KEGG and WIT.

- Gene Ontology

This scheme is being developed for the classification of the gene complements of both unicellular and multicellular organisms. The authors have separated different aspects of gene function by constructing three independent ontologies: biological process, molecular function and cellular component. The relationship of a gene product (protein) or gene product group to these three categories is one-to-many; a particular protein may function in several biological processes, contain domains that carry out diverse molecular functions, and participate in multiple interactions with other proteins, organelles or locations in the cell. Each ontology comprises a non-hierarchical network of nodes that can handle biological knowledge at different stages of completeness. Unlike hierarchical classifications (see text), this flexible structure allows for a particular function to be linked not to one, but to several more generalized categories of function, e.g. 'helicase' has two child nodes: 'DNA helicase' and 'ATP-dependent helicase'; 'ATP-dependent helicase' has two parent nodes: 'helicase' and 'adenosine triphosphatase'. All gene function data in Gene Ontology are attributed to the literature.

- Clusters of Orthologous Groups (COGs)

An entirely different classification approach has been taken in the compilation of Clusters of Orthologous Groups, a scheme that is both a sequence and function classification in nature. The COGs are derived by comparing genomes from phylogenetically distant lineages. Orthologs are equivalent genes in different species that evolved from a common ancestor by speciation. Consequently, they usually have the same function allowing the transfer of functional information from one COG member to all others. A pair of orthologs is expected to have the highest sequence similarity of any pair of genes in two genomes although the detection of many orthologous relationships is not so straightforward.

- Enzyme Commission

Of all of the functional classifications, the hierarchical Enzyme Commission (EC) scheme is the best developed and most widely used. A limitation of this scheme is that it is restricted to enzymes and their biochemical functions. It is a classification of enzyme reactions, and not enzymes. A reaction is assigned a four-digit EC number, where the first digit denotes the class of reaction (1, oxidoreductases; 2, transferases; 3, hydrolases; 4, lyases; 5, isomerases; 6, ligases). The meaning of subsequent levels in the hierarchy depends upon the primary EC number, e.g. the second level describes the substrate upon which the enzyme acts in oxidoreductase reactions, but the type of reorganization in isomerase reactions. Some enzymes catalyze two or more reactions. Accordingly, a protein may have more than one EC number assigned to it.

(continued)

Box 10.1 (*continued*)

● Sequence databases

SWISS-PROT and PIR both provide a valuable source of functional annotation of gene products. Each entry is manually curated and assigned functions have been experimentally characterized else carefully predicted by the curators. Beyond the provision of keywords from a controlled vocabulary, however, they do not formally classify sequences into functional categories. Other sequence databases such as Pfam, PROSITE and PRINTS do cluster sequences on the basis of functional motifs, and these provide functional descriptions of each family (see Chapter 5).

10.5.1 Limitations of functional schemes

10.5.1.1 Hierarchical structure

Most functional classification schemes are hierarchical or tree-like in nature. The top level categorizes function in very broad terms and subsequent levels or 'nodes' in the hierarchy become increasingly specific. For instance, 'enzyme' is a broad term, whereas 'alcohol dehydrogenase' is a more precise functional descriptor.

Although simple and easy to conceptualize, hierarchical schemes have their limitations. Certain functions can be involved in a number of more generalized functional classes, just as the ice-cream van referred to in *Table 10.1* can be classified under both 'vehicle' and 'food outlet'. The Gene Ontology team has taken a more complex, nevertheless more flexible, approach in the classification of function with a non-hierarchic structure, as outlined in *Box 10.1*.

10.5.1.2 Apples and oranges

Few schemes address the multidimensionality of protein function and instead they mix the 'apples and oranges' of function. For example, it is not uncommon to have the functional nodes 'cell regulation', a physiological function, and 'ion channel', a molecular function, included within a single classification.

It is important to be aware of these limitations, nevertheless they should not detract from the enormous usefulness of the many classification schemes in which they are not directly addressed.

10.6 From structure to function

This section outlines the functional information that can be derived from a protein structure itself and from the structural relationships it may share with other proteins of known fold.

10.6.1 Basic structure

The three-dimensional co-ordinates of all atoms within a protein structure are stored in a PDB file (see Chapter 6). The PDB file itself provides little functional information, although all PDB files give the name of the protein, when known. Some PDB files contain one or more 'SITE' records. These records provide a formatted list of residues of functional relevance, such as those involved in ligand-binding, metal-binding, catalysis or protein–protein interactions. From the co-ordinates themselves, however, we can derive the functional information summarized in *Figure 10.2*.

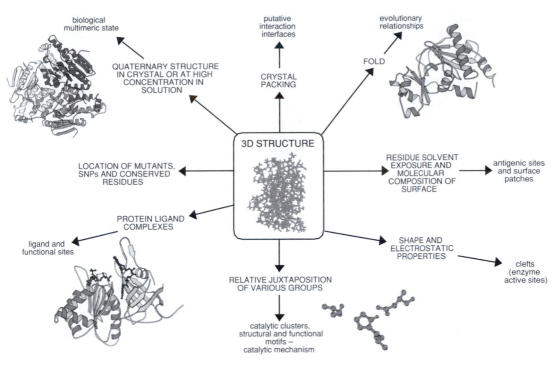

Figure 10.2

From structure to function: a summary of information that can be derived from three-dimensional structure, relating to biochemical function.

The structure reveals the overall organization of the polypeptide chain, in terms of its secondary structure elements, the packing arrangement of these elements and the connectivities between them. From this it is possible to determine those residues that are in the structural core and those that are in contact with solvent, as well as the shape and electrostatic properties of the protein surface. The quaternary structure of the protein in the crystal environment or in solution at high concentration often reveals the protein's biologically relevant oligomeric state.

Protein–ligand complexes are invaluable in terms of the functional information they provide. They reveal the location of the interaction site, the conformation of the ligand in its bound state and residues of functional significance. For enzymes, such complexes indicate the disposition of amino acids relative to each other and to the ligand within the active-site, from which one can postulate a catalytic mechanism. Typically, protein–ligand complexes are designed with a knowledge of the protein's function, e.g. commonly enzymes are crystallized with a transition-state analog. In structural genomics, however, the ligand is unknown, but the fortuitous inclusion of a ligand from the 'cloning' organism can provide a vital clue to function, examples of which have already been documented.

10.6.1.1 PDBsum

PDBsum is a Web-based database that provides detailed information on every protein in the PDB. The database provides a summary of the molecules in each PDB file (proteins, nucleic acids, metal ions, ligands, water molecules) together with various analyses of their structural features. The features summarized include secondary structure, disulfide bonds, protein–

ligand interactions and metal contacts. Additionally, PDBsum provides links to many other databases, such as SWISS-PROT, Pfam, PRINTS, ENZYME, KEGG and the structural classifications CATH and SCOP.

10.6.2 Structural class

The most gross level of structure, as defined by secondary structure content, cannot provide any specifics regarding the biochemical function of a protein. However, there are clear biases with respect to general function at the class level. There is a dominance of α/β folds in enzymes compared with non-enzymes, whilst the mainly-α class is under-represented in enzymes. This may be historical in origin, as well as having some physico-chemical basis.

Several common biological ligands have a notable bias towards certain protein classes defined by the stereochemical requirements for ligand-binding, e.g.

- Heme, α; the preferred binding mode is for the heme to slot between two or more helices; the hydrophobic faces are shielded from the solvent and they interact with non-polar sidechains in the helices.
- DNA, α and α/β; DNA recognition is dominated by the helix motif binding in the major groove.
- Nucleotides, α/β; many of the nucleotide-binding α/β folds correspond to three-layer αβα sandwiches with various topologies; the best known of these is the Rossmann fold, the first nucleotide-binding domain to be determined, and in this structure the nucleotide extends along the C-terminal ends of the strands in the parallel β-sheet.

In terms of general function, however, the nature of the fold can sometimes be a better guide. For example, almost all proteins having the TIM barrel fold are enzymes.

10.6.3 Global or local structural similarity

To infer function from structure, the knowledge-based approach which involves comparison of the protein fold, or structural motifs contained within it, to other structures in the PDB has the most potential. The amount of functional information which can be deduced in this way depends upon the nature of the relationship of the structure and its particular features to those already known.

Section 10.3 outlined the challenges presented by the complex relationships between structure and function. These are highlighted further by the seven pairs of proteins illustrated in *Figure 10.3*. The protein pairs are labeled as one of two types: homologs or analogs. Homologs, by definition, are derived from a common ancestor. Analogs are proteins that are similar in some way, yet show no evidence of a common ancestry. Structural analogs share the same fold, and functional analogs perform the same function.

10.6.3.1 Homologous relationship

Homologous relationships can be identified using one of the many protein structure comparison methods now available (see Chapter 6). The hemoglobins nicely illustrate the benefit of identifying homologs in this way for functional annotation (see *Figure 10.3c*). The sequences within this large protein family have diverged to such an extent that sequence analysis methods fail to detect their common origin, yet their structures are very well conserved. Hypothetically, if an uncharacterized hemoglobin family member was sequenced, through the determination of its structure one could assign to it a putative heme-binding oxygen transport function.

Considerable caution, however, must be exercised in the transfer of function between relatives. The assumption that homologous proteins have related functions appears to be largely applicable to orthologs of different species but the functions of paralogs, genes related

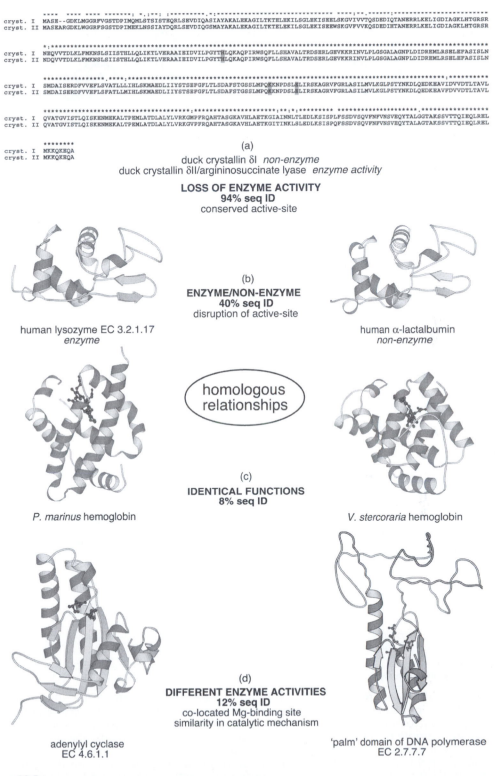

(a)

duck crystallin δI *non-enzyme*
duck crystallin δII/argininosuccinate lyase *enzyme activity*

LOSS OF ENZYME ACTIVITY
94% seq ID
conserved active-site

(b)

ENZYME/NON-ENZYME
40% seq ID
disruption of active-site

human lysozyme EC 3.2.1.17
enzyme

human α-lactalbumin
non-enzyme

homologous
relationships

(c)

IDENTICAL FUNCTIONS
8% seq ID

P. marinus hemoglobin

V. stercoraria hemoglobin

(d)

DIFFERENT ENZYME ACTIVITIES
12% seq ID
co-located Mg-binding site
similarity in catalytic mechanism

adenylyl cyclase
EC 4.6.1.1

'palm' domain of DNA polymerase
EC 2.7.7.7

Figure 10.3

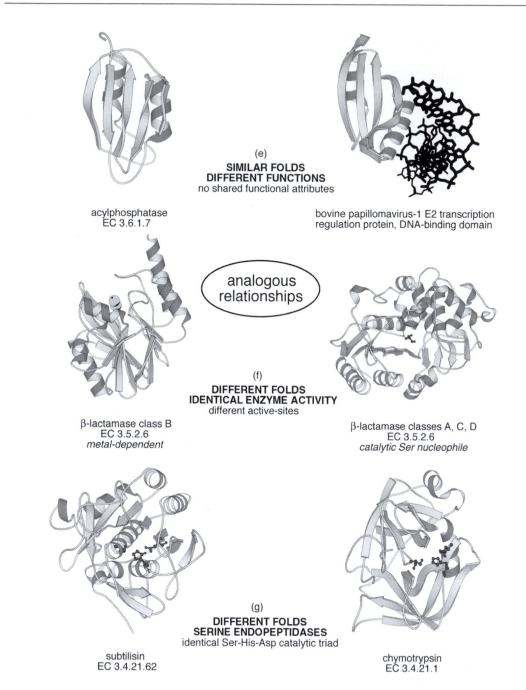

(e)
SIMILAR FOLDS
DIFFERENT FUNCTIONS
no shared functional attributes

acylphosphatase
EC 3.6.1.7

bovine papillomavirus-1 E2 transcription
regulation protein, DNA-binding domain

analogous
relationships

(f)
DIFFERENT FOLDS
IDENTICAL ENZYME ACTIVITY
different active-sites

β-lactamase class B
EC 3.5.2.6
metal-dependent

β-lactamase classes A, C, D
EC 3.5.2.6
catalytic Ser nucleophile

(g)
DIFFERENT FOLDS
SERINE ENDOPEPTIDASES
identical Ser-His-Asp catalytic triad

subtilisin
EC 3.4.21.62

chymotrypsin
EC 3.4.21.1

Figure 10.3 (*continued*)

Complexities of protein sequence, structure and function relationships. (a)–(d) and (e)–(g) illustrate homologous and analogous relationships, respectively. (a) Duck crystallins δI and δII function as structural proteins in the eye lens and δII also has argininosuccinate lyase activity, an example of gene sharing; this activity is lacking in δI despite the conservation of the catalytic residues (shaded in grey) (b) lysozyme functions as O-glycosyl hydrolase, but α-lactalbumin lacks this activity and instead regulates the substrate specificity of galactosyltransferase; the active-site is disrupted in α-lactalbumin (see Box 10.1 for an explanation of EC numbers) (c) despite their shared function, hemoglobins exhibit remarkable sequence diversity; the heme molecule bound to each structure is shown in

by duplication within a genome, may be quite different in divergent families. Even at high sequence identity, changes in function occur (see *Figure 10.3a*). Obviously the challenge for biologists is to differentiate sequence and structural variation that has specific functional consequences and variation that results from neutral drift. This requirement is highlighted further by *Figure 10.4*, which plots the sequence identity and structural alignment score for homologous domain pairs, where pairs with the same (circles) and different (squares) functions are distinguished. The relationship between global sequence and structural similarities and function is not straightforward.

A powerful method to infer function in homologous proteins is to assess the conservation of functional residues. An active-site architecture or functional interface may be conserved in the absence of significant global sequence similarity, and some or all components of function may be transferred to the hypothetical protein. In *Figure 10.4*, for example, all black points in the bottom left corner correspond to pairs of triacylglycerol lipases of the α/β hydrolase superfamily. This lipase function seems to be compatible with a wide variety of structures; the folds of these homologous proteins differ in β-sheet length (6–10 strands), topological connectivities close to the edges of the sheet and in the number and sizes of insertions within the sheet. Nevertheless, they all have a Ser-His-Asp triad that is crucial for catalysis.

10.6.3.2   Fold similarity and structural analogs

Structural similarity does not necessarily imply an evolutionary relationship (see Chapter 7). Evidence suggests that there is a limited number of folds and multiple superfamilies exist within a single fold group. Such fold groups are referred to as 'superfolds' or 'frequently occurring domains'. Therefore, proteins having a common structure can have completely unrelated functions (see *Figure 10.3e*). This illustrates the importance of establishing whether a common ancestry exists between an uncharacterized protein structure and its structural matches in the PDB.

Whilst functional prediction of an unknown protein belonging to a highly populated fold group may be difficult, for a few superfolds there are preferred binding-sites, termed 'supersites'. Therefore, in the absence of evolutionary homologs, a fold similarity can reveal a putative ligand-binding site. Although approximate, this information can guide biochemical experiments or aid ligand design. The TIM barrel and α/β Rossmann fold provide two good examples of supersites (see *Figure 10.5*). The existence of supersites may be associated with

Figure 10.3 (*continued*)

ball-and-stick representation (d) given their different sizes and insignificant sequence similarity, the evolutionary origin of adenylyl cyclase and the 'palm' domain of DNA polymerase was a subject of controversy until the identification of their co-located Mg^{2+}-binding sites; despite their different overall enzyme activities they share some similarities in their catalytic mechanism and act on chemically similar substrates; the 'fingers' domain of DNA polymerase interrupts the 'palm' domain at the top and it is represented by the long loops; Mg^{2+}-binding sites are shown in ball-and-stick representation (e) acylphosphatase and the DNA-binding domain of bovine papillomavirus-1 E2 both adopt the α/β-plait fold; they share no significant sequence similarity nor functional attributes to indicate an evolutionary relationship and so they belong to different homologous superfamilies; they are structural analogs (f) class B β-lactamases are metal-dependent, whilst the structurally distinct class A, C and D enzymes have a Ser nucleophile which is essential for catalysis and forms a covalent intermediate with the substrate; the two metal ions bound to class B β-lactamase are represented by the grey spheres, and the Ser nucleophile of the class A, C and D fold is shown in ball-and-stick representation (g) subtilisin and chymotrypsin have different structural scaffolds, yet they have the same Ser-His-Asp catalytic triad and both function as serine endoproteases via the same catalytic mechanism; the triad residues are shown in ball-and-stick representation.

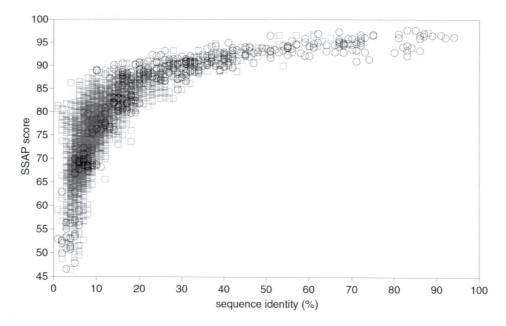

Figure 10.4

Correlation between sequence, structural and functional similarity in protein homologues. The graph plots pairwise sequence identity versus SSAP score (see chapter 6) for all unique homologous domain pairs within 31 structural superfamilies, with points coloured to distinguish pairs having identical (circles) and different (squares) functions. 99% of functionally distinct pairs share less than 40% sequence identity.

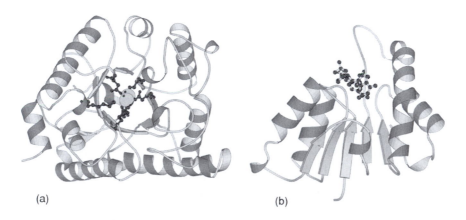

(a)   (b)

Figure 10.5

The location of (a) TIM barrel and (b) Rossmann supersites. In the TIM barrel, the enzyme active-site is invariably formed by the loops that connect the carboxy ends of the β-strands in the central β-barrel with the adjacent α-helices. In the doubly wound α/β Rossmann fold, ligands bind at the crossover point between the two identical halves of the structure, where the β-strand order is reversed.

the general principles of protein structure. For example, the dipole moment of a helix could assist a catalytic residue located at one end of it.

10.6.3.3 Structural motifs and functional analogs

As illustrated by the conserved catalytic triad in the triacylglycerol lipases, it is sometimes possible to identify local structural motifs that capture the essence of biochemical function and these can be used to assign function. Similar local three-dimensional sites may be identified in non-related proteins having different folds, as well as in evolutionary relatives. For example, subtilisin and chymotrypsin have different scaffolds, yet they have the same Ser-His-Asp catalytic triad and both function as serine endoproteases via the same catalytic mechanism (see *Figure 10.3g*). This same triad appears to be a useful framework for a number of activities, since it exists in other, non-proteolytic proteins, including the triacylglycerol lipases of the α/β hydrolase fold discussed above.

Given that catalytic activity depends upon the precise spatial orientation of only a few amino acids, it is perhaps not surprising that the same active-site architecture can recur in the contexts of different structural scaffolds. In addition, the concoction of the same enzyme mechanism more than once during evolution suggests that there may be a limited number of catalytic combinations of the 20 amino acids. Indeed, only a subset of residues has the required functionality for most catalytic tasks, such as proton abstraction. Local structural motifs are not restricted to enzyme catalytic sites and other well-known examples include the calcium-binding EF hand and the helix-turn-helix motif which binds to DNA.

Owing to the functional benefits of recognizing common local structural motifs in both related and unrelated folds, a bevy of new algorithms for their automatic identification has been developed. Supposing a cluster of functional atoms or residues is predefined in a 3-D template, the problem reduces to a search in 3-D to find the specified cluster in other proteins. Given the size of the PDB, however, a simple global calculation that involves the fitting of the template to all possible atom combinations within each structure is computationally prohibitive. Various approaches have been sought to improve the efficiency of the search. Most methods pre-process the co-ordinate files in some way and then use efficient search algorithms to scan the data. An example 3-D template is provided in *Figure 10.6*. The active-site of the aspartic proteases can be captured in an eight-atom template, and this can be used to identify all related enzymes in the PDB.

Usefully, given a set of protein structures, some methods identify common spatial motifs automatically and so they do not require any prior knowledge of functional sites nor a predefined atom cluster for the searches. *Box 10.2* briefly describes some of the methods currently available for the identification of structural motifs.

10.6.4 *Ab initio* prediction

In the absence of global or local structural similarities to a characterized protein (a situation that will be increasingly rare) *ab initio* prediction of function from structure is needed. That is, what functional information can be derived from the structure in isolation? There has been very little work in this area because in the past almost all new protein structures had been previously characterized. With the recent advent of structural genomics initiatives, this will no longer hold true.

Methods to detect a functional site must reflect the nature of that site. In enzymes, the active-site is usually situated in the largest cleft on the protein surface, and this cleft is often significantly larger than other clefts in the protein. A large cleft ensures that the number of complementary protein–substrate interactions is maximized and that the substrate is positioned precisely to facilitate catalysis. Thus, in most cases, the enzyme active-site can be identified using geometrical criteria alone. In contrast, protein–protein interaction sites are highly exposed and vary in character. Surface patches that constitute protein–protein bind-

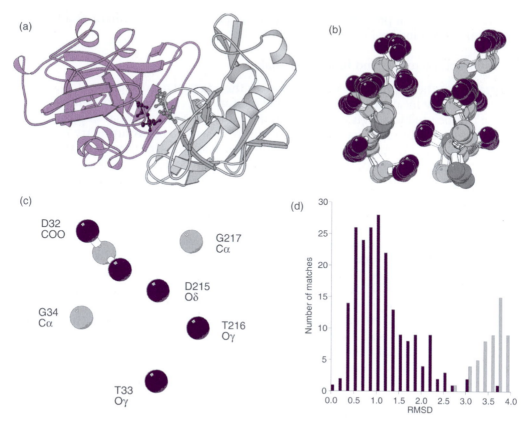

Figure 10.6

Three-dimensional template for the aspartic protease active-site. (a) The crystal structure of human pepsin (PDB code 1psn) consists of two domains (in pale purple and pale gray) which each contribute residues to the active-site (2xAsp-Thr-Gly; darker colours). (b) The active-site residues of 18 different families are superposed. (c) The atoms selected for the aspartic protease active-site template that yield the maximum specificity and sensitivity (using the pepsin 1psn numbering scheme): Asp 32 COO, Thr33 Oγ, Gly34 Cα, Asp215 Oδ, Thr216 Oγ, Gly217 Cα. (d) The matches found by a search against all entries in the PDB are plotted against the root mean square deviation (RMSD) in Å from the template shown in (c). True matches are shown in purple and false hits are in grey. The stray positive match at 3.69Å is proplasmepsin (PDB code 1pfz), an aspartic protease zymogen, a non-active protease.

ing sites can be identified if a series of physical and chemical parameters are considered; these include hydrophobicity, amino acid residue propensity and shape in terms of planarity, protrusion and size relative to the whole protein.

A new, alternative approach for the identification of enzyme active-sites uses theoretical microscopic titration curves. Ionizable residues having abnormal curves typically correspond to active-site residues. Therefore, given an uncharacterized structure, if those residues that have perturbed pK_as cluster together in the fold then a putative active-site is identified.

10.7 Evolution of protein function from a structural perspective

As outlined in Section 10.4, folds have been extensively combined, mixed and modulated to generate the multitude of functions necessary for life. A grasp of these evolutionary processes

---

Box 10.2 Methods to detect recurring structural motifs

● TESS, PROCAT

TEmplate Search and Superposition (TESS) has been used for the detection of enzyme-active-sites, but in theory it is applicable to any type of functional site. The search template is derived manually and it contains those atoms considered essential for the enzyme to perform its catalytic function. Functionally important residues are identified from experimental data available in the literature. TESS identifies those structures in the PDB having a similar constellation of atoms that is within a certain RMSD of the search template. All of the PDB files are pre-processed. This stage involves the construction of a quick 'look-up' table that contains geometrical information about the atoms within each protein structure. Once the table is generated, it may be used again and again for different TESS searches. PROCAT is an inventory of enzyme catalytic sites that have been derived using TESS.

● SPASM, RIGOR

The SPatial Arrangements of Sidechains and Mainchain (SPASM) program is similar to TESS in that it identifies matches to a user-defined motif. SPASM only treats residues, not individual atoms, and each residue is represented by its Cα atom and, for non-glycine residues, by a pseudo-atom located at the center of gravity of its sidechain atoms. This reduced representation ensures a faster search time, at the expense of sensitivity. The user may set a variety of options, e.g. conservation of sequence directionality of residues within the motif, user-defined set of allowed amino acid substitutions. RIGOR is a program that implements the reverse process of SPASM, that is, it compares a database of pre-defined motifs against a newly determined structure.

● FFF

Fuzzy Functional Forms (FFFs) are derived by the superposition of functionally significant residues in a few selected protein structures that have related functions, and the relative geometries (distances, angles) between the Cα atoms and the sidechain centers of mass are recorded. Using a search algorithm, residues that satisfy the constraints of the FFF may be identified in other experimentally determined structures in the PDB, as well as in low-to-moderate resolution models produced by *ab initio* folding or threading prediction programs (see Chapter 9). Therefore, given an uncharacterized protein sequence, one can predict its fold, identify a putative active-site that matches a FFF and then predict its function.

● Protein sidechain patterns

This method detects recurring 3-D sidechain patterns in protein structures without having any prior knowledge of the functional residues. A depth-first search algorithm detects all spatially adjacent groups of identical amino acids that are common to two proteins. To reduce the vast number of combinations, knowledge derived from multiple sequence alignments is used to concentrate on conserved polar residues, i.e. those that are likely to be involved in function. Matches are scored by RMSD, which is itself assessed by statistical significance, unlike the other methods listed above.

---

is essential not only if we are to benefit from the wealth of genome data but also for the successful design of proteins with novel functions.

In order to improve our understanding of the underlying molecular mechanisms of evolving new functions through sequence and structural changes, Todd, Orengo and Thornton at

University College London conducted a novel detailed, collective analysis of 31 functionally diverse enzyme superfamilies in the PDB. They analysed the conservation and variation of catalytic residues, reaction chemistry and substrate specificity, as well as changes in domain organization and quaternary structure. Most superfamilies are studied in isolation, but an analysis of this nature allows the identification of evolutionary trends, if any, in the evolution of new functions. Furthermore, one can assess whether the evolutionary strategies observed in one family are peculiar to that family or are a more general phenomenon observed in others. This section outlines the principal summaries of their work. The analysis and descriptions in the text are restricted to structural data, unless stated otherwise.

10.7.1   Substrate specificity

Of 28 superfamilies involved, at least in part, in substrate binding, in only one is the substrate absolutely conserved. Enzymes in six superfamilies bind to a common substrate type, such as DNA, sugars or phosphorylated proteins. However, in at least three of these families, variations within these ligand types may be extensive.

In as many as 19 superfamilies, substrate specificity is completely diverse, in that the substrates bound vary in their size, chemical properties and/or structural scaffolds (e.g. aromatic versus linear-chain hydrocarbons) illustrating the plasticity of protein structures with respect to ligand binding (see *Figure 10.7*). If any substrate similarity exists within these enzyme superfamilies, it is limited to a small chemical moiety such as a carbonyl group or peptide bond, as identified in nine superfamilies, and this is typically the center of reactivity during catalysis.

Substrate specificity implies diverse binding sites, achieved through structural variations and exploitation of the varying properties of the 20 amino acids. Surface loops commonly contain the structural determinants of substrate specificity, and this facilitates rapid evolutionary divergence and adaptation since the loops can vary whilst the structural integrity of the protein fold is maintained. For example, within some superfamilies of the TIM barrel fold, substrate differences are accommodated by loop excursions of varying lengths at the carboxyl termini of the central β-strands.

10.7.2   Reaction chemistry

Here there is a distinction between reaction chemistry and catalytic mechanism. Chemistry refers to the overall strategy of changing substrate into product, and the nature of the intermediates involved, whereas catalytic mechanism describes the roles played by specific residues in the active-site.

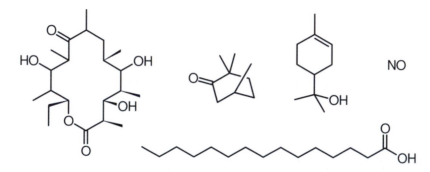

Figure 10.7

Diverse substrates bound by structural members of the cytochrome P450 superfamily.

10.7.2.1 Conserved chemistry

Far more common than the conservation of substrate binding is the conservation of reaction chemistry. There are 28 superfamilies involved directly in catalysis and for which some details of catalysis are known, and in two, the reaction chemistry is conserved. For example, catalysis by all members of the protein tyrosine phosphatase superfamily involves nucleophilic attack by a conserved Cys on the substrate and the thiol-phosphate intermediate formed is subsequently hydrolyzed.

10.7.2.2 Semi-conserved chemistry

In 21 superfamilies the chemistry is 'semi-conserved', in that enzyme members utilize a common chemical strategy in the contexts of different overall transformations. Typically, it is the initial catalytic step that is conserved, whilst the reaction paths that follow vary, sometimes extensively, just as several cars may approach a roundabout from the same direction but take different exits. For example, in the enolase superfamily the first reaction step invariably involves metal-assisted general base-catalyzed removal of a proton α to a carboxylate group. Overall, catalytic activities include β-elimination of water, racemization, cycloisomerization and epimerization (see *Figure 10.8*). Such superfamilies illustrate not only the versatility of protein folds with respect to function, but also the versatility of the chemistry involved, in that a single chemical step can be re-used in a number of contexts to provide completely different outcomes.

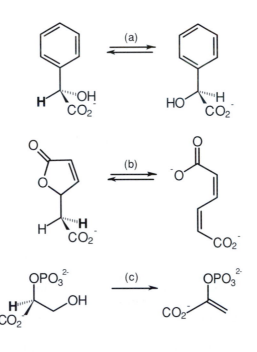

Figure 10.8

Reactions catalysed by three members of the enolase superfamily (a) mandelate racemase (b) muconate cycloisomerase I and (c) enolase. Their overall enzyme activities are quite different, but the initial step in each reaction involves metal-assisted general base-catalysed removal of a proton α to a carboxylate group. This proton is shown in bold in each enzyme substrate.

10.7.2.3   Poorly conserved chemistry

In three superfamilies the chemistry is 'poorly conserved' in that just one small aspect of the reactions catalyzed is conserved, such as a common intermediate, in the same way that several cars may approach and leave the same roundabout by completely different routes. The crotonase-like superfamily includes structural and sequence members having hydratase, dehydratase, dehalogenase, isomerase, esterase, decarboxylase and peptidase activities. Stabilization of an oxyanion intermediate by a conserved oxyanion hole, in which the intermediate is hydrogen-bonded to two backbone amide groups, is the only functional similarity conserved across all members, and in contrast to many of the 'semi-conserved' superfamilies, the reaction paths to this intermediate vary widely. They include proton abstraction, peptide bond addition and nucleophilic aromatic addition, and involve different catalytic residues within the active-site.

10.7.2.4   Variation in chemistry

In two superfamilies, reaction chemistry is completely dissimilar in at least one pair of enzymes. Typical members of the hexapeptide repeat protein superfamily are cofactor-independent acyltransferases and bind acyl-CoA or acylated-acyl carrier protein as a donor substrate. However, this enzyme family includes also an archaeal carbonic anhydrase which catalyzes reversible Zn-dependent hydration of carbon dioxide. Evidence for a common ancestry, in the absence of any functional similarity, is provided by the conservation of unusual sequence and structural features in all family members.

10.7.3   Catalytic residues

Two alternative situations with regard to the conservation or divergence of function and active-site architectures are observed. The same active-site framework may be used to catalyze a host of diverse activities, and conversely, different catalytic apparatus may exist in related proteins with very similar functions. The first situation occurs in the α/β hydrolase superfamily. The Ser-His-Asp triad of the triacylglycerol lipases referred to above is used by many other functionally distinct enzymes acting on diverse substrates, including various peptidases, lipases, a thioesterase and haloperoxidases.

One might expect that any similarity in reaction chemistry displayed by homologous enzymes is mediated by common functional groups conserved through evolution and so at least some aspects of the mechanisms of these enzymes would be identical. In half of the superfamilies in this analysis, however, there is poor positional conservation of residues which play equivalent catalytic roles in related proteins.

Although the ferritins of the di-iron carboxylate protein superfamily are not strictly enzymes, they have ferroxidase activity associated with a di-iron site and this is important for their iron storage function. In classical ferritins, the di-iron site is located within a four-helix bundle core, as in ribonucleotide reductase and other enzyme homologs. Ferritin from *L. innocua* is unusual in that despite having ferroxidase activity it lacks all of the iron-binding residues in the classical ferroxidase center. In this protein, the ferroxidase di-iron site is located at the interface between two subunits related by two-fold symmetry (see *Figure 10.9*). Despite the difference in location, this center is both chemically and structurally similar to that in classical ferritins.

The initial reaction step catalyzed by members of the phosphoglycerate mutase-like superfamily involves nucleophilic attack of a conserved His residue on the phospho group of the substrate to form a phosphohistidine intermediate. In this first step an acid catalyst protonates the oxygen of the scissile P–O bond in the substrate to make it a better leaving group. In all enzymes the acid catalyst is an active-site carboxylate: in the smaller proteins phosphoglycerate mutase and fructose-2,6-bisphosphatase it is a Glu located in a long loop after

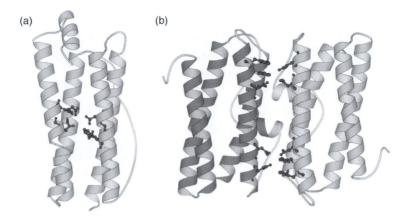

Figure 10.9

Di-iron sites in (a) classical ferritin and (b) *L. innocua* ferritin. Residues involved in iron-binding are shown in ball-and-stick representation. In (b), the two subunits are distinguished in different shades of grey.

the third strand of the central β-sheet, whilst in the acid phosphatases and 3-phytases of this superfamily it is an Asp located just after strand four. Nevertheless the relative disposition of the His nucleophile and the acid catalyst in all proteins is similar.

This variability is unexpected. It may reflect evolutionary optimization of the catalytic efficiency of these enzymes. 'New' functional groups in the active-site may take over the roles of old residues if they can do the job more efficiently. Secondly, the fortuitous appearance of alternative residues within the active-site, which have the necessary functionality, may itself be a method by which completely *new* activities have evolved. Lastly, examples of functional residue variability may indicate that the enzyme functions in question have evolved *completely independently*. That is, nature has arrived at two or more distinct solutions to the same catalytic conundrum within homologous structures. Evidence suggests that this has occurred in at least two superfamilies.

10.7.4 Domain enlargement

Holm and Sander introduced the concept of the minimal structural and functional core of related proteins. Provided the key structural elements of the functional site remain intact, large-scale diversions around this central core are permissible. Indeed, there is considerable structural variability in many of the superfamilies and in 11 superfamilies, the homologous domain size varies by at least two-fold (see *Figure 10.10*).

Size variations may involve the addition/loss of subdomains, variability in loop length, and/or changes to the structural core, such as β-sheet extension. These structural deviations are often associated with oligomeric contacts and variations in substrate specificity. Nature has probably embellished more simple ancestral folds in the evolution of protein complexity and functional specialization, through amino acid and intron insertions, whilst the evolutionary reduction in domain size is less common. In addition, many superfamilies contain members having insertions in their folds which are large enough to form separate structural domains.

(a)   (b)

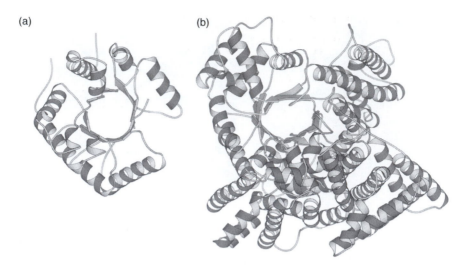

Figure 10.10

The phosphoenolpyruvate/pyruvate-binding TIM barrel domains of (a) pyruvate kinase and (b) phosphoenolpyruvate carboxylase differ by more than three-fold in size. Phosphoenolpyruvate carboxylase has a total of 40 α-helices, including the eight that pack around the central β-barrel. Most of the additional helices, introduced at different points throughout the primary sequence, are situated at the C-terminal end of the barrel, and in this cluster of helices lies the binding site for aspartate, an allosteric inhibitor. Extensive domain embellishment has provided the enzyme with a regulatory site and probably also an interface for stabilising its tetrameric structure.

10.7.5   Domain organization and subunit assembly

In 27 superfamilies, the domain organization varies between members, illustrating the importance of domain recruitment in functional evolution (see *Figure 10.11*). Additional modules fused to the catalytic domain of an enzyme may play a role in regulation, oligomerization, cofactor dependency, subcellular targeting or, commonly, substrate specificity. In at least 23 superfamilies, subunit assembly varies between members. For 12 of these, variation results only from a differing number of identical chains in the protein complex, whilst in the remainder, one or more members function in combination with different subunits, that is, as a hetero-oligomer.

10.7.6   Summary

This section has briefly illustrated the extent to which nature has adapted old folds for new functions, through changes at the atomic level to gross structural rearrangements. *Figure 10.12* summarizes the frequency with which particular types of changes occur and the extent of conservation of particular properties within the superfamilies analyzed. Some superfamilies have achieved their functional diversity by an extraordinarily wide variety of routes, including the loss/gain of catalytic metal sites, domain embellishments, internal domain duplications, domain rearrangements, fusion and oligomerization, whilst a small number have diversified considerably through modifications in the active-site alone, without the recruitment of other modules or subunits.

This analysis is restricted to the limited structural data currently available. With the growth in the PDB many more functions and structural variations in these superfamilies

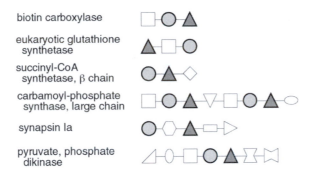

Figure 10.11

Domain organisation of members of the ATP-dependent carboxylate-amine/thiol ligase superfamily. Each shape corresponds to a structural domain and domains of the same shape are homologous. The two domains that together define the ATP-binding fold of this superfamily are highlighted in bold. In eukaryotic glutathione synthetase these domains occur in a different order owing to a genetic permutation event. In carbamoyl-phosphate synthase there are two copies of the ATP-binding fold and each copy has a distinct role in the reaction catalysed. This fold has been combined with a wide variety of other domains, including the ubiquitous Rossmann and TIM barrel folds, leading to the broad functional diversity of this superfamily.

are likely to be identified. Indeed, these structural families support over 200 enzyme and non-enzyme functions and with the inclusion of sequence data this number more than triples.

10.8   Structural genomics

Significant technological advances in recombinant DNA technology, high-level protein expression systems and structure determination methods have increased the rapidity with which protein structures can be solved. These advances have made viable the structural genomics projects that depend upon high-throughput structure determination on a genome-wide scale.

A group at the University of Toronto initiated a prototype structural genomics study of 424 proteins in the archaeon *Methanobacterium thermoautotrophicum*. These correspond to non-membrane proteins since membrane protein structure determination is not sufficiently advanced for high-throughput approaches. Of these 424 proteins, 20% was amenable to structural analysis. Progress in experimental techniques still needs to be made if these projects are to achieve their full potential.

10.8.1   From structure to function: specific examples

A few examples of efforts to determine function from structure are outlined below. These have been met with varying degrees of success. Clearly the structural data complement biochemical experiments and these in combination help to define biochemical and biological function.

10.8.1.1   Mj0577: putative ATP-mediated molecular switch

Mj0577, an ORF from the hyperthermophile *Methanococcus jannaschii*, had unknown function until structural determination of its gene product. The structure contained bound

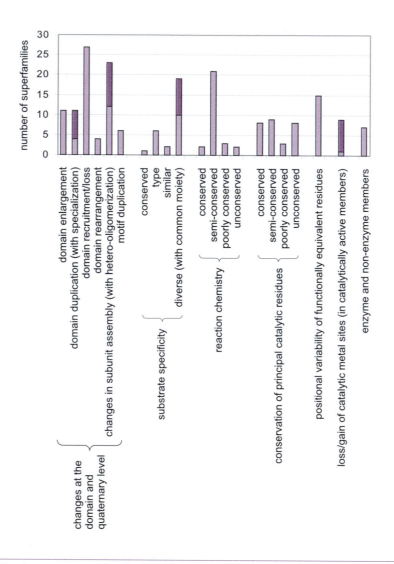

Figure 10.12

Summary of the frequency with which particular types of changes occur, as derived from the structural data only, and the extent of conservation of functional attributes within the 31 superfamilies analysed. Parts of histogram bars shaded in dark purple correspond to the x-axis label in brackets (e.g. the loss/gain of catalytic metal sites has occurred in nine superfamilies, and in eight of these nine families metal content varies in catalytically active domain members).

ATP, fortuitously scavenged from its *E. coli* host during over-expression, which suggested that the protein hydrolyzes ATP. Subsequent biochemical experiments showed that Mj0577 cannot hydrolyze ATP by itself, but it has ATPase activity only in the presence of *M. jannaschii* cell extract, and so the protein is likely to function as an ATP-mediated molecular switch.

10.8.1.2 Mj0266: putative pyrophosphate-releasing XTPase

Comparison of the three-dimensional structure of the uncharacterized Mj0266 gene product with the contents of the PDB revealed a structural similarity between this protein and the anticodon-binding domains of histidyl- and glycyl-tRNA synthetases, and the nucleotide-binding domains of several other proteins. Nucleotide-binding assays revealed the protein's ability to hydrolyze nucleotide triphosphates to monophosphates. Xanthine triphosphate was shown to be the best substrate by far and these results indicate strongly that the protein is a pyrophosphate-releasing XTPase.

10.8.1.3   *M. jannaschii* IMPase: bifunctional protein

Structure determination allowed the assignment of a second function to this protein of known function. Fructose-bisphosphate aldolase (FBPase) activity has been detected in crude protein extracts from *M. jannaschii* but sequence search tools have been unable to identify the FBPase enzyme in the genome. Inositol monophosphatase (IMPase), however, is present. Structure determination and analysis of this enzyme showed that the active-site is almost identical to that of human IMPase, but there are significant differences in five adjacent loop regions. Unexpectedly, the structure of *M. jannaschii* IMPase is similar also to pig kidney FBPase, and one of these five loops in the thermophilic IMPase is almost identical to a catalytic metal-binding loop in FBPase. The possibility that IMPase has FBPase activity also has been confirmed by experiment.

10.8.1.4   *E. coli* ycaC: bacterial hydrolase of unknown specificity

The structure of ycaC, and subsequent sequence studies, showed that this protein belongs to a superfamily of bacterial hydrolases with varying substrate specificities. Its fold bears a strong similarity to N-carbamoylsarcosine amidohydrolase (CSHase) despite a sequence identity of less than 20%. Conservation of amino acids in the active-site, including a Cys nucleophile and a *cis*-peptide bond involved in ligand-binding in CSHase, suggests that ycaC catalyzes a similar hydrolytic reaction. Biochemical experiments, however, have failed to identify a substrate.

10.8.1.5   *E. coli* yjgF

The structure of this hypothetical protein revealed a distant evolutionary relationship with chorismate mutase that was not apparent at the sequence level. A cavity in yjgF containing amino acids that are invariant in its close hypothetical sequence homologs was identified as the putative active-site. This functional site is in the same location as that of chorismate mutase and both proteins form homotrimers. However, their active-sites are completely unrelated in amino acid composition, which suggests they have different functions. Interestingly, whilst yjgF and protein tyrosine phosphatases share no global structural similarity, their active-sites have many features in common but they are unlikely to have the same function. Although the structure-to-function approach failed in this instance, the structure of yjgF has guided future site-directed mutagenesis experiments.

10.8.2   Implications for target discovery and drug design

The impact of structural genomics on target discovery and rational drug design is not yet clear. It is hoped that structures will provide new targets by revealing protein functions of relevance to therapy. Structures expose binding site characteristics, and so they reveal the suitability of a protein as a target for small molecule inhibitors (drugs). Knowledge of the distribution of proteins in species, gained through sequence and structural analyses, is important to identify species-specific targets and to thereby avoid toxicity problems. In addition, structures may be useful to understand and to predict cross-reactions of a drug with non-targeted proteins in the same organism, which can lead to side effects. This will help to prioritize candidate targets and to reduce the number of failures in the drug development process.

For structure-based drug design, as more structures become available it will become quite typical to incorporate knowledge of the target structure in drug development. Even today many of the newest drugs have been designed using combinatorial chemistry and structure-based design to bias libraries or to refine the lead molecule. With an increase in structural data, we can expect theoretical approaches to improve considerably thus making successful structure-based drug design *in silico* an achievable goal.

10.9   Conclusions

The three-dimensional structure of a protein is critical for a complete understanding of its biochemical function. Furthermore, structures allow one to identify distant evolutionary relationships that are hidden at the sequence level. These are two motivations behind the structural genomics projects. The increasing number of uncharacterized protein structures solved through these initiatives necessitates methods to infer function from structure, and the next few years will see the development of many more of these methods. However, the versatility of protein folds with respect to function presents a major challenge to their success. At present, it is difficult to obtain detailed functional information directly from structure, without complementary experimental work. Encouragingly, the knowledge-based approaches will be increasingly powerful as further protein data become available and we have a more complete catalog of protein sequences, structures and their functions. It is well to remember, however, that function is context-dependent, as exemplified by the moonlighting proteins, and typically structures provide clues only to biochemical function. This represents just the first stage in assigning cellular and phenotypic function.

References and further reading

Gerlt, J.A., and Babbitt, P.C. (1998) Mechanistically diverse enzyme superfamilies: the importance of chemistry in the evolution of catalysis. *Curr Op Chem Biol* **2**: 607–612.

Moult, J., and Melamud, E. (2000) From fold to function. *Curr Op Struct Biol* **10**: 384–389.

Murzin, A.G. (1998) How far divergent evolution goes in proteins. *Curr Op Struct Biol* **8**: 380–387.

Orengo, C.A., Todd, A.E., and Thornton, J.M. (1999) From protein structure to function. *Curr Op Struct Biol* **9**: 374–382.

Shapiro, L., and Harriss, T. (2000) Finding function through structural genomics. *Curr Op Biotech* **11**: 31–35.

Skolnick J., Fetrow J.S., and Kolinski A. (2000) Structural genomics and its importance for gene function analysis. *Nat Biotech* **18**: 283–287.

Teichmann, S.A., Murzin, A.G., and Chothia, C. (2001) Determination of protein function, evolution and interactions by structural genomics. *Curr Op Struct Biol* **11**: 354–363.

Thornton J.M., Orengo C.A., Todd A.E., and Pearl F.M.G. (1999) Protein folds, functions and evolution. *J Mol Biol* **293**: 333–342.

Thornton, J.M., Todd, A.E., Milburn, D., Borkakorti, N., and Orengo, C.A. (2000) From structure to function: approaches and limitations. *Nat Struct Biol Suppl* 991–994.

Todd A.E., Orengo C.A., and Thornton J.M. (1999) Evolution of protein function, from a structural perspective. *Curr Op Chem Biol* **3**: 548–556.

From structure-based genome annotation to understanding genes and proteins

11

Sarah A. Teichmann

- *Computational structural genomics*: large-scale structural annotation of the proteins encoded by genomes by identifying homologies between the genomic proteins and proteins of known structure.
- Methods and databases for computational structural genomics: complete proteome assignments to CATH domains in the *Gene3-D* database, SCOP domains in the *SUPERFAMILY* database.
- Insights into evolution of proteomes and *multidomain proteins* through structural annotation: a few large families[1] dominate proteomes, and there are many small families. At least two-thirds of the proteins in a proteome are multidomain proteins. In multidomain proteins, the large families have many types of domain neighbors, while the small families only occur next to one or two other types of domains.
- Insights into *enzyme* and *metabolic pathway* evolution through structural annotation: enzymes are assembled into metabolic pathways in a mosaic manner, with *domain* families tending to be distributed across different pathways.

11.1 Concepts

By identifying significant similarities between the sequences of proteins of known three-dimensional structure and the proteins encoded by genomes, the structures can be assigned to large fractions of *proteomes* at differing levels of accuracy. The procedure of *structural annotation* of genomes is fundamental to understanding and ordering the large-scale experimental sequence and structure data produced in the postgenomic era. Structural assignment of proteins of unknown structure has practical applications in terms of homology modeling and understanding functional properties of proteins. This chapter also describes how structural assignments to proteomes can help us understand the evolution of protein families in complete genomes, in multidomain proteins and in their functional context in cells such as in metabolic pathways.

11.2 Computational structural genomics: structural assignment of genome sequences

Knowing the three-dimensional structure of a protein provides information about the function of the protein and its protein–protein interactions. As explained in detail in

---

[1]In this chapter the term family is used in its broadest sense – that is to describe a set of homologous proteins related by a common evolutionary ancestor. This may also be described as a superfamily in other chapters.

Chapter 10, protein structure can provide functional clues in terms of active-sites identified in the structure, ligands that may be bound and functional motifs similar to those found in other structures. Chapters 12 and 13 discuss protein–protein interactions, and again the three-dimensional structure of a protein can be useful in giving insights into this aspect of protein function. The most obvious case is when a protein is crystallized as a complex with another protein, but in addition there are methods to predict protein interactions based on structure.

Much of the functional information inferred about a protein from its three-dimensional structure is based on homology to other proteins. Ideally, homology could be inferred by simply comparing protein sequences, but often sequences have diverged to the extent that their similarity is not detectable. Protein structure is much more conserved than sequence in evolution, so knowledge of the three-dimensional structure provides a powerful tool for determining even distant *homologies* between proteins. In this chapter, the later sections will describe insights we can get about the evolutionary history of proteins and genomes using structural information.

For any of the completely sequenced genomes, only a small fraction of the three-dimensional structures of gene products have been solved by X-ray crystallography or NMR spectroscopy. Therefore, in order to extend our knowledge of the protein structures of the proteins encoded by a genome, we have to perform structural predictions for the gene products. A reliable and commonly used method is to identify sequence similarities between proteins of known structure and the predicted protein sequences in genomes: thus three-dimensional structure is inferred by homology. This process of assigning structures to sequences by homology is termed 'computational structural genomics'. Since many sequence regions do not have a globular structure, annotation of sequences in terms of transmembrane regions, coiled coils, low complexity sequence and signal sequences can provide complementary information. An example of the extent of annotation of the predicted proteins of the small parasitic prokaryote *Mycoplasma genitalium* is shown as a pie chart in *Figure 11.1*.

In this chapter, first some common methods and databases for structural assignment to genome sequences will be introduced, and the connection to experimental structural genomics projects will be explained. Then insights into proteome and protein evolution gained using these methods will be described. Finally, a particular example of structural assignments to an entire molecular system will be discussed: the enzymes of small molecule metabolism in *Escherichia coli*. Structural assignments to almost the entire set of small molecule metabolism in this model organism have provided a better understanding of enzyme and pathway evolution.

11.3 Methods and data resources for computational structural genomics

11.3.1 Methods of assignment of structures to protein sequences

The simplest and most reliable, but not most powerful, way of identifying similarity between two protein sequences is by pairwise sequence comparison methods. BLAST is a pairwise sequence comparison method, as described in Chapter 3. The proteins in the GenBank database are linked to homologs of known three-dimensional structure if the relationship between the proteins can be found by BLAST.

Multiple sequence comparison methods capture the average characteristics of a family of protein sequences, and are considerably more powerful than pairwise sequence comparison methods. Therefore, more comprehensive structural assignments can be achieved using profile methods or hidden Markov model methods. As described in Chapter 5, both profiles and hidden Markov models are based on multiple alignments of a set of related protein sequences. Each column of the alignment represents one amino acid position and has a par-

Structural Annotation of the
Mycoplasma genitalium genome

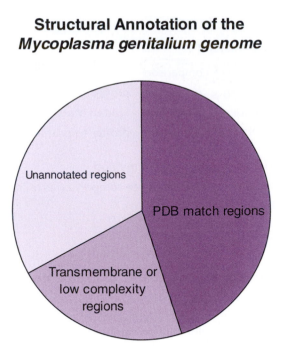

Unannotated regions

PDB match regions

Transmembrane or
low complexity
regions

Figure 11.1

Pie chart of structural assignments to the proteome of the bacterium *Mycoplasma genitalium*. Almost half of the amino acids (49%) in the *Mycoplasma genitalium* proteins have a structural annotation. In this case, the structural annotation was taken from the SUPERFAMILY database (version 1.59, September 2002), described in Section 11.3.2. Roughly one fifth of the proteome is predicted to be a transmembrane helix or low complexity region by the relevant computer programs. The remaining 30% of the proteome is unassigned.

ticular distribution of amino acids in the set of sequences at that position. The sequence of a query polypeptide chain is then compared to the profile or hidden Markov model and scored according to how well its amino acid sequence matches the distribution of amino acids at each position of the multiple alignment.

Other variants of multiple sequence comparison methods for structural assignment to genome sequences include profile-to-profile sequence comparison methods, where a profile including the sequence of known structure is compared to a profile including the query protein sequence. In addition, there are methods that include information about the three-dimensional structure, where a profile including the sequence of known structure and information about the secondary structure and solvent accessibility of each residue is compared to a profile including the query sequence and the predicted secondary structure of each amino acid in the query sequence.

Sometimes distant relationships can be detected by threading methods. Threading methods are based on the statistics of amino acid pairs contacting each other in the set of known structures, as described in Chapter 9, and threading can also be used for computational structural genomics. A summary of the different methods and representative programs is given in *Table 11.1*.

Table 11.1 Methods for structural assignment of proteomes

Name	Summary of Method	Reference
BLAST	Pairwise sequence comparison	Altschul *et al.* (1997)
PSI-BLAST	Multiple sequence comparison – profile	Altschul *et al.* (1997)
SAM-T99	Multiple sequence comparison – hidden Markov model	Karplus *et al.* (1998)
FFAS	Multiple sequence comparison – profile-to-profile	Rychlewski *et al.* (2000)
3-D-PSSM	Sequence and structure profile method	Kelley *et al.* (2000)
GenTHREADER	Threading and multiple sequence comparison profile	Jones (1999)
ProFIT	Threading	Sippl & Weitckus (1992)

11.3.2 Databases of structural assignments to proteomes

There are several databases that store the results of assignments of structures to the sets of proteins encoded by completely sequenced genomes by some of the methods mentioned above. All the three databases introduced here start from the domains of proteins of known structure instead of or in addition to the entire protein structure, as domains are the units of evolution and can occur in proteins in all different combinations, not just those observed in the proteins of known structure. (Please refer to Chapter 1 on Molecular Evolution for details.)

The databases differ in their domain definitions and sequence comparison methods and are listed in *Table 11.2*. Gough and colleagues developed the *SUPERFAMILY database*, which is associated with the SCOP database of Murzin and co-workers. SUPERFAMILY uses the domain definitions in the SCOP database. The assignments to the proteins of all the publicly available completely sequenced genomes in the SUPERFAMILY database are made using a hidden Markov model method. An example of the domains assigned to a *C. elegans* (worm) and *Drosophila* (fruitfly) protein in the SUPERFAMILY database is shown in *Figure 11.2*.

Structural assignments are one element of the database called *PartsList* compiled by Qian and co-workers, which contains many other types of information such as information on gene expression levels and protein–protein interactions. The structural assignments are also based on the domain definitions in the SCOP database and are made using PSI-BLAST. There are a growing number of databases such as PartsList, which could be termed 'comprehensive databases', in so far as they store large-scale datasets relating to completely sequenced genomes from many different sources, including aspects of sequences and structures, but also functional genomics data.

Table 11.2 Databases with structural assignments to proteomes

Name	Summary of Content	Reference
SUPERFAMILY	Assignments of SCOP domains with SAM-T99 hidden Markov models	Gough *et al.* (2001)
PartsList	Assignments of SCOP domains with PSI-BLAST	Qian *et al.* (2001)
Gene3-D	Assignments of CATH domains with PSI-BLAST and SAM-T99 Hidden Markov Models	Buchan *et al.* (2002)

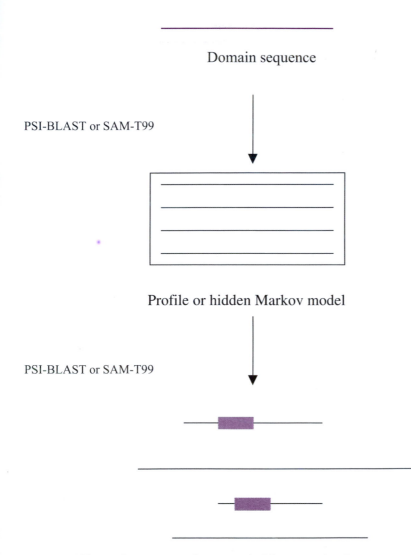

Domain sequence

PSI-BLAST or SAM-T99

Profile or hidden Markov model

PSI-BLAST or SAM-T99

All matches to proteins encoded by completely sequenced genomes in database

Figure 11.2

Flow chart of structural assignment procedure of the Gene 3-D, SUPERFAMILY and PartsList Databases. The annotation of complete proteomes by iterative multiple sequence comparison methods is carried out by creating a profile or hidden Markov model for each domain of known three-dimensional structure. Programs such as PSI-BLAST or SAM-T99 can create such models from a single query sequence and a large sequence database. The set of profiles or models is then scanned against all the protein sequences in the proteome, and regions of significant homology to the models are identified. Regions with a significant match are represented as violet rectangles.

As explained in Chapter 7, the SCOP database with its domain and family definitions is compiled entirely manually. In contrast, Orengo's CATH database is a semi-automatic structural classification scheme, and hence some of the domain and family classifications are different from SCOP. The *Gene3-D* database of Buchan and colleagues is linked to the CATH structural classification database, and takes the domain definitions from CATH to build its

Drosophila Lar phosphatase CT29268, 2037aa

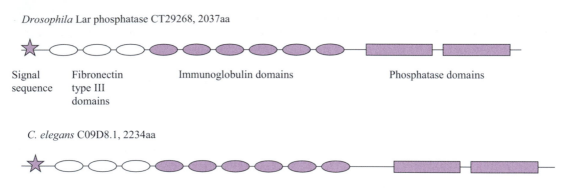

Signal   Fibronectin   Immunoglobulin domains   Phosphatase domains
sequence   type III
                domains

C. elegans C09D8.1, 2234aa

Figure 11.3

Domain assignments to the *C. elegans* (worm) and *Drosophila* (fruitfly) Lar phosphatase proteins. The domains in these two phosphatases have been assigned by the SUPERFAMILY database. The domain architecture of these homologous proteins is identical: three Immunoglobulin domains, nine Fibronectin type III domains and two Phosphatase domains.

profiles. The profiles are scanned against the proteome sequences. A generic scheme for the methods used to generate the structural assignments in the SUPERFAMILY, PartsList and Gene3-D databases is shown in *Figure 11.2*. For each protein in the SUPERFAMILY and Gene3-D databases, there is a graphical representation of the domain assignments in the spirit of *Figure 11.3*, as well as overall assignment statistics for each of the genomes processed in the database.

11.3.3 Computational and experimental structural genomics: Target selection

As mentioned in Chapter 7, numerous experimental *structural genomics initiatives* are underway. Their aim is to attain complete coverage of the structure of globular proteins and domains of proteomes. Because it is impossible to solve the structures of a whole proteome at once, proteins have to be ranked in terms of their importance or urgency for structural elucidation. The proteins that are picked for solution of their structure in the context of a structural genomics project are termed 'targets'.

Proteins that have a close homolog in the PDB are automatically low priority for structural elucidation, as their structure can be modeled to some degree of accuracy (depending on the level of sequence similarity between the two protein sequences). In fact, protein structures for complete proteomes have been modeled using an automatic comparative modeling method (see Chapter 8), Sali and Blundell's MODELLER, and are deposited in a database called *ModBase* constructed by Pieper and co-workers.

Even proteins that have distant homologs of known structure, as identified by profile methods in the Gene3-D or SUPERFAMILY databases, will not be high-priority targets for structural genomics projects that have as their aim novel protein folds. In addition to homolog detection via structural assignment, some structural genomics projects take into account families of sequence relatives, and only solve one structure from each of these families. Sequence families can be identified by sequence clustering methods or by homology to profiles of characterized and annotated sequence families, for which there are a variety of databases, as described in Chapter 5.

To keep track of the structural assignments to proteomes, sequence families and proteins whose structures are in the process of being solved, databases such as Brenner's *PRESAGE* database or the *SPINE* database set up by Bertone, Gerstein and co-workers. These databases provide a link between the computational structural genomics assignments and the experimental structural genomics projects, so that the computational structural assignments and sequence families can provide guidance in target selection.

11.4 Proteome and protein evolution by computational structural genomics

11.4.1 Protein families in complete genomes

The family membership of the assigned domains in a proteome can be inferred from the family membership of the homologous domains of known structure. Currently, about one-third to two-thirds of the predicted proteins of a completely sequenced genome can be assigned a homologous structure. Thus, for instance, the largest domain families in genomes can be compared across a range of genomes. The five largest domain families in four different genomes are shown in *Table 11.3*. The eubacterial genome is the well-studied bacterium *Escherichia coli*. Archaea are another type of prokaryote, which are similar to eukaryotes within the set of the proteins involved in information transfer, and the archaeal genome given is *Methanobacterium thermoautotrophicum*. The yeast *Saccharomyces cerevisiae* is a unicellular eukaryote, while the worm *Caenorhabditis elegans* is a multicellular eukaryote.

11.4.1.1 Common and specific domain families in the three kingdoms of life

Although bacteria diverged from archaea and eukaryotes roughly a billion years ago, one of the top two domain families is the P-loop nucleotide triphosphate hydrolase family in all four genomes. The domains belonging to this family hydrolyze nucleotides such as ATP or GTP and can thus either provide a phosphate group for alteration of small molecules or proteins, or the energy from ATP or GTP hydrolysis can be used to drive reactions, induce changes in protein conformation or induce protein motion. The functions carried out by the

Table 11.3 The largest protein families in proteomes from the three domains of life: The number of predicted proteins in each organism is given in brackets after the name of the organism. The structural assignments are from the SUPERFAMILY database, SCOP version 1.55. The number of domains in each protein family is given in brackets after the family name

Eubacterium	Archaeum	Unicellular Eukaryote	Multicellular Eukaryote
Escherichia coli (4289)	*Methanobacterium thermoautotrophicum* (1869)	*Saccharomyces cerevisiae* (yeast, 6343)	*Caenorhabditis elegans* (worm, 20100)
P-loop hydrolase (297)	P-loop hydrolase (145)	P-loop hydrolase (447)	Family A G protein-coupled receptor-like (1657)
Rossmann domains (124)	4Fe-4S Ferredoxins (56)	Protein kinase-like domains (131)	P-loop hydrolase (762)
Winged helix DNA-binding domain (114)	CBS domains (45)	Trp-Asp repeats (112)	Immunoglobulin (643)
Periplasmic binding protein II-like (95)	Rossmann domains (43)	Rossmann domains (104)	Protein kinase-like domains (529)
Homeodomain-like (86)	PYP-like sensor domains (27)	Armadillo repeats (104)	EGF/laminin domains (448)

domains of this family are thus useful in many different contexts in many proteins within a genome. The same is true of the NAD(P)-binding Rossmann fold domains, which bind NAD or NADP and provide oxidizing or reducing energy for many different reactions in cells.

Although many domain families are common to genomes from the three kingdoms of life, including some of the largest families, there are also families that are specific to particular kingdoms of life, assuming that they have not just gone undetected by the structural assignment methods. For instance, the eukaryotic protein kinases are among the largest domain families in the two eukaryote genomes (*Table 11.3*) and do not occur at all in prokaryotes. Several families that function in cell adhesion are only present in eukaryotes, such as the (non-enzymatic) immunoglobulin and EGF/laminin domains in *C. elegans* in *Table 11.3*. Thus new families were invented to carry out functions involved in signaling in eukaryotes, particularly multicellular eukaryotes.

11.4.1.2   The power law distribution of domain family sizes

Despite differences in the exact nature of the repertoire of domain families identified in each proteome, it is striking that the distribution of the number of families containing different numbers of domains follows a power law in all genomes. This means that there are a few families that are very large, and many families that are quite small in all the genomes, as shown in *Figure 11.4*.

Given this phenomenon, it is tempting to speculate that it is the result of the process of stochastic duplication of parts of genes in genomes. Such stochastic duplication can be simulated by starting with a single copy of a domain from each protein family, and then duplicating domains in the set at random. The domains that are duplicated early on in the simulation will result in large families, because once the initial domain is duplicated, the family has twice the probability of being duplicated and growing further. Such random duplication would result in a power law distribution like those in *Figure 11.4*. We must not forget, however, that the proteins in an organism are the result of selection of proteins that carry out effectively the functions that make the organism work. Therefore, it is more likely that the similarity in the domain family-size distributions across genomes is due to there being a few ubiquitously useful domain families and many small families that have specialized functions in all of the genomes.

11.4.2   Domain combinations in multidomain proteins

From the description of domains belonging to families in the previous section, it is clear that duplication is a major force in creating new proteins and in shaping proteomes. In addition to duplication, the genetic processes that lead to combinations of domains in proteins are also crucial, as explained in Chapter 1 on Molecular Evolution. From structural assignments to proteomes, we know that roughly two-thirds of the gene products in prokaryotes and three-quarters of the gene products in eukaryotes are multidomain proteins.

11.4.2.1   Selection for a small proportion of all possible combinations of domains

By analyzing the families of the domains that are adjacent to each other in proteins, it is also clear that only a small proportion of all possible domain combinations have been selected in functional proteins. Amongst the 1200 families that are observed in multidomain proteins with homologous structures, on the order of 2500 different types of pairwise combinations of domains are made, while the number of possible combinations is $1200^2 = 1,400,000$.

11.4.2.2   Large families have many types of neighboring domains

In multidomain proteins, the largest domain families also occur in the most different types of domain combinations. For example, the P-loop nucleotide triphosphate hydrolases and

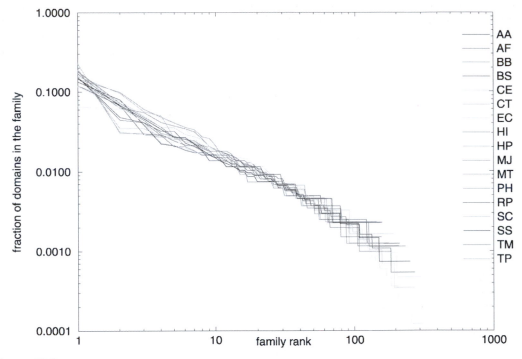

Figure 11.4

Superimposed distributions of family size. This graph shows the number of domains in each family on the y-axis as a fraction of all the domains in the proteome. The family rank on the x-axis is the number of the family when the families are sorted from largest to smallest. Family size decreases linearly with family rank in this log–log plot, meaning that there are a few large families in each proteome and many small families. The abbreviations for genomes are as follows: AA, *Aquifex aeolicus*; AF, *Archeoglobus fulgidus*; BB, *Borrelia burgdorferi*; BS, *Bacillus subtilis*; CE, *Caenorhabditis elegans*; CT, *Chlamydia trachomatis*; EC, *Escherichia coli*; HI, *Haemophilus influenzae*; HP, *Helicobacter pylori*; MJ, *Methanococcus jannaschii*; MT, *Methanobacterium thermoautotrophicum*; PH, *Pyrococcus horikoshii*; RP, *Rickettsia prowazekii*; SC, *Saccharomyces cerevisiae*; SS, *Synechocystis* sp.; TM, *Thermotoga maritima*; TP, *Treponema pallidum*.

Rossmann domains mentioned above and in *Table 11.3* have domain neighbors from many different families. All the smaller families have only one or two different types of neighbors. The correlation between the number of domains in a family and the number of different neighboring families is shown in *Figure 11.5*.

11.4.2.3 Conservation of N-to-C terminal orientation of domain pairs

The sequential order of *domain combinations* is strongly conserved in multidomain proteins. That is, the domains that belong to two particular families are almost always next to each other in one N-to-C terminal orientation. This could be either because all the domain pairs belonging to two particular families are related to each other by duplication, or because the function of a particular domain combination can only be achieved in one of the two orientations. Similarities in the linkers that connect the catalytic domains to Rossmann domains in PDB structure confirm homology of the complete unit of two domains. Therefore, conservation of domain order is due to duplication of whole domain combinations.

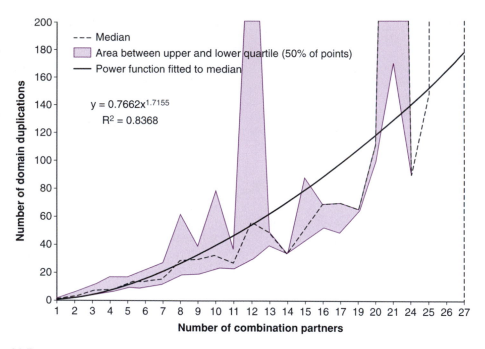

Figure 11.5

Number of domain partners in multidomain proteins as a function of family size. For each family in a set of 40 genomes, the family size in terms of number of domains is shown on the y-axis, and the number of families of the neighboring domains in multidomain proteins is given on the x-axis. This shows that a few large families are very versatile in terms of their domain neighbors, while most families are small and have few types of domain neighbors.

11.4.2.4 Many combinations are specific to one phylogenetic group

When one compares the repertoires of domain combinations in different phylogenetic groups (bacteria, archaea, eukaryotes), it becomes obvious that there are many domain combinations among common families that are specific to each of the phylogenetic groups. In fact, half of the domain combinations among the common families in eukaryotes and bacteria are specific to each kingdom, even though the families are present in all three kingdoms of course. Therefore, novel domain combinations have developed extensively since the divergence of the three kingdoms of life.

11.4.2.5 Multidomain proteins involved in cell adhesion

Studying particular families of multidomain proteins provides insights into the evolution of function through formation of proteins with new domain combinations. In multicellular eukaryotes, long multi-domain proteins with two to 50 domains belonging to specific families, such as immunoglobulin and cadherin domains, are involved in *cell adhesion* and in providing orientation to cells during their migration from one place to another in the developing embryo. In other words, these proteins are present on the outside surface of cells and provide cues and signals as to the location of the cell in the organism, or provide the glue that holds adjacent cells together.

Using structural assignments to the complete proteomes of the two multicellular eukaryotes *C. elegans* (worm) and *Drosophila* (fruitfly), a set of proteins containing immunoglobulin or cadherin domains can be retrieved and analyzed. One such study (Hill *et al.*, 2001) found that the repertoires of cadherin superfamily proteins in the fly (17 proteins) and worm (15 proteins) had only one protein in common with exactly identical domain architecture. Although worms and flies have cell types that are quite different, some basic cells and cell types are the same, and so it is surprising that there are so few proteins with identical domain architectures.

There are more immunoglobulin superfamily proteins than cadherin proteins in both the worm (about 100) and the fly (about 170), and the set of immunoglobulin proteins is about two-thirds larger in the fly than in the worm. Whether this is due to increased complexity in cell adhesion and the cell types of the fly is an unanswered question. Amongst the immunoglobulin superfamily proteins there are proteins that have identical domain architectures and are equivalent in function, such as the phosphatase shown in *Figure 11.3*. However, a large fraction of the repertoires in the worm and the fly are not shared with the other organism, and the hundred or so worm proteins are not simply a subset of the larger number of fly proteins. Nevertheless, there appear to be common motifs amongst these multi-domain proteins. For instance several *Cadherin* proteins in both *Drosophila* and *C. elegans* contain EGF and laminin domains too. Another example is the proteins in *Figure 11.6* that have several *immunoglobulin* domains followed by one fibronectin type III domain. Proteins with

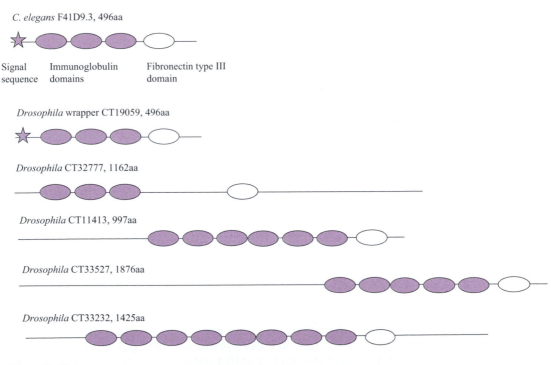

Figure 11.6

Immunoglobulin superfamily proteins with several Immunoglobulin domains and one Fibronectin type III domain in the fly and worm. The worm protein followed by five fly proteins at the bottom are all extracellular proteins that consist of several immunoglobulin domain proteins followed by one fibronectin type III domain. This motif of a number of immunoglobulin domains followed by one fibronectin type III domain may be characteristic of a particular feature of extracellular proteins. The worm protein has an identical domain architecture to the fly wrapper protein and the two proteins may therefore have equivalent functions in the two organisms.

these common motifs may have similar functions, for instance in their modes of binding to other cell adhesion proteins, even if they are not identical in the functions they perform. The class of multidomain proteins that comprise cell adhesion molecules, cell surface receptors and guidance molecules in development are an interesting test case for evolution of domain architecture coupled to emergence of new functions in multidomain proteins.

The example of cell adhesion proteins that belong to the immunoglobulin and cadherin families in fruitfly and worm shows that even amongst invertebrates, there are many domain architectures and proteins specific to each organism. Nevertheless, motifs of conserved sequences of a few domains can be identified across sets of proteins. However, when proteins carry out identical functions even in distantly related organisms, their domain architectures are extensively conserved: two-thirds of the enzymes of small molecule metabolism that catalyze identical reactions in *E. coli* and yeast have identical domain architectures.

11.5   Evolution of enzymes and metabolic pathways by structural annotation of genomes

We have seen above how structural assignments to protein sequences can provide us with insights into evolution of proteins and proteomes. Computational structural genomics also provides us with insights into evolution of the function of individual proteins such as enzymes, and the metabolic pathways they form. Here, the structural analysis of the enzymes of small molecule metabolism in the model organism *E. coli* will be presented. Historically, theories for the evolution of pathways have been formulated as early as 1945: Horowitz proposed a retrograde model of pathway evolution, in which enzymes evolve backwards from the protein that produces the final product. Subsequently, he suggested that this evolution occurred through gene duplications of the proteins within a pathway. Jensen (1976) showed that enzyme recruitment across pathways could occur by duplicated enzymes conserving their catalytic functions but evolving different substrate specificities. The analysis of *E. coli* metabolic pathways clarifies to what extent these theories hold.

11.5.1   Small molecule metabolism in *E. coli*: an enzyme mosaic

According to *The Concise Oxford Dictionary*, a mosaic is 'a picture...produced by an arrangement of small variously coloured pieces of glass or stone'. A mosaic, such as one showing a picture of Roman life in a Roman atrium, is analogous to small molecule metabolic pathways in several ways. In particular, the enzymes that form the metabolic pathways belong to a limited set of protein families, like the set of different colored pieces available to the artist to construct the mosaic. Furthermore, the picture of the mosaic as a whole is meaningful, even though there is no discernible repeating pattern in the way the pieces are arranged; instead, each piece has been selected to fill a space with the necessary color to make the mosaic picture.

We determined the 'colors' of the enzymes in the mosaic of *Escherichia coli* small molecule metabolic pathways by assigning the domains in each enzyme to a domain family. The protein families were derived mainly by structural assignment of SCOP domains to *E. coli* proteins using the assignment method of the SUPERFAMILY database mentioned above. Like the roughly hewn mosaic pieces of one color, the domains that belong to one family are not perfectly identical, but can be very divergent. The result of our domain assignments is a description of metabolic pathways and their enzymes like the diagram of glycogen catabolism shown in *Figure 11.7*. Clarifying the domain structure of enzymes gives us a picture of the structural anatomy of the individual enzymes in the metabolic pathways and allows us to investigate whether there are any patterns in duplicated enzyme domains within and across the metabolic pathways.

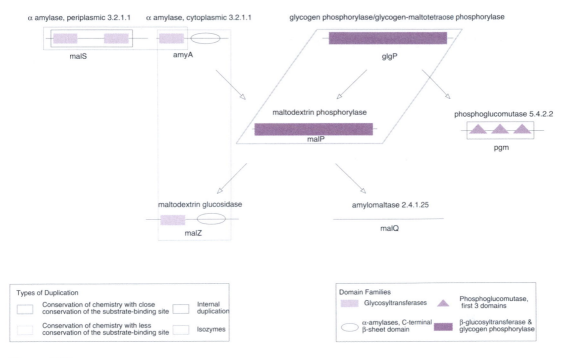

Figure 11.7

Glycogen catabolism pathway in *E. coli*. The enzymes in the glycogen catabolism pathway are represented by black lines and the structural domains in them by colored shapes in N-to-C-terminal order on the polypeptide chain. The arrows represent the flux of substrates and products through the pathway. There are two duplications with conservation of catalytic mechanism in this pathway. One is in consecutive enzymes (glgP/malP), so with a close conservation of substrate-binding site as well, whereas the other duplication occurs for enzymes one step apart (amyA/malZ), with less conservation of the substrate-binding site. There are also internal duplications, in which the same type of domain occurs several times in one polypeptide sequence (malS and pgm) and isozymes (malS and amyA).

The metabolic pathways in *Escherichia coli* are probably the most thoroughly studied of any organism. Although the details of the enzymes and metabolic pathways will differ from organism to organism, we expect the principles of the structure and evolution of metabolic pathways to hold across all organisms. Comprehensive information on small molecule metabolism in *E. coli* is contained in the EcoCyc database. We used the 106 pathways and their 581 enzymes described in this database. The results of the domain assignment procedure gave 722 domains in 213 families in 510 (88%) of the *E. coli* small molecule metabolism (SMM) enzymes. There are, on average, 3.4 domains per family, showing that even this fundamental set of pathways is the product of extensive duplication of domains and thus parts of genes. The distribution of family sizes of the 213 families is roughly exponential: there are 74 families with only one domain in *E. coli* SMM, and the largest family, the Rossmann domains, has 53 domains.

Half of the SMM enzymes are multidomain enzymes. This means that even proteins as fundamental as the SMM enzymes are not all simple single-domain enzymes but the product of extensive domain combinations. Therefore, either SMM enzymes developed by fusions and recombinations from a more basic set of proteins, which were single-domain proteins; or combinations of two or more domains occurred first, and then domains later split and recombined to crystallize as individual evolutionary units, the domains that we recognize today.

11.5.2 Types of conservation of domain duplications

After discussing the domain structures of the individual enzymes, we can now use this information to investigate aspects of the evolution of metabolic pathways. Of the 213 families of domains, 144 have members distributed across different pathways. The 69 families that are active in only one pathway are all small: 67 have one or two members, one has three members and one has four members. This distribution shows that the evolution of metabolic pathways involved widespread recruitment of enzymes to different pathways, supporting the Jensen model of evolution.

When discussing pathway evolution, it is useful to distinguish between different types of duplications of enzymes and their domains. Considering the glycogen catabolism pathway in *Figure 11.7*, we see that there are multiple copies of the four types of domains in the pathway. The glycosyltransferase domain family (yellow) and the phosphoglucomutase domains (green) recur within the individual proteins malS and pgm. This type of duplication is termed 'internal duplication' and can, by definition, only take place within pathways. Glycosyltransferase domains also occur in both the periplasmic and cytoplasmic versions of α-amylase. Duplication in enzymes that are isozymes can, by definition, also only occur within pathways, as *isozymes* are enzymes that have slightly different properties while catalyzing the same reaction in the same pathway. A glycosyltransferase domain also occurs in the maltodextrin glucosidase malZ. The duplication between α-amylase and maltodextrin glucosidase conserves catalytic mechanism, because both enzymes hydrolyze glucosidic linkages. Similarly, the two phosphorylase domains (purple rectangles) conserve reaction chemistry, because both glgP and malP are phosphorylases acting on different substrates. Recent studies have described this mechanism in detail and show how mutations in active-site residues produce new catalytic properties for enzymes.

There are two further types of duplication that do not occur in this pathway: duplication of cofactor- or minor substrate-binding domains such as Rossmann domains and duplication with conservation of the substrate-binding site but change in catalytic mechanism.

11.5.3 Duplications within and across pathways

As mentioned above, *recruitment* of domain families across pathways is common, and these duplications involve either conservation of reaction chemistry or conservation of a cofactor- or minor substrate-binding site. This supports the Jensen hypothesis of recruitment of families for their chemistry of catalysis. Of the different types of duplication listed above, internal duplication and duplication occurring in isozymes are frequent within pathways. Duplication with conservation of a cofactor- or minor substrate-binding site is also frequent within pathways.

However, within the entire set of almost 600 enzymes, there are only six examples of duplications in pathways with conservation of the major substrate-binding site and a change in the catalytic mechanism. This means that duplications in pathways are driven by similarity in catalytic mechanism much more than by similarity in the substrate-binding pocket, and speaks against Horowitz' model of retrograde evolution, in which it is suggested that enzymes within a pathway are related to each other.

Duplications within pathways do occur relatively frequently in situations such as that shown in *Figure 11.8*, in which fucA and rhaD are homologous. In this type of case, two enzymes are followed by the same enzyme(s) in a pathway and hence have the same or similar products. Alternatively, two enzymes can also be 'parallel' in this way when both have the same precursor enzyme in a pathway and thus have the same or similar substrates. About one-seventh of these 'parallel' enzymes in pathways have homologous domains.

As mentioned above, all the larger domain families in the metabolic pathways have members in more than one pathway; thus duplications across pathways are extremely common. However, it appears that little of this recruitment took place in an ordered fashion: cases of *serial recruitment*, where two enzymes in one pathway were both recruited to another pathway in the same order, such as fucK and fucA, rhaB and rhaD, and araB and araD in *Figure 11.7* are very rare. If duplication of segments of the bacterial chromosome took place, and all

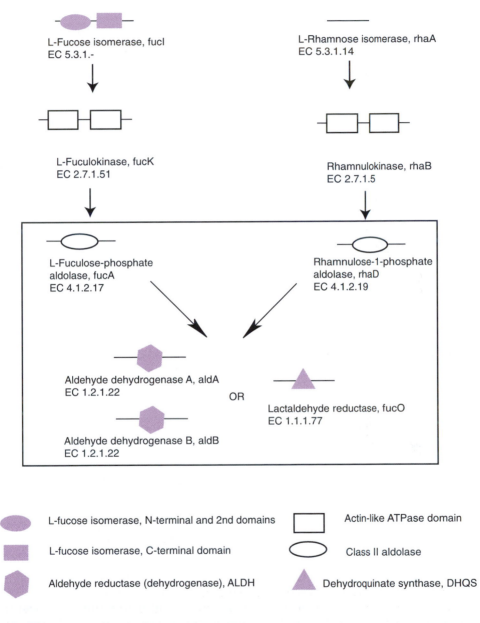

Figure 11.8(a)

Fucose and rhamnose catabolism. In fucose and rhamnose catabolism, there is an example of serial recruitment and an example of 'parallel' enzymes, which are boxed. Serial recruitment has occurred because fucK (L-fuculokinase) is homologous to rhaB (rhamnulokinase), and fucA (L-fuculose-phosphate aldolase) is homologous to rhaD (rhamnulose-1-phosphate aldolase). fucA and rhaD have the same product, and are both followed by aldA/aldB or fucO, and are thus 'parallel' enzymes. The enzyme classification (EC) numbers for each enzyme are given.

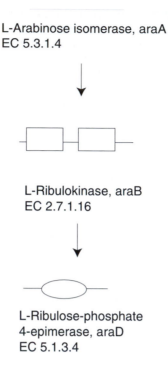

Figure 11.8(b)

Arabinose catabolism in *E. coli*. AraB is homologous to fucK and rhaB in (a), and araD is homologous to fucA and rhaD in (a). The three pairs of enzymes are an example of serial recruitment, as supported by their similar positions on the *E. coli* chromosome: the genes in each pair are divided by one gene on the chromosome. The key is the same as for part (a).

the genes in a duplicated segment were used to form a new pathway, serial recruitment would be expected. In fact, only 89 out of 26,341 (0.3%) possible pairs of enzymes are homologous in both the first and second enzymes. Only seven of these 89 pairs of doublets of enzymes have both doublets close to each other on the chromosome, providing evidence that the two initial enzymes were duplicated as one segment. The three kinase and aldolase/epimerase pairs of enzymes involved in sugar catabolism (*Figure 11.8(a)* and *(b)*) are a good example of this rare situation: all three pairs are one gene apart on the *E. coli* chromosome.

11.5.4 Conclusions from structural assignments to *E. coli* enzymes

The above description of how a relatively small repertoire of 213 domain families constitutes 90% of the enzymes in the *E. coli* small molecule metabolic pathways is paradoxical to some extent. Although the SMM enzymes have arisen by extensive duplication, with an average of 3.4 members per SMM family, the distribution of families within and across pathways is complex in that there is little repetition of domains in consecutive steps of pathways and little serial homology across pathways. Together with the analysis of the chromosomal locations of genes, it is evident that metabolic pathways have, in general, not arisen by duplication of chunks of the *E. coli* chromosome, either to extend a pathway or make a new pathway. There are obvious exceptions to this, such as the enzymes discussed in the fucose, rhamnose and arabinose catabolic pathways. With a few important exceptions, some of which are well-studied, duplication of enzymes for the purpose of conserving a substrate-binding site has not occurred much either, or the fraction of consecutive homologous enzymes would be larger.

The main selective pressure for enzymes in pathways appears to be their catalytic mechanism. This pattern of evolution has resulted in a mosaic of enzyme domains optimized for smooth-functioning small molecule metabolism in *E. coli*, with little order in the pattern of domains with respect to position within or between pathways. The comprehensive structural assignments to 90% of the enzymes in all *E. coli* small molecule metabolic pathways confirm that pathways are constructed by recruitment for the sake of catalytic mechanism, with few instances of duplication of enzymes within a pathway or serial recruitment across pathways.

11.6 Summary and outlook

Computational structural genomics is fundamental to understanding and ordering the large-scale experimental data produced in the postgenomic era. Structural assignment of proteins of unknown structure has practical applications in terms of homology modeling and understanding functional properties of proteins. This chapter has shown how structural assignments to proteomes can help us understand the evolution of protein families in complete genomes, in multidomain proteins and in their functional context in cells such as in metabolic pathways.

Structural assignments to genome sequences provide a guide for experimental structural genomics projects which are underway. Once these initiatives have provided extensive coverage of the domains in proteomes, a complete overview of the structural domains in each proteome will be possible by the combination of actual structures and structural assignments.

References

Altschul, S.F., Madden, T.L., Schaffer, A.A., Zhang, J., Zhang, Z., Miller, W., and Lipman, D.J. (1997) Gapped BLAST and PSI-BLAST: a new generation of protein database search programs. *Nucleic Acids Res* **25**: 3389–3402.

Bertone, P., Kluger, Y., Lan, N., Zheng, D., Christendat, D., Yee, A., *et al.* (2001) SPINE: an integrated tracking database and data mining approach for identifying feasible targets in high-throughput structural proteomics. *Nucleic Acids Res* **29**: 2884–2898.

Brenner, S.E. (2001) A tour of structural genomics. *Nat Rev Genet* **2**: 801–809.

Buchan, D., Pearl, F., Lee, D., Shepherd, A., Rison, S., Thornton, J.M., and Orengo, C. (2002) Gene3-D: structural assignments for whole genes and genomes using the CATH domain structure database. *Genomes Res* **12**: 503–514.

Gough, J., Karplus, K., Hughey, R., and Chothia, C. (2001) Assignment of homology to genome Sequences using a library of Hidden Markov Models that represent all proteins of known structure. *J Mol Biol* **313**: 903–991.

Jones, D.T. (1999) GenTHREADER: an efficient and reliable protein fold recognition method for genomic sequences. *J Mol Biol* **287**: 797–815.

Karplus, K., Barrett, C., and Hughey, R. (1998) Hidden Markov models for detecting remote protein homologies. *Bioinformatics* **14**: 846–856.

Kelley, L.A., MacCallum, R.M., and Sternberg, M.J. (2000) Enhanced genome annotation using structural profiles in the program 3-D-PSSM. *J Mol Biol* **299**: 499–520.

Qian, J., Stenger, B., Wilson, C.A., Lin, J., Jansen, R., Teichmann, S.A., *et al.* (2001) PartsList: a web-based system for dynamically ranking protein folds based on disparate attributes, including whole-genome expression and interaction information. *Nucleic Acids Res* **29**: 1750–1764.

Rychlewski, L., Jaroszewski, L., Li, W., and Godzik, A. (2000) Comparison of sequence profiles. Strategies for structural predictions using sequence information. *Protein Sci* **9**: 232–241.

Sippl, M.J., and Weitckus, S. (1992) Detection of native-like models for amino acid sequence of unknown three-dimensional structure in a database of known protein conformations. *Proteins* **13**: 258–271.

Teichmann, S.A., Chothia, C., and Gerstein, M. (1999) Advances in structural genomics. *Curr Op Struc Biol* **9**: 390–399.

Teichmann, S.A., Rison, S.C.G., Thornton, J.M., Riley, M., Gough, J., and Chothia, C. (2001) Small molecule metabolism: an enzyme mosaic. *Trends Biotech* **19**: 482–486.

Global approaches for studying protein–protein interactions

12

Sarah A. Teichmann

- There are three types of interactions between domains: the interactions between the domains in multidomain proteins, between the domains of proteins in stable complexes and of proteins in transient interactions.
- The proteins in stable complexes tend to be more closely co-expressed and more closely conserved compared to proteins in transient interactions and monomeric proteins.
- Two types of large-scale experimental approaches are introduced here: *yeast-two-hybrid screening* and routine purification of complexes followed by *mass spectrometry*. Both approaches have been applied to the *Saccharomyces cerevisiae* proteome on a genomewide scale in efforts towards determining the complete network of protein interactions in this organism.
- Theoretical analyses of structural domains have mapped the pattern of *domain family*[1] *interactions* and established that homologous proteins interacting have homologous interfaces and conserved *geometry of domain interactions*.
- Genomewide prediction methods for physical and functional protein interactions have been developed based on *gene fusion* and *conservation of gene order*. These most frequently predict proteins that physically interact, and less frequently enzymes part of the same metabolic pathway.

12.1   Concepts

Interactions between domains occur in *multidomain proteins*, in *stable complexes* and in *transient interactions* between proteins that also exist independently. Therefore, the physical contacts between domains are crucial for the functioning of the cellular machinery. Experimental approaches for the large-scale determination of protein interactions are emerging. Theoretical analyses based on protein structures have unraveled some of the overall principles and features of the way domains evolved to interact with each other. Computational methods for genomewide *prediction of protein–protein interactions* and functional associations between proteins are introduced here. These methods are based on *conservation of gene order*, as well as absence or presence of genes in genomes.

12.2   Protein–protein interactions

Proteins interact in many pathways and molecular processes, and are fundamental to the workings of cells and organisms. The physical interactions between protein domains occur in three different contexts: in multidomain proteins, in stable complexes and in transient interactions. These three different types of interactions between domains are shown schemati-

---

[1]In this chapter the term family is used in its broadest sense – that is to describe a set of homologous proteins related by a common evolutionary ancestor. This may also be described as a superfamily in other chapters.

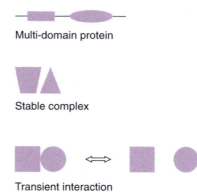

Multi-domain protein

Stable complex

Transient interaction

Figure 12.1

Three types of interactions between domains. Interactions between domains determine the structure of multidomain proteins, in which there are several domains on one polypeptide chain. Given that all proteins consist of domains, interactions between domains also occur between the proteins that are permanently associated in stable complexes and proteins that interact transiently, but also exist independently of each other.

cally in *Figure 12.1*. Domain interactions in multidomain proteins are the contacts that occur between the different domains on one polypeptide chain. Stable complexes consist of proteins that are permanently associated with each other, like many oligomeric proteins for instance. Well-known stable complexes include the histone octamer, the ribosome and DNA and RNA polymerases. Transient interactions on the other hand are all those protein–protein interactions that occur between proteins that also exist independently.

Sets of proteins that are part of stable complexes and sets of proteins involved in transient interactions differ in terms of the similarity in gene expression among the set of proteins. Proteins permanently associated in a stable complex need to be present or absent in the cell at the same time. Analysis of microarray data by Gerstein and co-workers by methods along the lines of those described in Chapters 14 and 15, has shown that the members of stable complexes in the yeast *Saccharomyces cerevisiae* have highly correlated gene expression patterns. In contrast, correlation of gene expression for pairs of transiently interacting proteins is only marginally significant compared to randomly chosen pairs of proteins. In prokaryotes, genes are co-regulated if they are a member of the same operon, and many proteins that are members of the same stable complex are part of the same operon. For instance, Ouzounis and Karp determined that over 90% of the enzymes that are in stable complexes in *E. coli* metabolic pathways are adjacent on the *E. coli* chromosome.

Membership in a stable complex also differs from transient interaction in terms of evolutionary constraints upon sequence divergence. Thus the proteins in stable complexes are more similar across species, having higher sequence identity between orthologs, than the proteins in transient interactions. A calculation by Teichmann showed that there are significant differences between the average values for sequence identities between *S. cerevisiae* and *S. pombe* orthologs in stable complexes, transient interactions and monomers. For proteins in stable complexes the average sequence identity is 46%, while for proteins in transient interactions it is 41%. (Proteins not known to be involved in any type of interaction have an average sequence identity of 38%.) One of the main reasons for this is the surface area involved in interfaces of stable complexes which is larger than in transient complexes. Sequence divergence may be slower in order to conserve these extensive interfaces.

12.3 Experimental approaches for large-scale determination of protein–protein interactions

In the postgenomic era and the spirit of large-scale analyses of genes and proteins, one of the most interesting and important challenges is to understand the network of all protein interactions in cells. *Proteomics* efforts are trying to do this experimentally. Examples of this are large-scale yeast-two-hybrid screens as well as complex purification with mass spectrometry.

12.3.1 Yeast-two-hybrid screens

The yeast-two-hybrid system uses the transcription of a reporter gene driven by the Gal4 transcription factor to monitor whether or not two proteins are interacting. As shown in *Figure 12.2a*, if the interaction between two proteins, A and B, is being tested, one of their genes would be fused to the DNA-binding domain of the *Gal4 transcription factor* (Gal4-DBD) while the other would be fused to the activation domain (Gal4-AD). The DNA-binding domain chimeric protein will bind upstream of the reporter gene. If the activation domain chimeric protein interacts with the DNA-binding domain chimeric protein, the reporter gene will be transcribed.

This experiment can be carried out hundreds or even thousands of times on array plates, as in the case of the study by Uetz and colleagues on *S. cerevisiae* (yeast) interactions.

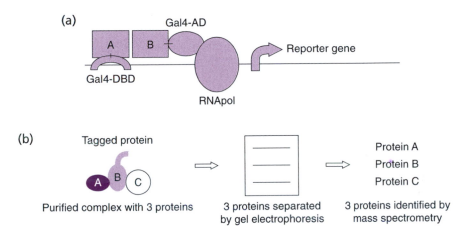

Figure 12.2

Large-scale experimental approaches for determining protein–protein interactions. The principle of yeast-two-hybrid assays and purification of complexes followed by mass spectrometry is shown here (a). In a yeast-two-hybrid assay, two chimeric proteins are made. One of these consists of the 'bait' protein, protein A here, fused to the DNA-binding domain of the yeast Gal4 transcription factor (Gal4-DBD), as indicated by the black line connecting the two proteins. The Gal4-DBD will bind in the promoter region of a reporter gene. The second chimeric protein consists of a 'prey' protein, protein B here, fused to the activation domain of the Gal4 transcription factor (Gal4-AD); the fusion is again indicated by a black line connecting the two proteins. If proteins A and B interact, the Gal4-AD will be recruited to the promoter of the reporter gene as well, and will activate the RNA polymerase and thus stimulate transcription of the reporter gene. Thus the interaction between proteins A and B can be monitored by expression of the reporter (b). In the large-scale purification of protein complexes, fusion proteins are created as well. These consist of the protein of interest fused to a tag, that is a peptide that allows affinity purification, represented by the hook at the top of the tagged protein in this diagram. Thus when the protein is purified through its tag, associated proteins will also be retrieved. The identity of these proteins can then be determined by mass spectrometry.

Each array element on these plates contains yeast cells transformed with a particular combination of two plasmids, one carrying the DNA-binding domain chimeric protein and the other the activation domain chimeric protein. Disadvantages of the method are that only pairwise interactions are tested, and not interactions that can only take place when multiple proteins come together, as well as a high false-positive rate. The false-positive rate is due to several factors, including bridging proteins that are mediating the interaction between two proteins that are not actually in physical contact, the detection of interactions between proteins that are not in the same cellular compartment under physiological conditions and interactions that arise by chance and are not reproducible.

12.3.2 Purification of protein complexes followed by mass spectrometry

Isolating protein complexes from cells allows identification of interactions between ensembles of proteins instead of just pairs. Systematic purification of complexes on a large scale is done by tagging hundreds of genes with an epitope. Like in the yeast-two-hybrid assay, this is done by making chimeric genes that are introduced into cells. As shown in *Figure 12.2b*, affinity purification based on the epitope will then extract all the proteins attached to the bait protein from cell lysates. The proteins in the purified complex are separated by gel electrophoresis and each one is characterized using mass spectroscopy. The principle of mass spectrometric identification of proteins is that the protein is chopped into fragments by tryptic digestion, and the mass of each fragment is measured by matrix-assisted laser desorption/ionization-time-of-flight mass spectrometry (MALDI-TOF MS). This measurement is so accurate that the combination of amino acids in each fragment can be calculated and compared to a database of all the proteins in the proteome of the organism in order to find the correct one. Gavin and colleagues have carried out such an analysis for hundreds of complexes in the yeast *S. cerevisiae* in 2002.

Purification and identification of the components of complexes in this way allows detection of interaction of proteins in stable complexes as well as transient interactions, as do yeast-two-hybrid screens. Because purification of complexes does not itself provide information on which components of the complex are in physical contact, pairwise interaction information, for example from yeast-two-hybrid experiments, can potentially complement data from complex purification.

12.4 Structural analyses of domain interactions

Many entries in the Protein DataBank (PDB) are three-dimensional structures of multiple domains. These structures provide experimental information about interactions between domains at atomic detail. Unfortunately, there are comparatively few three-dimensional structures compared to the amount of data available from the lower resolution large-scale experiments described above. Nevertheless, analysis of structures consisting of multiple domains has uncovered some of the principles of domain interactions in three dimensions. This information can therefore be complementary to the experimental data on protein interactions described above and to the predicted interactions described in the later sections of this chapter.

12.4.1 Interaction map of domain families

To study the large-scale patterns and evolution of interactions between protein domains, we can summarize the interactions in terms of the domain families. Thus the interactions of one family represent the sum of all the interactions of domains in that family. Precise information about contacts between individual domains can be extracted by analysis of PDB entries.

The result of the known interactions between members of structural protein families is a graph of connections between families like that shown in *Figure 12.3*, where the nodes are

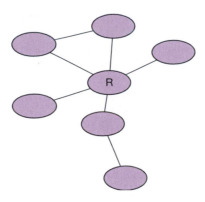

Figure 12.3

Schematic protein domain family interaction map. Each node in this graph represents a protein domain family. The physical contacts of the domains in different families are represented by the lines between the nodes. There are a few families that are hubs in the network: these are large families that are functionally versatile, such as Rossmann domains indicated by an 'R' here. Most families engage in only one or two types of interactions, such as the unlabelled families shown here.

protein families and the edges represent an interaction between at least one of the domains from each of the two families. Most domain families only interact with one or two other families, while a few families are extremely versatile in their interactions and are connected to many families. This pattern is observed at the level of individual proteins as well, as similar networks can be constructed for the individual proteins in the yeast proteome, for instance. Almost half of all known families engage in interactions with domains from their own family when one includes oligomeric proteins. Such symmetrical interactions appear to be particularly favorable.

12.4.2   The geometry of domain combinations

The network of domain family interactions is a purely two-dimensional map: it lays out the connections between families but does not provide information on the three-dimensional geometry of interactions. In order to understand the geometry of domain combinations, different structures of homologous pairs of domains must be studied. This is important, because though the methods for structure prediction of individual domains are well established (Chapter 9), much less is known about assemblies of domains.

There has been one investigation of this question in multidomain proteins and one in transient interactions. For 20 pairs of domain families engaged in transient interactions, the different entries for each pair of families was analyzed by Aloy and Russell. They found that on average, 70% of the residues in contact at the interface were conserved across the different structures of a particular pair of families.

The investigation of domain combinations in multidomain proteins by Bashton and Chothia focuses on two-domain proteins belonging to the Rossmann domain family. These proteins generally consist of one Rossmann domain and one catalytic domain. As for the analysis of transient interactions, all the proteins belonging to one family of catalytic domains form the same type of interface to the Rossmann domains, as shown in *Plate 4*. Furthermore, the linkers between the catalytic domain and the Rossmann domain were conserved in each family. This means that interface conservation within one catalytic family is a result of the direct evolutionary relationship between the proteins that have a particular pair of domains. In other words, each set of Rossmann domain proteins with a particular catalytic domain has descended from one common ancestral recombination event.

Across the different types of catalytic families, the position of the two domains with respect to one another varied, but only within a range of about 90°. This is the result of a functional constraint in these enzymes: the catalytic domain can only take up a variety of positions, as the substrate needs to be held sufficiently close to the NAD(P) cofactor of the Rossmann domain. In other multidomain proteins where there is no such strict functional constraint, the domain interfaces of one domain family to other families may well be more variable.

12.5 The use of gene order to predict protein–protein interactions

The structural analyses discussed above are useful for understanding protein–protein interactions and potentially for modeling them in three dimensions, but not for predicting interactions between proteins. Computational methods for predicting protein–protein interactions are desirable as experimental determination is time-consuming and expensive. Several prediction methods are based on the observation that if two proteins are part of the same complex, it is favorable for the two interaction partners to be co-expressed and co-regulated. This is particularly true of stable complexes as opposed to transient interactions, as described above.

Thus interactions between proteins can be predicted computationally by looking for sets of genes that occur as a single gene in at least one genome, or for prokaryotic genes that have conserved adjacent gene order across several genomes. Methods for prediction of protein interactions based on gene structure or gene order are described in this section.

12.5.1 Conservation of gene order

In prokaryotes, genes are organized into operons of co-regulated and co-expressed groups of genes. Genes that are consistently part of the same operon across different, distantly related genomes are likely to be part of the same protein complex or functional process across all species, because they have been selected to remain as a co-regulated unit throughout the extensive shuffling of gene order that takes place in prokaryote genomes. Thus conservation of gene order across different, distantly related genomes has been used as a method for predicting protein interactions. This principle is illustrated schematically in *Figure 12.4*.

When comparing pairs of genes or sets of genes in different genomes for this purpose, it is important to ensure that the genes are truly equivalent, in other words that they are orthologs, as opposed to merely similar genes. This is frequently done by only accepting a

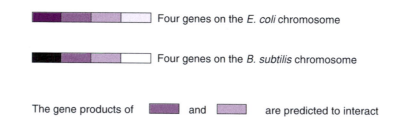

Four genes on the *E. coli* chromosome

Four genes on the *B. subtilis* chromosome

The gene products of [ ] and [ ] are predicted to interact

Figure 12.4

Conservation of gene order for predicting protein interactions. Genes of the same color are orthologs, in other words equivalent genes in the two genomes. If a pair of genes is consistently adjacent across a range of distantly related prokaryote genomes, one can infer that they need to be co-regulated in order to function properly. Thus the two genes are likely to interact physically or functionally. It is estimated that in about two-thirds to four-fifths of all such cases the interaction is a physical interaction.

pair of proteins as orthologs if they are 'bi-directional best hits'. This means that both proteins are the best match to each other when searching against the other proteome. The other extreme would be to consider proteins as equivalent if they share just one of many domains, for instance.

Members of a stable complex are often co-regulated and thus will be detected by this method, but proteins which are part of the same metabolic pathway or part of the same biological process can also be co-regulated. Therefore, although some of the genes detected as interacting by this method physically interact, others are just functionally associated with each other. Thus in a quantitative assessment of this method using the genome of the parasitic organism *Mycoplasma genitalium* as a benchmark, Huynen and colleagues found that two-thirds to four-fifths of the general interactions detected correspond to physical interactions and another 13% correspond to a metabolic or non-metabolic pathway. Genes that are physically associated with each other while being regulated and expressed individually will not be detected by this method.

Conservation of gene order due to operon structure is not applicable to eukaryote genomes, so interactions of proteins specific to eukaryotes cannot be detected by this method. Co-regulation of genes in eukaryotes can be inferred by similarity in the expression patterns of genes, as described in Chapter 14. A comparison of co-expressed yeast and worm genes showed that 90% of those pairs of genes with conserved co-expression are members of stable complexes (Teichmann and Babu, 2002). Thus, the principle of conserved co-regulation across distantly related organisms applies to stable complexes in both prokaryotes and eukaryotes, and can be used as a prediction tool in both.

12.5.2 Gene fusions

Another approach for predicting protein interactions is to look for cases across a set of genomes where two or more orthologs are part of the same gene in one genome, presumably as a result of gene fusion. The prediction is then that the orthologs that are on separate genes in the other genomes interact with each other, as shown in *Figure 12.5*. In the case of gene fusion, the fused proteins are not only co-regulated, as in conservation of gene order described above, but also permanently co-localized in the cell. The additional requirement of co-localization beyond just co-regulation poses a further limitation on the prediction method.

Domains that are part of a multidomain protein are automatically co-regulated and co-localized. Therefore, members of stable complexes as well as consecutive enzymes may be involved in gene fusions. However elements of signal transduction chains are seldom part of the same gene, for instance, as it is an essential part of their function that they can be regulated and localized independently. Due to the requirement for co-regulation as well as co-localization, the method is mostly limited to certain classes of protein–protein interactions:

Single multi-domain polypeptide chain in yeast

Separate polypeptide chains in *E. coli*

Figure 12.5

Gene fusions for predicting protein interactions. The arrows connect orthologous genes. Two proteins encoded by separate genes in *E. coli* are shown here. In yeast, the orthologs to these genes are encoded by a single gene. The two separate gene products in *E. coli* are thus predicted to interact. In about two-thirds of such cases, the unfused versions of the proteins interact physically, and in a further 15% of such cases the proteins are part of the same metabolic pathway.

members of the same stable complex and proteins in the same metabolic pathway. In their assessment of computational methods for prediction of all types of protein interactions, Huynen and colleagues found that two-thirds of the interactions detected in this way were between proteins that physically interact and another 15% between proteins part of the same metabolic pathway. The remaining interactions involved hypothetical proteins of unknown function.

12.6   The use of phylogeny to predict protein–protein interactions

There are a few other methods of predicting protein–protein interactions computationally that are independent of the assumptions discussed above. A method that appears to reliably predict a loose functional correlation between proteins is the phylogenetic profile method developed by Pellegrini and colleagues. The phylogenetic profile method relies on detection of orthologs (or homologs, in a variation of the method) in a set of genomes. If the pattern of ortholog presence or absence is the same in a group of proteins, then these proteins are clustered together as belonging to the same functional class, as shown in *Figure 12.6*. The phylogenetic patterns of clusters of orthologous groups of proteins deposited in the COG database of Koonin and co-workers could in principle be used for prediction in the same way. In the assessment of methods to predict protein–protein interactions mentioned above, one-third of such pairs were found to physically interact, and an additional third to belong to the same metabolic pathway or functional process.

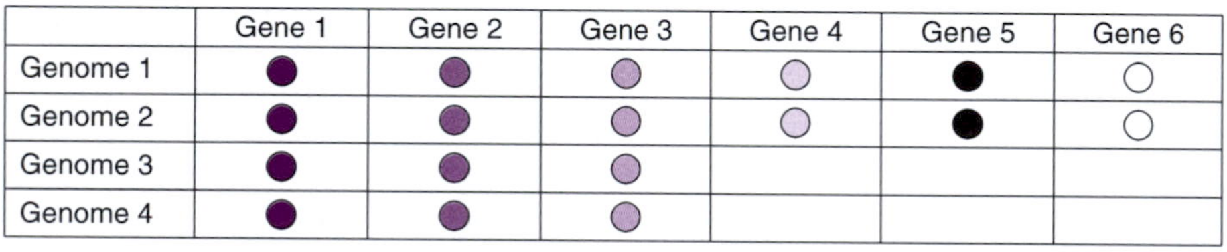

	Gene 1	Gene 2	Gene 3	Gene 4	Gene 5	Gene 6
Genome 1	●	●	○	○	●	○
Genome 2	●	●	○	○	●	○
Genome 3	●	●	○			
Genome 4	●	●	○			

Figure 12.6

Phylogenetic profiles. In the phylogenetic profile method for predicting protein interaction, presence or absence of orthologous genes is scored across a variety of genomes. This is represented by presence or absence of a dot in the row corresponding to one of four genomes here. Genes that have the same pattern of presence or absence across genomes are predicted to interact. Thus the gene products of genes 4, 5 and 6 would be predicted to interact in the example here.

12.7   Summary and outlook

There are three different types of interactions between domains: interactions in multidomain proteins, stable complexes and transient protein–protein interactions. Members of stable complexes are particularly tightly co-regulated and conserved in sequence.

Structural analyses on small sets of multi-domain proteins and transient complexes have shown that the domains from a pair of families often bind to each other with the same geometry. Most domain families engage in interactions with one or two other types of families, but a few families are very versatile and interact with many families. These versatile families are ubiquitously useful families such as P-loop nucleotide triphosphate hydrolases and Rossmann domains.

In order to predict protein–protein interactions, computational methods based on the use of gene order and phylogeny have been developed. These are particularly effective at predicting the members of stable complexes and consecutive enzymes in metabolic pathways. Experimental approaches for the global determination of pairwise protein interactions as well as complexes have yielded thousands of interactions in the yeast *S. cerevisiae*.

The most detailed experimental information about protein–protein interactions comes from three-dimensional structures. It is likely that structures of protein complexes will be solved more frequently in the wake of the structural genomics projects.

References and further reading

Aloy, P., and Russell, R.B. (2002) Interrogating protein interaction networks through structural biology. *Proc Natl Acad Sci USA* **99**: 5896–5901.

Bashton, M., and Chothia, C. (2002) The geometry of domain combinations. *J Mol Biol* **315**: 927–939.

Gavin, A.C., Bosche, M., Krause, R., *et al*. (2002) Functional organization of the yeast proteome by systematic analysis of protein complexes. *Nature* **415**: 141–147.

Huynen, M., Snel, B., Lathe, W., and Bork, P. (2000) Predicting protein function by genomic context: quantitative evaluation and qualitative inferences. *Genome Res* **10**: 1204–1210.

Jansen, R., Greenbaum, D., and Gerstein M. (2001) Relating whole-genome expression data with protein–protein interactions. *Genome Res* **12**: 37–46.

Pellegrini, M., Marcotte, E.M., Thompson, M.J., Eisenberg, D., and Yeates, O. (1999) Assigning protein functions by comparative genome analysis: protein phylogetic profiles. *Proc Natl Acad Sci USA* **96**: 4285–4288.

Tatusov, R.L., Koonin, E.V., and Lipman, D.J. (1997) A genomic perspective on protein families. *Science* **278**: 631–637.

Teichmann, S.A., and Babu, M.M. (2002) Conservation of gene co-regulation in prokaryotes and eukaryotes. *Trends Biotech* **20**: 407–410.

Teichmann, S.A., Murzin, A., and Chothia, C. (2001) Determination of protein function, evolution and interactions by structural genomics. *Curr Op Struc Biol* **11**: 354.

Uetz, P., Giot, L., Cagney, G., *et al*. (2000) A comprehensive analysis of protein–protein interactions in *Saccharomyces cerevisiae*. *Nature* **403**: 623–627.

Predicting the structure of protein–biomolecular interactions

13

Richard M. Jackson

- Predicting the three-dimensional structure of protein–biomolecule (protein, DNA, ligand) complexes; concepts, the docking problem and practical considerations.
- Molecular complementarity; using shape, surface properties, molecular mechanics and knowledge-based force fields and experimental constraints.
- Search problem; local and global search.
- Conformational flexibility; ligand and protein.
- Evaluation of models and molecular visualization.

13.1   Concepts

Interactions are key to understanding biological processes as they are involved in all regulatory and metabolic processes. Determining the structure of the interactions of biomolecules (protein–protein, protein–DNA and protein–small molecule) with other molecules is important in understanding biological processes at the molecular level. The knowledge of three-dimensional structure allows the possibility of intervention and manipulation of molecular interactions, via structure-based drug design, site-directed mutagenesis and protein engineering. Therefore, docking methods and other prediction methods have the potential to add to insights obtained from structure determination methods such as X-ray crystallography, nuclear magnetic resonance (NMR) and electron microscopy (EM).

At the center of predicting the structure of molecular interactions is 'the docking problem'. It can be described very succinctly. Given two biological molecules of known structure that are known to interact can we determine their three-dimensional structure when in a complex? The molecules might be protein, DNA/RNA, or small molecules (substrate, cofactor or inhibitor). Whilst the problem can be easily stated, searching three-dimensional space and determining the energetically most favorable interaction is a computationally complex problem.

13.2   Why predict molecular interactions?

At the start of 2002 there are more than 15,000 protein structures in the Brookhaven structure databank (PDB) but only ~700 non-redundant coordinate sets for protein–protein complexes. This large difference is similar in nature to the gap between the protein sequence databank and the structure databank and stresses the need to develop methods both experimental and theoretical that can bridge this gap. With the completion of the human genome an unprecedented wealth of information is becoming available about proteins. Structural and functional genomics are giving rise to a wealth of structural and protein interaction data. For example yeast two-hybrid screening (as well as other experimental methods) is providing information on possible protein–protein interactions. A comprehensive analysis of protein–protein interactions in *Saccharomyces cerevisiae* using large-scale yeast two-hybrid screens has already been published. Increasingly these proteins will have structures already solved or homologs present in the structural database. We are therefore entering a period in which there is a growing need for methods that can utilize this information. Finding which of these interactions are biologically relevant and the structural basis of these interactions is an area in which bioinformatics will have a significant impact.

13.3   Practical considerations

For the purpose of simplification, distinction is often made between the general treatment of *Protein–Protein/DNA docking* and *Protein–Ligand* (small molecule) *docking*. This involves issues relating to *molecular complementarity* (section 13.4), *the search problem* (section 13.5) and *conformational flexibility* (section 13.6). Some general distinctions can be made.

13.3.1   Protein–protein docking

- A simplified description of the protein (or protein surface) is used instead of atomic level detail.
- Some form of simple surface area complementarity measure is used to score the 'fitness' of different solution complexes.
- Both of the protein molecules are usually considered rigid.
- Search problem is restricted to 6° of freedom (three translational and three rotational).
- Treatment of protein flexibility is generally restricted to a post-processing or further refinement stage following the initial rigid-body docking.

13.3.2   Protein–ligand docking

- A detailed scoring function is used involving an atom level force-field.
- The search usually focuses on exploring an area (or volume) of likely complementarity such as a known binding site.
- Descriptor-based methods are often used. These descriptors are used as a grid for mapping the ligand into the cavity.
- Alternatively, a search of conformational space is carried out using Monte-Carlo, simulated annealing or molecular dynamics methods.
- The protein is usually considered rigid whilst the ligand is treated as flexible or as a large number (ensemble) of rigid structures.

13.3.3   What is needed?

- Atomic resolution structures of the protein(s), DNA/RNA or small molecules.
- A computer workstation preferably with high-resolution graphics.
- An executable version of a docking program.

An atomic resolution structure implies atomic coordinates in PDB format determined experimentally by either X-ray crystallography or NMR. For small molecules the structure may be built according to stereochemical rules (determining bond lengths, angle, and torsion angles) using molecular modeling software. A computer with high-resolution graphics will aid in viewing the results and manipulation of protein structure. There are a number of docking programs available, many of which are freely (or at nominal cost) available to academic users subject to a license agreement being signed and returned to the program authors.

13.4   Molecular complementarity

At its most elementary the problem can be thought of as a three-dimensional jigsaw puzzle. We are trying to find the best fit between two objects that happen to be biomolecules. Indeed researchers have constructed physical plastic or plasticine models of two proteins and tried to find the best fit by manual manipulation. Alternatively we can use virtual representations of the molecules and use interactive graphics to manually explore how the two molecules might fit together. However, this can also be automated allowing the computer to search for a best fit. The problem is how do we know when we have the right answer? The scoring function used to determine the level of complementarity between two biomolecules is a key component of any docking method. In many cases this will determine success or failure in recognizing the correct complex from amongst a large number of possible alternative solutions.

13.4.1 Shape complementarity

The complementarity between two proteins or a protein and ligand can be described by surface contact or overlap, i.e. the number of adjacent surface points (atoms or residues) or the overall buried surface area of two molecules in contact (*Figure 13.1*). For example simple atom neighbor counting and the simple surface contact scores are of this type. Whilst these measures are easily calculated they also have some physical basis for being effective scoring functions. The principal driving force for protein folding and binding is the hydrophobic effect,

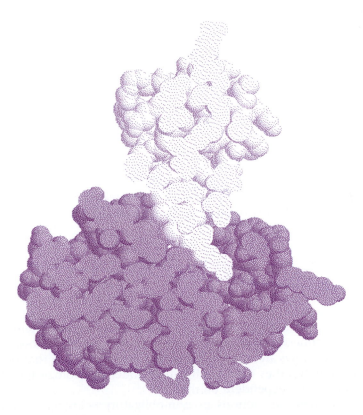

Figure 13.1

Cross-section through a Corey-Pauling-Koltun (CPK) or spacefill model where each atom is represented by a solid van der Waals sphere. The protein trypsin (purple) is in complex with bovine pancreatic trypsin inhibitor (pale purple). Note that several amino acid residues in the inhibitor bind into a cleft (which is the substrate-binding site) on trypsin suggesting atomic level shape complementarity.

which involves the free energy gain of removing non-polar surface area from water. The burial of surface area (or the maximization of surface contact) is an approximation of this effect. However, since no distinction is generally made between polar and non-polar surface area this is an approximation (since burial of the former is unfavorable or neutral).

13.4.1.1 Grid representation

To speed up the matching process the topology of the protein can be simplified from atomic level detail to a series of cubic elements by *discretizing* the 3-dimensional space using a grid. This allows very fast computer matching using search methods such as Fourier transform (see section 13.5.2). The shape of a molecule is described by mapping it to a 3-D grid of uniformly spaced points. Clearly the level of detail is controlled by the grid spacing. The larger the grid spacing the cruder the representation with respect to the atomic level (*Figure 13.2*).

In a translational scan the mobile molecule B moves through the grid representing the static molecule A and a signal describing shape complementarity, f_C, is generated for each mapping. Mathematically the correlation function, f_C of f_A and f_B is given by:

$$f_C = \sum_{i=1}^{N} \sum_{j=1}^{N} \sum_{k=1}^{N} \text{of } f_{A\,i,j,k} * f_{B\,i+\alpha,j+\beta,k+\gamma}$$

where N is the number of grid points along the cubic axes i, j, and k and α, β, and γ are the translational vectors of the mobile molecule B relative to the static one A. The overlap between points representing the surface of the molecules is scored favorably, however, overlap with points representing the core of the static molecule are scored unfavorably. Zero correlation score is given to two molecules not in contact. Negative scores represent surface overlap with the core region of molecule A. The highest score represents the best surface complementarity for a given translational scan. Note that the mobile molecule must be mapped differently to the grid for each rotational change applied to the molecule.

13.4.2 Property-based measures

In many cases, displaying physical properties on the molecular surface of molecules can help to guide molecular docking. Alternatively sequence conservation might also help, particularly where a homologous family of proteins maintain a specific binding partner.

13.4.2.1 Hydrophobicity

The hydrophobic effect plays a dominant role in the folding of proteins. Hydrophobic residues aggregate away from contact with water to form hydrophobic cores with more polar residues forming the solvent accessible surface and allowing solubility of the protein. Clearly a hydrophobic interface will drive the formation of protein–protein or protein–ligand interactions. It has been noted that hydrophobicity is fairly common at protein–protein interfaces particularly in homodimers (two identical protein monomers that associate) and oligomeric proteins. (*Plate 5*) Oligomers are often obligate complexes meaning that the free-energy cost of dissociation is high and they exist as oligomers under physiological conditions. In some cases biological function is dependent on this.

However, many protein interactions are non-obligatory being made/broken according to their environment. These proteins must be independently stable in solution. These are commonly heterodimeric complexes including enzyme–inhibitor and antibody–antigen complexes as well as a host of other casual interacting proteins. The hydrophobic effect is often much less dominant and interfaces are more polar in nature partly because of issues relating to protein stability and aggregation. Therefore, hydrophobicity is useful as a guide to molecular complementarity in selective protein–protein interactions.

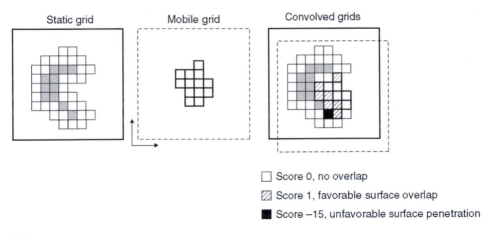

Static grid   Mobile grid   Convolved grids

☐ Score 0, no overlap

▨ Score 1, favorable surface overlap

■ Score −15, unfavorable surface penetration

Figure 13.2

This simplified representation of two molecules (shown here in 2-D) allows them to be matched. The molecules are discretized differently. The static molecule (usually the larger of the two molecules) is described by repulsive core points (in gray) and attractive surface layer points (in white). Overlap of points with the mobile molecule score differently. A 2-D representation of the convolved grids for one point in the translation scan is shown. The favorable surface overlaps are spoilt by one overlap with a core point.

13.4.2.2   Electrostatic complementarity

The electrostatic properties of biomolecules often play an important role in determining interactions. The burial of charged residues at protein–protein/DNA interfaces is thought to be generally net destabilizing with the hydrophobic effect being the primary driving force. However, charged groups involved in the biomolecular interface are often stabilized by other polar or oppositely charged groups on the interacting molecule. Therefore charge complementarity can play an important role in determining the specificity of the interaction. In many cases in biology a protein must recognize a highly charged molecule such as polyanions like DNA or RNA. In order to make a close approach the protein must have charged residues that complement the negative charges present on the phosphate backbone. Electrostatic complementarity is also important in many protein–protein and protein–ligand interactions.

EcoRV is a clearly defined case where large areas of positive and negative surface potential map to one another (*Plate 6*). Visually, most protein–protein and many protein–DNA interactions are less clearly defined than this.

13.4.2.3   Amino acid conservation

It has been known for some time that conservation of residues at the surface of a protein family is often related to function. This may be an enzyme active-site or binding site. Unlike hydrophobicity or electrostatic potential, displaying residue conservation on the molecular surface has no physical or chemical basis. However, the evolutionary information can sometimes delineate a functional epitope allowing residues to be identified that are important for binding. In order to infer structure–function relationships the proteins must be structurally related, preferably a large family of proteins with related function (*Plate 7*).

13.4.3 Molecular mechanics and knowledge-based force fields

These scoring functions require atomic level detail. Biomolecules and ligands are defined by their heavy atoms in addition to which hydrogen atoms are often added according to stereochemical rules. An archetypal potential function is given in Chapter 8 (*Box 8.1*). You are encouraged to re-read this. The parameters used for equilibrium bond lengths (b_0), bond angles (θ_0), torsion angles (ϕ) and their associated force constants (k), the van der Waals parameters (A_{ii}, B_{ii}) and point charges (q_i, q_j) are defined by molecular mechanics force-field parameters. There are a large number of parameters and these are defined from experimental observation or by quantum mechanics. Calculation of interaction energies between proteins or proteins and ligands often involves energy minimization (EM) or molecular dynamics (MD); these techniques are discussed in Chapter 8. Alternatively, energy minimization can be used in combination with other search methods to refine or explore the molecular interaction. These are discussed in section 13.6.

In a similar way knowledge-based or rule-based force fields apply a set of predefined parameters to describe the energy (or score) of the system. These parameters are derived from statistical analysis of known structures of protein–protein or protein–ligand complexes. For example, the spatial distributions of atomic contacts derived from a non-redundant set of protein complexes are used to derive atom contact preferences. These rules may incorporate directional properties such as preferred hydrogen-bonding geometries. Like conventional molecular mechanics force fields the idea is to be able to calculate an energy or score from the 3-D atomic coordinates. Alternatively, knowledge-based force fields may describe contact energies based on spatial distributions of residue contacts derived from protein–protein contacts in a similar way to scoring functions derived for fold recognition and threading (see Chapter 9). As with a conventional molecular mechanics force field optimization or search methods can then be used to try to find optimal interaction geometries.

13.4.4 Experimental and knowledge-based constraints

Although not technically a scoring function, experimental constraints constitute very valuable information. At the simplest level this may be information about a known binding site or epitope. For example, most protein or small molecule inhibitors of enzymes bind so as to obstruct the catalytic/substrate binding site (see *Figure 13.1*; trypsin in complex with bovine pancreatic trypsin inhibitor). Simply knowing these sites can aid in screening false-positive solutions or allow docking to focus on a specific area. A similar situation arises with antibody Complementary Determining Regions (CDRs), which are involved in antigen molecular recognition. The search for viable solutions can be restricted to interactions involving these functional epitopes.

Site-directed mutagenesis can also act as an important experimental constraint. Knowing that the mutation of a surface residue strongly influences binding (whilst still maintaining structural integrity of the protein) implicates this location in binding. However, in certain circumstances residue mutations at sites distant from the interface can influence binding through cooperative effects or fold destabilization.

Although not as definitive as experimental information, functional prediction of a protein belonging to a well-populated protein family can also act as a constraint in predicting biomolecular interactions. Conserved interactions in protein families (homologs) are fairly common so knowledge of both the structure and common function may be enough to infer catalytic/binding site location. In certain highly populated protein folds called the *Superfolds* there is conservation of binding-site location even in the absence of homology, these have been termed *Supersites* (see Chapter 10). They may act as a constraint in predicting interactions.

13.5 The search problem

Determining the structure of two interacting proteins still remains a formidable problem. The problem is a six-dimensional search problem with three degrees of translational freedom (translate along x, y and z axes in Cartesian space) and three degrees of rotational freedom (rotate around the x, y and z axes) for the mobile molecule assuming one molecule is kept stationary. This also makes the assumption that the two molecules can be treated as rigid bodies and clearly this is not always a sensible approximation. A complete six-dimensional search was originally considered computationally prohibitive for large molecules like proteins.

The first approach was to use interactive graphics and manually explore how two molecules might fit together. This can be very time consuming. Early computational methods had to use highly simplified representations of the protein molecular surface. Clearly, for the problem to be tractable the search must be either directed and focus on areas of likely complementarity using descriptor-based methods, or use a simplified description of the molecular surface and search space and perform a complete search. A comprehensive summary of all methods is beyond the scope of this chapter. Two widely used methods are described here in more detail. Firstly, a constraint-based method (section 13.5.1) and secondly a complete search of conformational space (section 13.5.2). For descriptions of other docking methods you are encouraged to read the reviews given at the end of this chapter.

13.5.1 Constraint-based methods

Constraint-based methods form the basis of many protein–ligand (small molecule) docking algorithms. The idea is that the binding site of the receptor molecule can be described by a series of points or vectors that represent favorable interaction points for the ligand atoms. The search of ligand conformational space is constrained to these points, i.e. a restricted area/volume of the receptor-binding site. These points/vectors are determined by either geometric or force-field complementarity including knowledge-based scoring functions. They are used as a guide for mapping the ligand into the cavity and work by trying to match a ligand molecule (whose shape is defined by atoms or spheres) onto a defined set of points or vectors that represent a positive image of the protein cavity. The DOCK algorithm is one of the first and most widely used methods developed by Kuntz and co-workers.

13.5.1.1 DOCK algorithm

- Define a limited surface area for exploration (might contain one or several cavities).
- Use clustered spheres to represent cavities on the receptor.
- Spheres are computed for every pair of surface points (i and j) that do not intersect the protein surface (*Figure 13.3*). However only the smallest surface sphere is retained for every surface point i. For example, four pairs of surface points are shown below. Surface points are defined by the Molecular Surface (the solvent excluded surface defined by the point of contact of a solvent probe sphere rolled over the surface of the protein).
- A set of overlapping spheres defines a cluster (e.g. above i_1, i_2, i_3). Typically a protein will have 10–20 such clusters describing different surface cavities.
- The molecule/ligand to be docked is characterized by spheres in a similar way to the cavity or by its atom centers.

13.5.1.2 Matching

Matching is performed based on similarity in distances between receptor centers i->j and ligand centers k->l. Sets match if all the internal distances match (within a tolerance value). Four pairs are found to uniquely determine the rotation/translation matrix for placing the

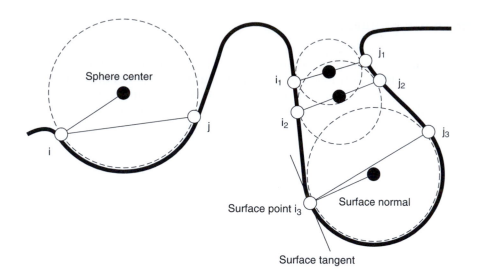

Figure 13.3

Two-dimensional representation of the protein molecular surface (bold line) and four pairs of surface points (i, j). The spheres cannot intersect the surface and sphere centers are on the surface normal from point i.

ligand in the receptor site. This placement is performed by a least-squares fitting procedure (see Chapter 6). The number of matches generated (and hence the computational run time) for a given ligand in a given cavity is a function of the size of both the cavity and the ligand. At its most simple scoring involves mapping the ligand placement onto a contact grid. More complex calculation of the interaction energy using force-field evaluation is also possible.

13.5.1.3   Applications

The DOCK program and other constraint-based methods are particularly suited to small molecule docking where there is a knowledge of the ligand-binding site and it can focus the search on a specific cavity, or region. For example, where the cavity is known to be the substrate or co-factor binding site in an enzyme. Such methods have been applied to a lesser degree where the search is intended to be global because the search space is less constrained and more difficult to define using this approach.

13.5.2   Complete search of space

There are relatively few methods that perform a systematic search of six-dimensional space. A brute force calculation which involves computing all possible configurations at atomic level detail is not computationally tractable for protein-sized molecules. Firstly, some level of simplification of the protein 3-D representation is essential. Secondly, a computationally efficient search method must be used. Possibly one of the most extensively used methods which confronts these two problems is the Fourier transform docking method proposed by Katchalski-Katzir and co-workers. The method has been adopted by several groups particularly in applications to protein–protein docking.

13.5.2.1   The Fourier transform method

The Fourier correlation theory allows the complexity of the translational scan to be reduced considerably (see *Figure 13.4*). The method uses the discretized representation of protein

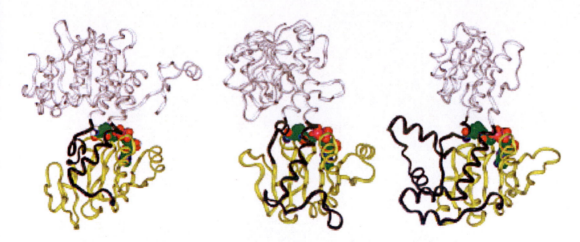

Plate 4

The geometry of domain combinations. Three different two-domain proteins are shown here, all in the same orientation with respect to their yellow Rossmann domains. The NAD cofactor is shown as a space-filling model. In all three proteins, the black linker regions and grey catalytic domains of the formate/glycerate dehydrogenase-like family are homologous and the geometry of the catalytic domain towards the Rossmann domain is conserved. (See Chapter 12, Section 12.4.2)

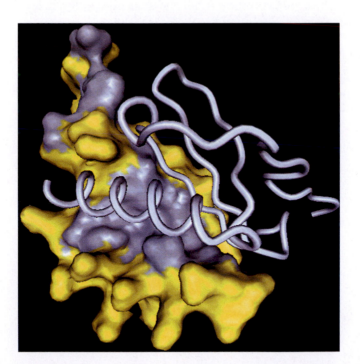

Plate 5

Homodimer interaction between two subunits of interleukin. One subunit is shown as a ribbon, the other by its solvent accessible molecular surface. The surface is color coded according to residue hydrophobicity. The following amino acids are considered hydrophobic (G, A, P, C, V, I, L, M, F, W) and colored grey, other residue types are considered polar and colored yellow. It can be seen that the dimer interface is predominantly hydrophobic in nature. (See Chapter 13, Section 13.4.2.1)

Plate 6

Type II restriction enzyme EcoRV (DNA bound conformation) and its cognate DNA fragment, which has been translated ~40Å along the x-axis out of the hollow binding groove. The blue regions show areas of positive electrostatic potential at the molecular surface and the red show negative potential. The enzyme binding grove is predominantly positive, the DNA negative. (See Chapter 13, Section 13.4.2.2)

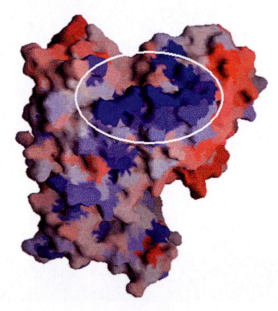

Plate 7

Sequence conservation of Rab GDP-dissociation inhibitor (GDI), a family of conserved proteins involved in cellular transport. Ten GDI sequences were used in construction of the residue conservation pattern. Blue represents a high level of residue conservation, with white intermediate and red high variability. The highlighted area of high conservation corresponds to a region identified by site-directed mutagenesis as the Rab-binding platform. (See Chapter 13, Section 13.4.2.3)

> **Box 13.1   Virtual screening and structure-based drug design**
>
> Computer-aided strategies have become an increasingly important component of modern drug design. Given an experimental protein structure for a protein of pharmaceutical interest the aim is to identify small molecules that bind and inhibit or otherwise alter activity of the target. Current computational methods involve database screening of small molecule libraries by molecular docking or the *de novo* design of ligands by assembly of molecules from smaller fragments that interact favorably with the target. The number of possible compounds in molecular space is too large to screen experimentally. The aim is to produce a manageable subset that can be screened experimentally with high-throughput methods.

---

structures described in section 13.4.1 as its input. Practically, this model could be applied using a systematic search in 3-D space (involving N^3 calculations where N^3 is the number of grid points). However, Since f_A and f_B are discrete functions representing the discretized molecules A and B (see section 13.4.1.1) it is possible to calculate f_C more quickly by fast Fourier transform requiring only $\log_e(N^3)$ calculations as opposed to N^3 for a systematic search. This is a considerable computational saving (try calculating the difference for large values of N on a scientific calculator). After, each translational scan molecule B is rotated about one of its x,y,z axes and the translational scan repeated until all rotational space has been completely scanned. Typically angular sampling is in steps of 15–20°. Finer angular sampling is computationally demanding, for example decreasing it from 15° to 10° would increase the number of orientations by over three-fold. The ability to control the fineness of the search by both grid resolution and angular sampling allows the user to limit the search space and therefore reduce the complexity of the problem. Clearly, there is a balance between run time and level of molecular detail that is critical in achieving a successful outcome.

13.5.2.2   Scoring

See the Grid representation method in section 13.4.1.1 for scoring molecular complementarity.

13.5.2.3   Applications

The method is most suited to macromolecular docking such as protein–protein docking or situations where dealing with atomic level detail is computationally prohibitive. The inability of the method to include a flexible description of the ligand molecule means it has been applied to a lesser degree to protein–ligand docking.

13.6   Conformational flexibility

Proteins are dynamic entities that undergo both limited conformational change of amino acid side-chains and fluctuation of flexible loop regions about equilibrium positions when in solution. Such flexibility can often be adequately treated by 'soft' potentials or limited conformational flexibility and/or refinement of side-chains on docking. However, proteins often undergo more extensive conformational changes which may involve large-scale motions of domains relative to one another or possibly conformational change involving order–disorder transitions. Clearly, these types of motions will be poorly treated by a rigid body model for docking. Again distinction is often made between the general treatment of *protein–protein docking* and *protein–ligand docking*.

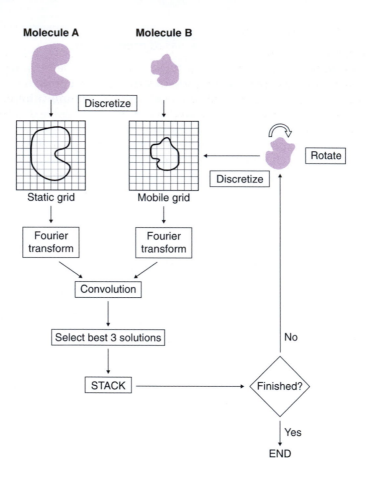

Figure 13.4

The two molecules are discretized as described in section 13.4.1.1. Molecule A is static and is described by repulsive core and attractive surface layer points. Molecule B is the mobile molecule and no distinction is made between surface and core elements. These representations are Fourier transformed. A translational scan is then performed in Fourier space and a signal describing shape complementarity, f_c, is generated for each mapping. Only the best three solutions are stored to a stack. The mobile molecule B is then rotated and a new translational scan performed. This process is repeated until all rotational combinations have been tried. (Note: the process is carried out in 3-D but shown here in 2-D for simplicity.)

13.6.1 Protein–ligand docking

Small molecule ligands are often highly flexible adapting their shape to fit the receptor-binding pocket. The degree of conformational flexibility of a small molecule ligand (substrate, cofactor or inhibitor) can be considerable particularly where there are multiple torsion angles. This presents a major challenge in protein–ligand docking and several different approaches have been used to solve this problem. In general the protein is still treated as a rigid body in these simulations. The treatment of protein flexibility will be addressed later. Here we review the concepts and methods of three approaches for treating ligand flexibility. The interested reader is encouraged to read the reviews given at the end of this chapter.

13.6.1.1   Multiple conformation rigid-body method

The simplest approach is to assume that the ligand can be treaded as a rigid body. In this case a ligand is assumed to be able to adopt a number (N) of different low-energy conformations that are computed prior to the ligand being docked into the receptor. These, N, low-energy conformations are then docked individually into the receptor assuming a rigid conformation using a descriptor-based approach. The scoring function is used to determine which of the resulting solutions is optimal.

The advantage of this approach is that the search can be restricted to a smaller number of relevant ligand conformations, the disadvantage is that the active conformation may be missed as the result of a minor structural difference not considered in the N ligand conformations.

13.6.1.2   Stochastic search methods

Stochastic processes use a random sampling procedure to search conformational space. This includes methods such as Monte Carlo simulation, simulated annealing, Tabu search, genetic algorithms and evolutionary programming. The ligand molecule performs a random walk in space in the receptor cavity. Usually, the ligand is placed in a random orientation in the receptor cavity. Then at each step a small displacement is made in any of the degrees of freedom of the ligand molecule (translation, rotation or torsion angle).

In a Monte Carlo simulation the score (or energy) is calculated at each step and compared to the previous step. If the new energy is lower the step is accepted, otherwise the result is treated probabilistically by a Boltzmann mechanism. The probability of accepting the step is given by

$$P(\Delta E) = \exp(-\Delta E/k_B T)$$

where ΔE is the difference in energy, k_B is Boltzmann's constant and T the temperature. If $P(\Delta E)$ is greater than a random number generated between 0 and 1 then the step is accepted. The higher the temperature (or the smaller ΔE at a given T) the higher the likelihood the step is accepted. A conventional Monte Carlo simulation proceeds at constant temperature, whilst in simulated annealing the temperature is gradually cooled during the simulation in an attempt to locate a globally optimal solution. In simulated annealing the computer stores a single solution and generates a new solution randomly.

Another method called Tabu search performs a local search in a subspace of the total search space to generate a new guess at the optimal solution. In both methods only the previous solution is used to generate a new solution, whilst the best solution to date is stored in the computer memory.

Genetic methods (genetic algorithms and evolutionary programming) store multiple solutions. These solutions form a *population* of *members*. Each member has an associated score or *fitness*. During the search for the global optimal solution successive new populations are created by a procedure involving selection of the fittest members. These members then have *offspring* to create a new population. Differences arise in how the methods generate offspring. In a genetic algorithm two solutions are *mated* to form a new offspring solution. In evolutionary programming each member of the population generates an offspring by *mutation*.

The advantages of stochastic methods are that the ligand is able to explore conformational space in a relatively unconstrained way, frequently leading to the globally optimal solution. The disadvantages are that it cannot guarantee reaching a global optimal solution and the methods are computationally costly in comparison to the other methods described here.

13.6.1.3 Combinatorial search methods

Methods have been developed to systematically search conformational space given the initial placement of a (rigid) fragment of the ligand in the receptor site. The ligand is divided up into different rigid-body fragments. These are docked independently of each other and determine a limited number of *anchor* fragments in the receptor-binding site. Once a set of placements has been generated remaining fragments (making up the rest of the ligand) are reattached. This is done in a stepwise fashion so fragments connected to the anchor are attached first and a conformational search is performed often in a stepwise manner (to limit the degree of sampling) to generate all possible configurations. These are then *pruned* so only the N best energetic placements are taken into the next iteration. This is known as a *greedy strategy*. The next set of fragments is then connected and the process is repeated. This is sometimes called *incremental construction*. The advantage of this greedy strategy is that the number of possible conformations is controlled choosing the most energetically favorable solutions at each stage. This prevents a possible combinatorial explosion. The method is focused and therefore very fast. The disadvantage is that sub-optimal placements might be pruned such that they disallow the next fragment in the process to bind in an optimal way. Therefore the global optimal placement is not guaranteed.

In all the above methods there is a balance between speed and ability to find the global optimal solution. Different methods will be most appropriate depending on the application. Fast methods are preferred for looking at large numbers of small molecules within reasonable time frames, such as docking a large number of ligands to a given protein (small molecule database screening), however, a more extensive search of conformational space is more appropriate where reliance is placed on finding the global optimum solution. In no case is finding the global optimum solution guaranteed.

13.6.2 Protein flexibility

The inclusion of protein flexibility in docking applications is still not widespread. However, methods have been described for the introduction of side-chain flexibility to both protein–ligand and protein–protein docking. A method that has been applied to both these docking applications uses the Mean Field principle. The Mean Field approach is one type of what are called *bounded search* methods. Others include the Dead-end-elimination theorem and the A* algorithm. These methods use different approaches to find a solution and a detailed discussion is beyond the scope of this chapter. However, they use a multiple copy representation of protein side chains built using a *rotamer* library as discussed in Chapter 8. In bounded searches the conformational degrees of freedom are finite, i.e. defined by a limited set of rotamers. Therefore a self-consistent solution to the problem exists, and a global optimal solution exists within this definition. The inclusion of protein flexibility is achieved at greater computational cost but this allows side-chain flexibility to be included, potentially allowing for an induced fit on protein binding.

13.7 Evaluation of models

In a similar way to other structure prediction methods models can be evaluated using RMSD to measure the similarity between two molecular complexes (see Chapter 6). This is only the case if an experimental structure already exists. Several docking methods have been evaluated at 'Critical Assessment of Structure Prediction 2' (CASP2) for both protein–ligand and protein–protein interactions (see Dixon 1997). Prior to this protein–protein docking was evaluated in an international challenge to the docking community by Mike James and co-workers (see Strynadka *et al.* 1996). The level of success is dependent on the system under study. For the protein–protein docking evaluation several of the same docking methodologies were used in both docking challenges. In the first challenge involving a protein inhibitor all the groups successfully predicted the complex. However, there was little degree of success in docking an antibody–antigen complex in the second challenge. This may be because modeling molecular recognition is in general more difficult in antibody–antigen than protein–inhibitor systems. In the protein–ligand docking challenge at CASP2 there was generally a good level of success, however, again certain targets proved to be problematic. The lesson from these assessments is that the level of success depends on the system under study.

Clearly there is an ongoing need for assessment of methods. The 'Critical Assessment of Prediction of Interactions' (CAPRI) is an ongoing community-wide experiment on the comparative evaluation of protein–protein docking for structure prediction (see References for details). There is currently no one universal method or scoring function that will work on all occasions. Therefore, an understanding of the importance of different factors (shape, hydrophobicity, electrostatics, evolutionary relationships and conformational flexibility) in a particular interaction is important if we are to have confidence in the results. This information is not always available and the search for a universally applicable scoring function as well as an adequate treatment of conformational flexibility is ongoing. Having discussed the caveats many individual successes have been reported in the literature. The technique of virtual screening of small molecule drugs by computer has become commonplace and a very useful tool in narrowing down the very large number of drugs that might need to be screened by experimental methods. In one study virtual screening provided lead compounds where conventional experimental random screening had failed.

In practical circumstances an experimental structure of the complex does not already exist. Clearly, some other evaluation criteria are needed. This is by necessity experimental validation. The hypothetical protein–biomolecular complex predicts a mode of interaction between the two molecules that can be tested experimentally. In the case of an enzyme and a small molecule inhibitor initial test of activity will indicate if the molecule inhibits the enzyme. For a protein–protein interaction site directed mutagenesis experiments of residues in the proposed interface may confirm or draw into question the validity of the model.

13.8   Visualization methods

Visualization methods are very important in viewing molecular properties on molecules. Of particular note is the rendering of molecular surfaces according to their various properties (that can be expressed numerically). This approach has been popularized by the GRASP program. Also *Virtual Reality Modeling Language* (VRML) viewers can provide similar displays. These programs were used to generate the color figures in this chapter. The popular program RasMol can also be made to view molecular properties by assigning those properties to the temperature factor column of the PDB file in question. The molecules are then best viewed in *Spacefill* mode, however, the color scheme used is not flexible. The GRASP and VRLM have been incorporated into the GRASS server. This allows a Web-based interactive exploration of molecules in the PDB allowing the molecular properties to be viewed on the molecular surface. There are also several other popular molecular graphics programs that allow a similar visualization.

References and further reading

Abagyan R., and Totrov M. (2001) High-throughput docking for lead generation. *Curr Opin Chem Biol* **5**: 375–382.

Critical Assessment of Prediction of Interactions (CAPRI) see http://capri.ebi.ac.uk/

Dixon, J.S. (1997) Evaluation of the CASP2 docking section. *Proteins* **29** (Suppl 1): 198–204.

Halperin, I., Ma, B., Wolfson, H., and Nussinov, R. (2002) Principles of docking: An overview of search algorithms and a guide to scoring functions. *Proteins* **47**: 409–443.

Leach A.R. (2001) *Molecular Modelling: Principles and Applications.* Prentice Hall, Harlow, UK.

Second Meeting on the Critical Assessment of Techniques for Protein Structure Prediction (CASP2) see http://predictioncenter.llnl.gov/casp2/

Smith G.R., and Sternberg M.J. (2002) Prediction of protein–protein interactions by docking methods. *Curr Opin Struct Biol* **12**: 28–35.

Strynadka, N.C., Eisenstein, M., Katchalski-Katzir, E., Shoichet, B.K., Kuntz, I.D., Abagyan, R., et al. (1996) Molecular docking programs successfully predict the binding of a beta-lactamase inhibitory protein to TEM-1 beta-lactamase. *Nat Struct Biol* **3**: 233–239.

Experimental use of DNA arrays

Paul Kellam and Xiaohui Liu

- The principles of DNA microarray experiments are discussed. How such global methods of measuring transcription can help understand complex biological systems is outlined and technical aspects of performing two-color DNA microarrays are illustrated.
- The methods of DNA microarray data capture and pre-processing are illustrated. This includes the entire process from image scanning to a normalized gene expression data matrix.
- Current standards and types of DNA microarray data management are outlined. In particular, the principles of the Minimum Information About a Microarray Experiment (MIAME) are described.
- This chapter provides the starting point for more advanced gene expression analysis discussed in Chapter 15.

14.1   Concepts

With the massive increase in worldwide DNA sequencing capacity we are now in the era where the complete DNA sequence of an organism's genome has become a minimum level of information for further study. However for any given organism, a genome is a static description of its genetic potential. The genome needs to be transcribed and translated into effector molecules (generally proteins) which then act together to shape the organism. This relationship is captured in the 'central dogma' of information flow in the cell (*Figure 14.1a*).

To date the study of biological processes during health and diseases has often followed the reductionist method of identifying single gene and encoded protein products that define an observed phenotype. In many cases this has led to a greater understanding of the processes involved within a cell and how these processes are dynamically linked. However, such reductionist approaches often over-simplify complex systems and are less useful when multi-gene/multiprotein processes are involved. An alternative strategy, facilitated by the availability of complete organism genome sequences is to use high-throughput functional genomics methods to investigate multiple events at the same time in a cell or tissue that together define a phenotype. DNA microarrays are one such methodology that allows the simultaneous determination of mRNA abundance for many thousands of genes in a single experiment. However, when thought of in terms of all the transcripts and proteins that are present in a given cell, the total complexity of the system increases dramatically (*Figure 14.1b*).

Therefore the emerging functional genomics research methods also require computational biology to extract and interpret such information. The detection and listing of individual components is also the first step to modeling how the components functionally interact in a '*systems biology*' framework, reviewed by Kitano. Therefore high-throughput post-genomic biology is integral to detailed biological understanding.

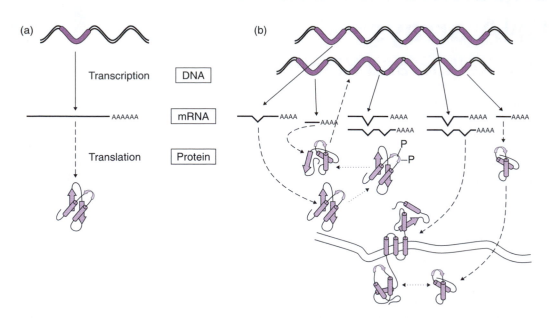

Figure 14.1

(a) The central dogma of information flow in the cell. Information encoded within the genome is transcribed into messenger RNA and translated into proteins, the effector biological molecules. (b) A dramatic increase in complexity involving the central dogma when more than one gene (including splice variants) and protein are produced. Here the proteins act together in a functional module. Solid arrows represent transcription, dashed arrows represent translation and movement of proteins within the cell and dotted arrows represent protein interactions.

14.1.1 The cellular transcriptome

The central dogma of information flow in the cell identifies potential cellular components that can be measured. An obvious part of the information flow that has the lowest additional complexity is at the level of the transcribed messenger RNA (mRNA), collectively known as the *transcriptome*. However, to detect, quantify and interpret a transcriptome requires a careful blend of molecular and computational biology.

The core assumption of transcriptome studies is that a cellular phenotype is linked to the types and expression levels of transcribed genes. Therefore the transcriptome defines the phenotype by measuring a surrogate marker for the eventual translated proteins. In reality, this is only part of the story, as a gene can have multiple different transcript *splice variants* encoding alternative proteins, mRNA can be sequestered untranslated in the nucleus and, once translated, a given protein can be modified in different ways resulting in alternative functions. Nevertheless, proteins cannot be synthesized without an mRNA, therefore defining a transcriptome can give valuable insight into what information the genome is currently using. In this chapter we will illustrate methods and the initial stages of data analysis using DNA microarrays. In Chapter 15 we will illustrate more advanced array analysis methods.

14.2 Methods for large-scale analysis of gene expression

Traditionally *Northern blots* have been used to determine the level of a particular transcript in a given cell type or tissue. Northern blots make use of hybridizing a labeled DNA or RNA probe specific for a gene of interest to a solid support that has had RNA from the cells or tissues immobilized onto it. However, Northern blots are not very scalable. Therefore other methods have been developed to allow high-throughput determination of the presence of multiple different transcripts. These methods come in many forms, from random sequencing of short stretches of cloned complementary DNAs (cDNAs, produced *in vitro* from mRNA) followed by counting the abundance of each different transcript cDNA, known as *Serial Analysis of Gene Expression* (SAGE), to 96-well-plate-based quantitative *reverse transcriptase-polymerase chain reaction* (RT-PCR).

However, the technique that has found greatest acceptance for large-scale gene expression analysis is *DNA arrays*. Arrays are effectively the Northern blot method in reverse. This time DNA probes for many genes are immobilized on a solid support (such as nylon membranes or glass slides) at defined, fixed positions. RNA from cells or a tissue are labeled and hybridized to the gene probe array, with each transcript finding and binding to its respective probe, with hybridized signal intensity being relative to the abundance of the transcript. For each type of array, the probes are always located in the same position making multiple experiments easily cross comparable.

Arrays come in a variety of forms from cDNA or PCR products spotted onto nylon membranes (*macroarrays*) or glass slides (*microarrays*) to gene-specific oligonucleotides synthesized *in situ* onto a glass slide such as *Affymetrix* chip arrays. The design and production of arrays has now become a small industry. However, it is still possible to produce arrays in standard molecular biology laboratories with the right equipment. Array production methods are still evolving but the basic principles have been written about in great detail (The Chipping Forecast, 1999).

14.3   Using microarrays

14.3.1   Performing a microarray experiment

As different array types already exist, so the methods for performing specific array experiments can also differ. Nevertheless, all rely on the same principle of labeling mRNA isolated from cells with a detectable nucleotide, such as fluorescent nucleotides and detection of probe-bound labeled nucleic acids on the array. One of the most common types of array protocols is two-color microarrays. In this case, mRNA from the sample of interest is converted to cDNA in a reaction that incorporates one specific fluorescent dye (i.e. *Cy5* – red). mRNA is also labeled as cDNA with a different fluorescent dye (i.e. *Cy3* – green) from a control or reference RNA source. The two different labeled samples are then mixed and hybridized to the microarray where labeled cDNA sequences bind to their respective probes on the array (*Plate 8*).

Since two differently labeled cDNAs are applied to one array, a local competitive hybridization occurs for each probe. If a gene is expressed at a high level in the experiment sample but at a low level in the control/reference sample, the spot on the array will be predominantly red. If the gene is expressed at a low level in the experimental sample but a high level in the control/reference sample, the spot on the array will be predominantly green. Equal levels of gene expression in the sample and reference results in equal amounts of red and green hybridization and therefore a yellow spot on the array.

The level of hybridization of labeled cDNA to each array probe is determined by laser scanning the array for fluorescence at the Cy5 and Cy3 wavelengths thereby measuring the amount of bound probe. This results in a series of numerical values for each spot on the array. By using a common labeled reference cDNA for a series of arrays with different sample cDNAs it is possible to cross-compare all the different gene expression values by comparing the ratio of sample hybridization signal to reference hybridization signal across all arrays. This competitive hybridization and ratio measurement also has the effect of minimizing variations that exist between one printed array and the next. For example, if a gene is equally expressed in the sample and reference, but due to array printing or normal experimental variation the sample/reference hybridizes twice as well to array 1 as it does to array 2, then although the absolute gene expression levels for both channels will be different, the ratio will be the same.

14.3.2 Scanning of microarrays

One aspect of microarray analysis that is often only commented on in commercial array scanner manuals, is the specific type of data produced when an array is scanned. It is essential that when array experiments are undertaken that researchers understand the type of scanner they have and the range of data fields that are produced (*Figure 14.2*). Modern array scanners will scan a slide at a resolution of 5 or 10 µm, meaning the laser and photo multiplier tube will excite and measure fluorescence of an area of 5 µm × 5 µm (25 µm²) or 10 µm × 10 µm (100 µm²), respectively. These areas are often called 'pixels'. The entire hybridized array is therefore scanned at a given resolution.

Following scanning, a spot-finding algorithm is used to define the location and edge boundaries of each array spot. The algorithm also defines an area around the spot to measure

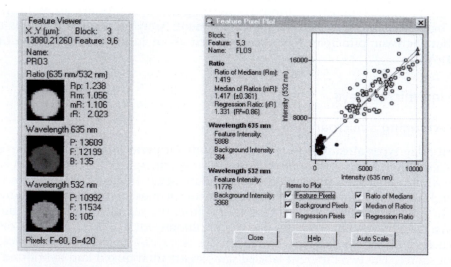

Name	F635 Median	F635 Mean	F635 SD	B635 Median	B635 Mean	B635 SD	F532 Median
Proto-oncogene c-cot (protein-serine/thr	4914	4732	1940	701	748	381	13773
KIAA0308	2070	2141	871	703	760	575	3062
CDW52 antigen (CAMPATH-1 antigen)	3267	3304	1443	712	777	572	8400
RACK-like protein PRKCBP1	2483	2496	859	711	783	410	2650
Homologue of mouse tumor rejection antig	31095	28608	11727	740	1036	1865	34448
aflatoxin aldehyde reductase AFAR	2998	3019	1337	740	994	1773	4837

Name	F532 Mean	Ratio of Medians	Ratio of Means	Median of Ratios	Mean of Ratios	Log Ratio
Proto-oncogene c-cot (protein-serine/thr	12369	0.326	0.35	0.352	0.339	-1.617
KIAA0308	3070	0.599	0.628	0.627	0.577	-0.74
CDW52 antigen (CAMPATH-1 antigen)	8135	0.336	0.353	0.348	0.335	-1.575
RACK-like protein PRKCBP1	2627	0.953	0.972	0.97	0.928	-0.069
Homologue of mouse tumor rejection antig	31491	0.902	0.908	0.914	0.891	-0.149
aflatoxin aldehyde reductase AFAR	4792	0.561	0.573	0.571	0.552	-0.833

Figure 14.2

An example of the graphical user interface and selection of data that is produced per spot following laser scanning of a microarray. These data are from an Axon 4000B GenePix scanner and software.

the local background hybridization. The size of the array spots varies between array-printing machines and the printing methods used, but on a given print run of arrays the spot sizes will generally be consistent. The average spot size on an array can be about 100 μm in diameter. A 100 μm diameter spot has an area of 7854 μm^2 and therefore can be divided up into about 78 squares (pixels) when scanned at 10 μm.

The fluorescent measurements for all pixels at each wavelength, the defined spots and the background values are all measured. However, rather than exporting every 100 μm^2 pixel measurement associated with a spot and its background, the data are reported as the mean and median of the pixels from a given spot and the spot's local mean and median background for both the Cy5 and Cy3 channels along with standard deviation of the values (*Figure 14.2*). Other exported data are derived from these values such as the ratio of the medians and ratio of the means of the Cy3 to Cy5 channels. In addition other comparisons are performed. For example, each pixel in a spot can have a Cy3 and Cy5 ratio, which can be used to calculate the mean or median of the pixel ratios (median of the ratios). Depending on the scanner software different aspects of the measured data are reported.

The amount of data from a single array can therefore rapidly become overwhelming and, as yet, the robust analysis and use of all scanned array data is not routinely undertaken. Simple data filters can be used to extract good-quality array spot measurements. For example, if the ratio of the medians, median of the ratios and regression ratio of Cy5 compared with Cy3 for all pixels of a given spot produce similar values, the spot could be considered good quality (*Figure 14.2*). In addition the signal to noise ratio (SNR = signal/background) can be calculated and used to exclude spots with a low SNR as they have either low spot intensity relative to the background, or high background intensity relative to the spot.

However, there is still no universally accepted way of assessing spot measurement reliability. Ultimately the best assessment of spot reliability is to replicate array experiments and determine the standard error of the replicated spots together with spot quality assessments. Once good-quality scanned data are identified from a set of arrays and quality control filters have been applied, the next stage of analysis can be undertaken.

14.4   Properties and processing of array data

When using and transforming gene expression data it is preferable to work with log intensities and/or log ratios of intensities. One reason for this is that raw intensity values are not normally distributed but log-transformed intensities approach a normal distribution. This can be seen by plotting the raw or log-intensity ratios of Cy3 and Cy5 gene expression ratio data (*Figure 14.3a, b*).

In *Figure 14.3b* two sets of array data are plotted. Although they are clearly different, with different mean and median values, the actual shape and spread of the data in the histogram is very similar, i.e. the two data sets have similar standard deviations. In fact, the standard deviation (a measure of expression ratio data spread) in a normal distribution is related to the population mean. Therefore log transformation aids in comparisons between array and the array normalization processes (see section 14.5) as discussed below.

Log ratios also facilitate the interpretation of relative increases or decreases in gene expression from different samples. If a common reference mRNA (normally Cy3) is used for all arrays, then the reference gene expression in effect becomes a gene-specific constant across all arrays. Logarithms to base 2 are used rather than natural or decimal logarithms (log to base 10) (*Plate 9*). This allows the intuitive interpretation of gene expression ratio changes between genes from different samples. For example, if a gene is up-regulated by a factor of two relative to the reference sample, then the gene has a ratio of 2 (15,000 Cy5/7500 Cy3), whereas those down-regulated by a factor of two have a ratio of 0.5 (6000 Cy5/12,000 Cy3). Down-regulated genes therefore only have ratios between 0 and 1. However, a gene

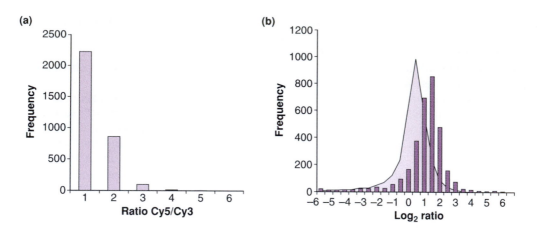

(a) Frequency histogram produced by plotting the distribution of the intensity ratios of the Cy5 and Cy3 channels per spot. **(b)** The same Cy5 and Cy3 spot ratio data (shaded distribution) compared to a second set of array data (bar distribution). Both sets of data are converted to $\log_2$.

up-regulated by a factor of 2 has a $\log_2$ ratio of 1 and a gene down-regulated by a factor of 2 has a $\log_2$ (ratio) of –1. A gene whose expression value is unchanged i.e. ratio of 1 has a $\log_2$ ratio of 0. This gives a much more intuitive view of gene expression as the magnitudes for under and over expression are identical whilst also allowing a log transformation, into an almost normal distribution of the expression data.

One important question is how variable array experiments and the derived gene expression ratios are. The easiest way of seeing inherent variations in array data is when the same mRNA sample is divided into two, labeled with Cy5 and Cy3 and hybridized together to an array. In this case the perfect array would have a y(Cy3) = x(Cy5) or $\log_2$y(Cy3) = $\log_2$x(Cy5) relationship between the intensities. One way of viewing data from a single slide is to plot the background subtracted $\log_2$ (Cy5) intensities against the $\log_2$ (Cy3) intensities (*Figure 14.4a*).

However, this can make interesting features of the data harder to see. Therefore, an alternative data plot called the MA plot has been proposed, where M=$\log_2$Cy5/Cy3, and A=$\log_2\sqrt{}$(Cy5Cy3) (see Yang *et al.* 2002). These plots allow the easier identification of spot artefacts and intensity-dependent patterns (the A axis) in the $\log_2$ ratios (*Figure 14.4b*). From the MA plot it can be seen that there are intensity-dependent differences in $\log_2$ ratios with more variation in $\log_2$ ratios at low signal intensities. However, it is important to realize in this array of the same mRNA labeled with Cy5 and Cy3 that 90% of the data in the plot falls between a $\log_2$ ratio of ±0.5 (ratio of 0.7–1.4) (*Figure 14.4b*, dark data points), indicating the robust nature of array experiments.

This does create one problem in that as the data are not all y = x, how can you determine low-level gene expression changes from normal array variation (for example the 10% of spots with $\log_2$ ratio of >±0.5, *Figure 14.5b*). As with other biological experiments, the only way you can be sure of the variation in the data is to repeat the arrays to produce at least three data points per gene, per sample. This allows the calculation of standard deviations and therefore the ability to test if a gene expression ratio change is significantly above the normal experimental variation.

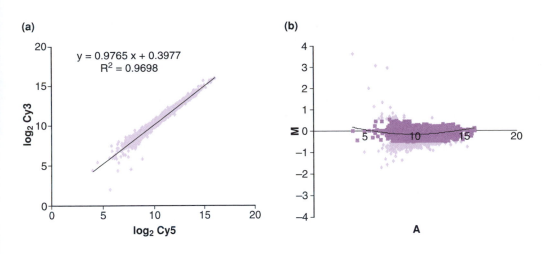

Figure 14.4

(a) Graph showing the relationship between the background subtracted $\log_2$ intensity values for Cy5 and Cy3 labelled mRNA from the same cell. (b) The same data plotted as an MA graph (see text). The dark spots representing 90% of the data are within $\pm 0.5 \log_2$ (Cy5/Cy3).

14.5   Data normalization

14.5.1   Normalizing arrays

As outlined above, like all other biological experiments, gene expression measurements using microarrays are subject to both random and systematic variations. For example, experimental differences in the absolute amounts of mRNA used for each array and differences in array hybridization and washing can occur. Systematic differences between the different fluorescent dyes are also known which are linked to physical properties of the dyes such as light sensitivity and excitation and emission half lives. The process of array normalization is used to minimize such systematic variations in the gene expression levels of the two samples hybridized to the array and allows the comparison of gene expression levels across multiple slides.

Many methods have been developed for normalization and are based on certain assumptions about the underlying data. For more information, see the review by Quackenbush. Normalization methods take advantage of the fact that when log transformed, gene expression data approach a normal distribution and therefore the entire gene expression data distribution can be shifted to the left or right of the population mean or median without changing the standard deviation or relationships within the data. One of the most common array normalization methods, 'Global normalization' assumes that the Cy5 and Cy3 intensities on a slide are related by a constant factor, (Cy5 = kCy3), and that the expression ratio of the average (mean) or median gene of the population is zero (i.e. the Cy5/Cy3 ratio for an average gene is 1). The assumption about the 'average gene' is based on the observation that the majority of genes do not change their expression levels when comparing two mRNA populations. Therefore

$$\log_2 (\text{Cy5}/(k\text{Cy3})) = \log_2 (\text{Cy5}/\text{Cy3}) - a$$

where, as indicated above, a common choice for 'a' is the median or mean of the $\log_2$ ratios, (i.e. $a = \log_2 k$). The effect of this normalization process is shown in *Figure 14.5a and b*. After normalization the data set are centered around a $\log_2$ ratio of 0.

More advanced normalization methods are now being introduced. For example, when looking at data from an individual slide (*Figure 14.4b*), there is not an absolute linear relationship between $\log_2$ ratio values and intensity. The scatter plot smoothing method, LOWESS (Locally Weighted Least Squares) is now used to produce location-specific estimates of the $\log_2$ ratios for various intensities and uses these as distinct local normalization values dependent on overall intensity. Essentially LOWESS divides the scatter-plotted values into subgroups and produces a low-degree polynomial to this subset of the data. The local polynomials fit to each subset of data are almost always of first or second degree, that is, either locally linear (in the straight line sense) or locally quadratic.

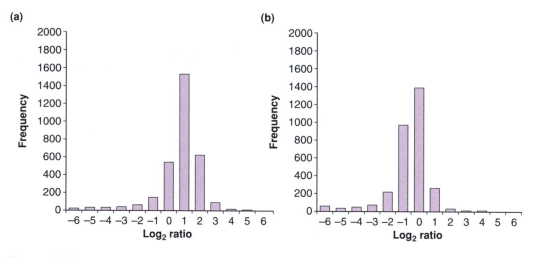

Figure 14.5

The frequency histogram of one microarray dataset $\log_2$ ratio values before (a) and after (b) global normalization to a mean ratio of 1 ($\log_2$ ratio 0).

14.5.2 Normalizing genes

Normalization processes, when applied to all arrays in an experimental set, allow all data to be centered around 0. This makes each array cross-comparable. As mentioned in section 14.3, in two-color microarrays a common reference RNA labeled with Cy3 should be used for all arrays in an experimental set. Whilst this is advantageous in allowing the derivation of log ratios, it can produce the confounding effect of two identically expressed genes across a set of samples in terms of Cy5 intensities having different $\log_2$ ratios (*Figure 14.6*).

The graph in *Figure 14.7* readily shows a confounding effect of the reference Cy3 RNA especially in the raw intensity ratios. This effect is due to the reference RNA having a higher level of Gene A expression compared to Gene B. This is simply because the reference RNA is like any other RNA sample, where different genes have different absolute expression levels. However, unlike the Cy5-labeled RNA samples which are different on each array the reference Cy3-labeled RNA is the same for each array, therefore the denominator of each Cy5/Cy3 ratio should be roughly equivalent for a given gene on a set of arrays.

		Sample/Arrays				
		1	2	3	4	5
Cy5 Intensity (sample)	Gene A	100	200	400	800	1600
	Gene B	100	200	400	800	1600
Cy3 Intensity (ref)	Gene A	100	100	100	100	100
	Gene B	10	10	10	10	10
Raw Intensity ratio	Gene A	100/100	200/100	400/100	800/100	1600/100
	Gene B	100/10	200/10	400/10	800/10	16/10
Intensity ratio	Gene A	1	2	4	8	16
	Gene B	10	20	40	80	160
Log_2 ratios	Gene A	0	1	2	3	4
	Gene B	3.32	4.32	5.32	6.32	7.32

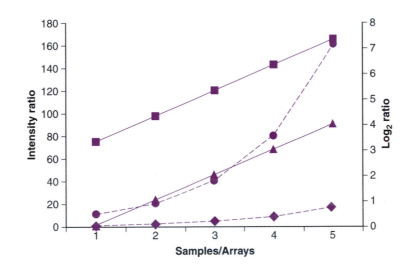

Figure 14.6

Table and figures of Cy5 (sample) and Cy3 (reference) raw intensity values and their respective ratios and log_2 ratios for two genes A and B determined for five samples on five different arrays. On each array the same reference RNA is used hence the same Cy3 value appears across all five arrays for genes A and B. However, the absolute level of gene A expression in the reference sample is 10 times greater than gene B. On each array different samples are hybridized with the common reference. As can be seen from the raw Cy5 intensities both gene A and B are expressed at the same level in each sample. The raw intensity ratios represent division of Cy5 values by Cy3 resulting in the final Intensity ratios and log_2 ratios. The same ratio data is plotted with the raw intensity ratios being dotted lines, gene A diamonds and gene B filled circles and the log_2 ratios values being solid lines, gene A triangles, gene B filled squares. Given that gene A and B are expressed in an identical pattern in the five samples the differences in the ratio between the genes A and B is caused solely by the reference RNA.

The effect of the reference on the log_2 ratio values has the effect of shifting all values by a constant. This fact allows the removal of the reference effect on each gene thereby making the gene expression ratios independent of the reference sample. This is achieved by adjusting the values of each gene to reflect their variation from some property of the series of observed values. Such a property is often the median gene expression value for the gene. Essentially the genes can be centered by subtracting the row-wise median from all the values in the row thereby making the median value of the row 0. This is performed for each row in the gene expression matrix (*Figure 14.7*).

Following these processes, from experimental design to data pre-processing, produces an expression data set ready for discovering new biological knowledge. Methods used for this analysis are found in Chapter 15 with an example of the final gene expression matrix. However, all the data produced up to this point need to be stored and curated in a way to allow future analysis. This has resulted in new database systems and ways of databasing biological experiments.

14.6 Microarray standards and databases

Array experiments require the ability to manage large amounts of data both before and after an array experiment. In addition, for array experiments from different people to be comparable, there needs to be consistency in experimental design and recording of experimental

		Sample				
		1	2	3	4	5
Log_2 ratios	Gene A	0	1	2	3	4
	Gene B	3.32	4.32	5.32	6.32	7.32
Median value	Gene A			2		
	Gene B			5.32		
Median Centered	Gene A	−2	−1	0	1	2
	Gene B	−2	−1	0	1	2

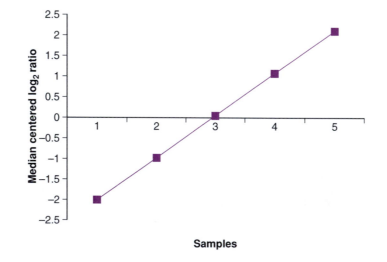

Figure 14.7

Data from Figure 14.7 median centered for both genes and plotted. The median gene expression value is now zero. Median centering removes the effect of the reference RNA having 10 times greater gene A expression compared to gene B. After median centering the log_2 ratios for genes A and B are the same reflecting the observed fact that the raw Cy5 intensities both gene A and B are expressed at the same level in each sample.

and array details. An international consortium of array groups has defined a Minimum Information About a Microarray Experiment (MIAME) that provides a framework for defining the type of data that should be stored and dividing the large amount of array data types into defined groups for database implementation. This allows microarray labs to design their own databases, which will be compatible with ArrayExpress, the proposed international microarray database.

There are six parts to MIAME:

1. Experimental design: the set of array hybridizations that comprise the whole experiment.
2. Array design: the description of the array used in the experiment, for example spots on the array.
3. Samples: the samples used and the extraction and labeling of the RNA.
4. Hybridization: procedures and parameters of each hybridized array.
5. Measurements: images and spot quantifications.
6. Normalization: types and specifications.

Each of these parts is defined in great detail with the type of information that should be included. For example, a table called Experiment within the Experimental design section should include:

Field Name	Field type	Field properties
experiment_id	NUMBER	NOT NULL
name	VARCHAR2	NOT NULL
experiment_type	VARCHAR2	
experimental_factors	VARCHAR2	
ordered	CHAR	
serial	CHAR	
series_type	VARCHAR2	
grouping	CHAR	
grouping_type	VARCHAR2	
experiment_date	DATE	
submission_date	DATE	
release_date	DATE	
description	VARCHAR2	
publication	VARCHAR2	

where CHAR, VARCHAR, DATE and NUMBER represent database field types. Each field type is further defined by MIAME, for example, experimental factors are further defined below.

Experimental factors, i.e. parameters or conditions tested, for instance:

1. time
2. dose
3. genetic variation
4. response to a treatment or compound.

All information on MIAME and other aspects of array data management and databasing can be found at http://www.mged.org/ and http://www.ebi.ac.uk/microarray/index.html.

Having now described the underlying methods, pre-processing and data storage for microarray experiments we can now consider more advanced array analysis methods and how you can turn the large investment in producing gene expression data into biological knowledge.

References

Brazma, A., Hingamp, P., Quackenbush, J. *et al.* (2001) Minimum information about a microarray experiment (MIAME) – toward standards fro microarray data. *Nature Genet* **29**: 365–371.

Kitano, H. (2002) Systems biology: a brief overview. *Science* **295**: 1662–1664.

Quackenbush, J. (2001) Computational genetics: computational analysis of microarray data. *Nature Review Genet* **2**: 418–427.

The Chipping Forecast. *Nature Genet* 1999; **21**: 1–60. http://www.nature.com/ng/chips_interstitial.html.

Yang, Y.H., Dudoit, S., Luu, P., Lin, D.M., Peng, V., Ngai, J., and Speed, T.P. (2002) Normalization for cDNA microarray data: a robust composite method addressing single and multiple slide systematic variation. *Nucleic Acids Res* **30**(4): e15.

Mining gene expression data

Xiaohui Liu and Paul Kellam

- Key issues in data mining are discussed in the context of modern developments in computing and statistics.
- *Data mining* methods that have been commonly used in gene expression analysis are presented with simple examples.
- Current and future research issues are outlined.

15.1 Concepts

DNA microarray technology has enabled biologists to study all the genes within an entire organism to obtain a global view of gene interaction and regulation. This technology has great potential in obtaining a deep understanding of biological processes. However, to realize such potential requires the use of advanced computational methods. In the last chapter, experimental protocols and technical challenges associated with the collection of gene expression data were presented, the nature of the data was highlighted, and the data standards for storing expression data were discussed. In this chapter, we shall introduce some of the most common data mining methods that are being applied to the analysis of microarray data and discuss the likely future directions.

Data mining has been defined as the process of discovering knowledge or patterns hidden in (often large) datasets. The data mining process is often semi-automatic and interactive, but there is a great desire for it to be automatic. The knowledge or patterns discovered should be meaningful and interesting from a particular point of view. For example, loyalty patterns of previous customers should allow a company to detect among the existing customers those likely to defect to its competitors. In the context of DNA microarray data, the patterns detected should inform us of the likely functions of genes, how genes may be regulated, and how they interact with each other in health and disease process. A comprehensive treatment of various issues involved in building data mining systems can be found in Hand (2001).

15.1.1 Data analysis

The job of a data analyst typically involves problem formulation, advice on data collection (though it is not uncommon for the analyst to be asked to analyze data which have already been collected, especially in the context of data mining), effective data analysis, and interpretation and reporting of the findings. Data analysis is about the extraction of useful information from data and is often performed by an iterative process in which 'exploratory analysis' and 'confirmatory analysis' are the two principal components.

Exploratory data analysis, or data exploration, resemble the job of a detective: understanding the evidence collected, looking for clues, applying relevant background knowledge and pursuing and checking the possibilities that clues suggest. This initial phase of data examination typically involves a number of steps, namely:

1. *Data quality checking*, processes for determining the levels and types of noise (random variation) in the system, processes for dealing with extreme outlying values and missing values.
2. *Data modification*, processes for transforming data values (such as the $\log_2$ transformation seen in Chapter 14) and processes for error correction.
3. *Data summary*, defined methods for producing tables and visualization of summary statistics for the data.
4. *Data dimensionality reduction*, methods for reducing the high dimensionality of the data to allow patterns to be identified or other data mining tools to be used.
5. *Feature selection and extraction* methods that allow defining features of the data to be identified and significance values applied to such features.
6. *Clustering methods*, such as hierarchical clustering, which identify potential structures in the data that share certain properties.

Data exploration is not only useful for data understanding, but also helpful in generating possibly interesting hypotheses for further investigation, a more 'confirmatory' procedure for analyzing data. Such procedures often assume a potential model structure for the data, and may involve estimating the model parameters and testing hypotheses about the model.

15.1.2   New challenges and opportunities

Over the last 15 years, we have witnessed two phenomena which have affected the work of modern data analysts more than any others. Firstly, the size of machine-readable data sets has increased and the problem of 'data explosion' has become apparent. Many analysts no longer have the luxury of focusing on problems with a manageable number of variables and cases (say a dozen variables and a few thousand cases); and problems involving thousands of variables and millions of cases are quite common, as is evidenced in biology throughout this book.

Secondly, data analysts, armed with traditional statistical packages for data exploration, model building, and hypothesis testing, now require more advanced analysis capabilities. Computational methods for extracting information from large quantities of data, or '*data mining*' methods, are maturing, e.g. artificial neural networks, Bayesian networks, decision trees, genetic algorithms, statistical pattern recognition, support vector machines, and visualization. This combined with improvements in computing such as increased computer processing power and larger data storage device capacity both at cheaper costs, coupled with networks providing more bandwidth and higher reliability, and On-Line Analytic Processing (OLAP) all allow vastly improved analysis capabilities.

These improvements are timely as functional genomics data provide us with a problem where traditional mathematical or computational methods are not really suited for modeling gene expression data: such data are noisy and high-dimensional with up to thousands of variables (genes) but a very limited number of observations (experiments).

In this chapter we will present some of the most commonly used methods for gene expression data exploration, including hierarchical *clustering*, K-means, and self-organizing maps (SOM). Also, support vector machines (SVM) have become popular for classifying expression data, and we will describe the basic concepts of SVM.

15.2   Data mining methods for gene expression analysis

To date two main categories of data mining methods have been used to analyze gene expression data: clustering and classification. *Clustering* is about the organization of a collection of unlabeled patterns (data vectors) into clusters based on similarity, so that patterns within the same cluster are more similar to each other than they are to a pattern belonging to a different cluster. It is important for exploratory data analysis since it will often reveal interesting structures of the data, thereby allowing formulation of useful hypotheses to test. In the context of gene expression data analysis, clustering methods have been used to find clusters of co-expressed/co-regulated genes, which can be used to distinguish between diseases that a standard pathology might find it difficult to tell apart as reported by Alizadeh and colleagues in *Nature* (2000).

As early DNA microarray experiments have shown that genes of similar function yield similar expression patterns, work on the use of *classification* techniques to assign genes to functions of a known class has increased. Instead of learning functional classifications of genes in an unsupervised fashion like clustering, classification techniques start with a collection of labeled (pre-classified) expression patterns; the goal being to learn a classification model that will be able to classify a new expression pattern. Classification has also been extensively used to distinguish (classify) different samples, for example, breast cancer samples into distinct groups based on their gene expression patterns. We will now explore these methods in more detail.

15.3   Clustering

Given a gene expression matrix, clustering methods are often used as the first stage in data analysis. The goal of *clustering* here is to group sets of genes together that share similar gene expression patterns across a dataset.

There are three essential steps in a typical pattern clustering activity:

1. *Pattern representation* refers to things such as the number of available patterns, the number of features available to the clustering algorithm, and for some clustering methods the desired number of clusters. Suppose we have a gene expression data set with 1000 genes, whose expression levels have been measured for 10 time points or 10 different samples. In this case, we would have 1000 available patterns and 10 features available for each pattern. One clustering aim would therefore be to see how we could group these 1000 patterns using these 10 features into a number of *natural* clusters.

2. *Pattern proximity* is usually measured by a *distance (dissimilarity) metric* or a similarity function defined on pairs of patterns. The Euclidean distance is probably the most commonly used dissimilarity measure in the context of gene expression profiling. On the other hand, a similarity measure such as Pearson's correlation or Spearman's rank correlation is also used. For a variety of commonly used proximity measures see *Box 15.1*. Other distance metrics such as Mahalanobis distance, Chebyshev distance and Canberra metric are described in Webb (1999).

3. *Pattern grouping* or clustering methods may be grouped into the following two categories: hierarchical and non-hierarchical clustering (Jain, 1999). A hierarchical clustering procedure involves the construction of a hierarchy or tree-like structure, which is basically a nested sequence of partitions, while non-hierarchical or partitional procedures end up with a particular number of clusters at a single step. Commonly used non-hierarchical clustering *algorithms* include the K-means algorithm, graph-theoretic approaches via the use of minimum spanning trees, evolutionary clustering algorithms, simulated annealing based methods as well as competitive neural networks such as self-organizing maps.

To explain how clustering methods work we will use an artificial dataset of simple numbers represented as a distance matrix (*Table 15.1*). In addition, we will use a simple eight-gene and six-sample log$_2$ ratio gene expression matrix as described in Chapter 14 (*Figure 15.1a*) to show how clustering methods may work on DNA array data. This gene expression matrix contains three easily identifiable gene expression patterns relative to the reference mRNA source namely: (i) non changing; (ii) over-expressed to under-expressed; and (iii) under-expressed to over-expressed (*Figure 15.1b*).

Box 15.1 Measures of dissimilarity (distance) between two individual patterns

A gene expression pattern, $\mathbf{x}$, is represented by a vector of measurements $[x_1, x_2, ..., x_p]$ (expression levels at p different time points). Many of the clustering techniques require some measure of dissimilarity or distance between $\mathbf{x}$ and $\mathbf{y}$, the two p-dimensional pattern vectors, so it is useful to summarize several commonly used distances.

Euclidean distance

$$d_e = \sqrt{\sum_{i=1}^{p} (x_i - y_i)^2}$$

Euclidean distance appears to be the most commonly used measure for a variety of applications, including gene expression profiling. Nevertheless it does have the undesirable property of giving greater emphasis to larger differences on a single variable.

City-Block or Manhattan distance

$$d_{cd} = \sum_{i=1}^{p} |x_i - y_i|$$

This metric uses a distance measure that would be suitable for finding the distances between points in a city consisting of a grid of intersecting thoroughfares (hence the name). It is a little cheaper computationally than Euclidean distance, so it may be a good candidate for applications demanding speedy solutions.

Minkowski distance

$$d_m = \left(\sum_{i=1}^{p} |x_i - y_i|^m \right)^{1/m}$$

The Minkowski distance is a general form of the Euclidean and city block distances: the Minkowski distance of the first order is the city-block distance, while the Minkowski distance of the second order is the same as the Euclidean distance. The choice of m depends on the amount of emphasis placed on the larger differences $|x_i - y_i|$.

Table 15.1 A dissimilarity matrix for five individuals

	1	2	3	4	5
1	0	1	5	9	8
2		0	4	8	7
3			0	3	4
4				0	2
5					0

15.3.1 Hierarchical clustering

The concept of the hierarchical representation of a data set was primarily developed in biology. Therefore it is no surprise that many biologists have found it easier to use hierarchical methods to cluster gene expression data. Basically a hierarchical algorithm will produce a

(a)

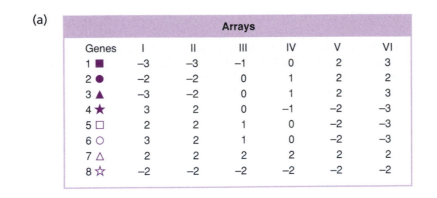

Arrays						
Genes	I	II	III	IV	V	VI
1 ■	−3	−3	−1	0	2	3
2 ●	−2	−2	0	1	2	2
3 ▲	−3	−2	0	1	2	3
4 ★	3	2	0	−1	−2	−3
5 □	2	2	1	0	−2	−3
6 ○	3	2	1	0	−2	−3
7 △	2	2	2	2	2	2
8 ☆	−2	−2	−2	−2	−2	−2

(b)

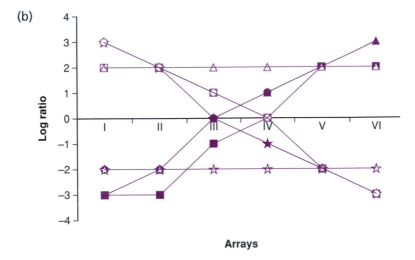

Figure 15.1

(a) Gene names are in column 1 (gene 1 to 8) and each gene is coded per row, with respect to (b). Arrays (I to VI) are in columns (2–7) with each feature being a gene expression $\log_2$ ratio value (see Chapter 14). (b) A graphical representation of the data for eight genes measured across six arrays corresponding the gene matrix in (a).

hierarchical tree or *dendrogram* representing a nested set of partitions. Sectioning the tree at a particular level leads to a partition with a number of disjoint clusters. A numerical value, essentially a measure of the distance between two merged clusters, is associated with each position up the tree where branches join.

There are two main classes of *algorithm* for producing a hierarchical tree. An *agglomerative* algorithm begins with placing each of the *n* available gene expression values across a set of arrays (patterns, also known as the gene expression vector) into an individual cluster. The algorithm proceeds to merge the two most similar groups to form a new cluster, thereby reducing the number of clusters by one. The algorithm continues until all the data fall within a single cluster. A *divisive* algorithm works the other way around: it operates by successively dividing groups beginning with a single group containing all the patterns, and continuing until there are *n* groups, each of a single individual. Generally speaking, divisive algorithms are computationally less efficient.

A typical hierarchical agglomerative clustering algorithm is outlined below:

1. Place each pattern (gene expression vector) in a separate cluster.

2. Compute the proximity matrix of all the inter-pattern distances for all distinct pairs of patterns.
3. Find the most similar pair of clusters using the matrix. Merge these two clusters into one, decrement number of clusters by one and update the proximity matrix to reflect this merge operation.
4. If all patterns are in one cluster, stop. Otherwise, go to the above step 2.

The output of such an algorithm is a nested hierarchy of trees that can be cut at a desired dissimilarity level forming a partition. Hierarchical agglomerative clustering algorithms differ primarily in the way they measure the distance or similarity of two clusters where a cluster may consist of only a single object at a time. The most commonly used inter-cluster measures are:

$$d_{AB} = \min_{\substack{i \in A \\ j \in B}} (d_{ij}) \tag{15.1}$$

$$d_{AB} = \max_{\substack{i \in A \\ j \in B}} (d_{ij}) \tag{15.2}$$

$$d_{AB} = \frac{1}{n_A n_B} \sum_{i \in A} \sum_{j \in B} d_{ij} \tag{15.3}$$

where d_{AB} is the dissimilarity between two clusters A and B, d_{ij} is the dissimilarity between two individual patterns i and j, n_A and n_B are the number of individuals in clusters A and B, respectively. These three inter-cluster dissimilarity measures are the basis of three of the most popular hierarchical clustering algorithms. The *single-linkage* algorithm uses Equation 15.1: the *minimum* of the distances between all pairs of patterns drawn from the two clusters (one pattern from each cluster). The *complete-linkage* algorithm uses Equation 15.2: the *maximum* of all pairwise distances between patterns in the two clusters. The *group-average* algorithm uses Equation 15.3: the average of the distances between all pairs of individuals that are made up of one individual from each cluster. To see how these algorithms work, we shall apply them to the artificial data set first, and then to the illustrative gene expression data set.

15.3.1.1 Single linkage clustering

From *Table 15.1* it is clear that the closest two clusters (which contain a single object each at this stage) are those containing the individuals 1 and 2. These are merged to form a new cluster {1,2}. The distances between this new cluster and the three remaining clusters are calculated according to equation 15.1 as follows:

$$d_{(12)3} = \min(d_{13}, d_{23}) = 4$$
$$d_{(12)4} = \min(d_{14}, d_{24}) = 8$$
$$d_{(12)5} = \min(d_{15}, d_{25}) = 7$$

This gives the new dissimilarity matrix:

	{1,2}	3	4	5
{1,2}	0	4	8	7
3		0	3	4
4			0	2
5				0

The closest two clusters are now those containing individuals 4 and 5, so these are merged to form a new cluster {4,5}. We now have three clusters: {1,2}, 3, {4,5}. The distances between the new cluster and the other two clusters are then calculated:

$$d_{(45)3} = min(d_{43}, d_{53}) = 3$$

$$d_{(45)(12)} = min(d_{14}, d_{24}, d_{15}, d_{25}) = 7$$

These lead to the following new dissimilarity matrix:

	{1,2}	3	{4,5}
{1,2}	0	4	7
3		0	3
{4,5}			0

The closest two clusters are those containing 3 and {4,5}, so these are merged to form a new cluster {3,4,5}. The distance between the new cluster and the other remaining cluster {1,2} is calculated to be 4, which results in the following new dissimilarity matrix:

	{1,2}	{3,4,5}
{1,2}	0	4
{3,4,5}		0

Finally, merge of the two clusters at this stage will form a single group containing all five individuals. The dendrogram illustrating this series of mergers is given in *Figure 15.2a*.

15.3.1.2 Complete linkage clustering

As with single linkage this clustering approach begins by merging clusters containing individuals 1 and 2 into a new cluster {1,2}. The dissimilarities between this cluster and the three remaining clusters are calculated according to equation 15.2 as follows:

$$d_{(12)3} = max(d_{13}, d_{23}) = 5$$

$$d_{(12)4} = max(d_{14}, d_{24}) = 9$$

$$d_{(12)5} = max(d_{15}, d_{25}) = 8$$

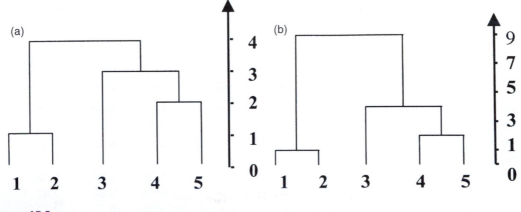

Figure 15.2

Dendrograms from single-linkage (a) and complete-linkage clustering (b)

This gives the different dissimilarity matrix:

	{1,2}	3	4	5
{1,2}	0	5	9	8
3		0	3	4
4			0	2
5				0

The similar process takes place as described above for the single linkage algorithm. The final complete-link dendrogram is shown in *Figure 15.2b*. In this example the tree structure is the same regardless of the linkage method used. The readers are encouraged to apply the group-average algorithm to this data set and work out the corresponding dendrogram.

15.3.1.3 Clustering gene expression data

Here we would like to show how clustering algorithms such as single-linkage may be applied to the gene expression data set as described in *Figures 15.1a,b*. First, we calculate the Euclidean distances between each gene pair using the gene expression profiles in *Figure 15.1a* as follows:

Gene Pair	Array I	Array II	Array III	Array IV	Array V	Array VI	Σ of values	$\sqrt{}$ of Σ = distance (d_e)
(gene 1–2)^2	1	1	1	1	0	1	5	2.36
(gene 2–3)^2	1	0	0	0	0	1	2	1.41
(gene 3–4)^2	36	16	0	4	16	36	108	10.39
•								
•								
(gene 1–3)^2	0	1	1	1	0	0	3	1.73
(gene 2–4)^2	25	16	0	4	16	25	88	9.27
•								
•								
(gene 1–7)^2	25	25	9	4	0	1	64	8
(gene 2–8)^2	0	0	4	9	16	16	45	6.7
(gene 1–8)^2	1	1	1	4	16	48	48	6.93

The corresponding distance matrix is then shown in *Table 15.2*.

Table 15.2 Gene expression distance metrix

	Gene 1	Gene 2	Gene 3	Gene 4	Gene 5	Gene 6	Gene 7	Gene 8
Gene 1	0	2.36	1.73	10.72	10.29	10.81	8.00	6.92
Gene 2		0	1.41	9.27	8.66	9.16	6.08	6.70
Gene 3			0	10.39	9.75	10.30	6.86	7.41
Gene 4				0	1.73	1.41	7.41	6.86
Gene 5					0	1.00	6.86	6.78
Gene 6						0	6.86	7.42
Gene 7							0	2.72
Gene 8								0

Now let us see how the single-linkage method may be applied to this case. The closest two clusters (which contain a single object each at this stage) are those containing genes 5 and 6. These are merged to form a new cluster {5, 6}. The distances between this new cluster and the six remaining clusters are calculated according to Equation 15.1 as follows:

$$d_{(56)1} = min(d_{15}, d_{16}) = \textbf{10.29}$$

$$d_{(56)2} = min(d_{25}, d_{26}) = \textbf{8.66}$$

. . .

$$d_{(56)8} = min(d_{58}, d_{68}) = \textbf{6.78}$$

This gives the new dissimilarity matrix below where genes 5 and 6 are merged and the distances recalculated as per the single linkage metric:

	Gene 1	Gene 2	Gene 3	Gene 4	Gene {5,6}	Gene 7	Gene 8
Gene 1	0	2.36	1.73	10.72	10.29	8.00	6.92
Gene 2		0	1.41	9.27	8.66	6.08	6.70
Gene 3			0	10.39	9.75	6.86	7.41
Gene 4				0	1.41	7.41	6.86
Gene {5,6}					0	6.86	6.78
Gene 7						0	9.80
Gene 8							0

The process of merging the two least dissimilar gene vectors and recalculating the distance matrix is continued until all genes are merged and plotted as a dendrogram.

Interpretation of the dendrogram in *Plate 10* is greatly aided by displaying the log_2 ratio gene expression values as a colored block diagram, where each row represents a gene across the columns of samples. The color of each row/column block (feature) is related to the actual log_2 ratio as shown by the color range at the bottom of *Figure 15.3*. The data are taken from *Figure 15.1a* and represent in shades of green and red the same information as plotted in *Figure 15.1b*. From this it is relatively easy to see the gene expression patterns that lead to the dendrogram structure derived from the single linkage of the gene expression Euclidean distances. Consistent with the known gene expression patterns in *Figure 15.1b* the eight genes split into three groups in the dendrogram.

A challenging issue with hierarchical clustering is how to decide the *optimal* partition from the hierarchy, i.e. what is the best number of groups? One approach is to select a partition that best fits the data in some sense, and there are many methods that have been suggested in the literature (Everitt, 1993). It has also been found that the single-link algorithm tends to exhibit the so-called *chaining* effect: it has a tendency to cluster together at a relatively low level objects linked by chains of intermediates. As such, the method is appropriate if one is looking for 'optimally' connected clusters rather than for homogeneous spherical clusters. The complete-link algorithm, on the other hand, tends to produce clusters that tightly bound or compact, and has been found to produce more useful hierarchies in many applications than the single-link algorithm (Jain, 1999). The group-average algorithm is also widely used. Detailed discussion and practical examples of how these algorithms work can be found in Jain (1999) and Webb (1999).

15.3.2 K-means

A non-hierarchical or partitional clustering algorithm produces a single partition of the data instead of a clustering structure such as the dendrogram. This type of algorithm has advantages in applications involving large data sets where the construction of dendrograms is computationally prohibitive. These algorithms usually obtain clusters by optimizing a criterion function, of which the square error is the most commonly used. K-means is the best-known partitional algorithm employing such a criterion where the square error for each cluster j, [j= 1,2, ...K], is the sum of the squared Euclidean distances between each pattern $x_i^{(j)}$ in the cluster j and its center, or mean vector, of the cluster, $m^{(j)}$

$$E_j = \sum_{i=1}^{n_j} d_{x,m^{(j)}}^2 \text{ where } m^{(j)} = \frac{\sum_{i=1}^{n_j} x_i^{(j)}}{n_j}$$
(15.4)

where E_j is referred to as the *within-cluster variation or sum of squares* for cluster j, n_j is the number of patterns within cluster j, and is the Euclidean distance from pattern x_i to the center of the cluster to which it is assigned. Therefore, the total square error for the entire clustering with K clusters is the sum of the within-cluster sum of squares:

$$E = \sum_{j=1}^{K} E_j$$
(15.5)

Basically the K-means algorithm starts with an initial partition with a fixed number of clusters and cluster centers, and proceeds with assigning each pattern to its closest cluster center so as to reduce the square error between them. This is repeated until convergence is achieved, e.g. there is no reassignment of any pattern from one cluster to another, or the squared error ceases to decrease significantly after certain number of iterations.

Here is a view of the K-means algorithm:

1. Choose K cluster centers to coincide with K randomly chosen patterns. Repeat the following steps until the cluster membership is stable.
2. Assign each pattern to its closest cluster center.
3. Compute the new cluster centers using the new cluster memberships.
4. Repeat the above two steps until a convergence criterion is satisfied.
5. If desirable, adjust the number of clusters by merging and splitting existing clusters.

It should be noted that the last step is not an essential part of the K-means algorithm, but it has been included in some of the most well-known systems, e.g, in ISODATA (Webb, 1999). The adjustment of the number of clusters may be desirable if certain conditions are met. For example, if a cluster has too many patterns and an unusually large variance along the feature with the largest spread, it is probably a good idea to split the cluster. On the other hand, if two cluster centers are sufficiently close, they should perhaps be merged.

Let us have a look at how this algorithm works with a simple example. Consider the two-dimensional data as shown in *Figure 15.3* where there are six patterns. Suppose we set K (the number of clusters) to 2 and choose two patterns from the data set as initial cluster center vectors. Those selected are points 5 and 6. We now apply the algorithm to the data set and allocate individuals to clusters A and B represented by the initial vectors 5 and 6, respectively. It is clear that individuals 1, 2, 3, 4, 5 are allocated to cluster A and individual 6 to cluster B. New centers (means) are then calculated with (2,2) for cluster A and (4,3) for cluster B, and the sum of within-cluster sum-of-squares is calculated using Equation 15.5, which gives 6:

$$[(1\text{-}2)^2 + (1\text{-}2)^2] + [(2\text{-}2)^2 + (1\text{-}2)^2] + [(2\text{-}2)^2 + (2\text{-}2)^2] + [(2\text{-}2)^2 + (3\text{-}2)^2] +$$
$$[(3\text{-}2)^2 + (3\text{-}2)^2] + [(4\text{-}4)^2 + (3\text{-}3)^2] = 2 + 1 + 0 + 1 + 2 + 0 = 6.$$

The results of this iteration are summarized in *Table 15.3* in Step 1. This process is then repeated, using the new cluster centers for re-grouping individuals. This leads to the assignment of individuals 1, 2, 3, and 4 to cluster A and 5 and 6 to cluster B. The sum of within-cluster sum-of-squares is now reduced to 4. A third iteration produces no change in the cluster membership or the within-cluster sum-of-squares. The algorithm terminates here.

The K-means algorithm is popular because it is easy to understand, easy to implement, and has a good time complexity. However, the algorithm can still take a considerable time if the number of patterns involved is very large and the number of clusters is substantial as with some of the large-scale gene expression applications. Another problem is that it is sensitive to the initial partition – the selection of the initial patterns, and may converge to a *local minimum* of the criterion function value if the initial partition is not properly chosen. A possible remedy is to run the algorithm with a number of different initial partitions. If they all lead to the same final partition, this implies that the global minimum of the square error has been achieved. However, this can be time-consuming, and may not always work.

15.3.3 Self-organizing maps

The self-organizing map (SOM) algorithm was proposed by Teuvo Kohonen in 1981. Apart from being used in a wide variety of fields, the SOM offers an interesting tool for exploratory data analysis, particularly for partitional clustering and visualization. It is capable of representing high-dimensional data in a low-dimensional space (often a two- or one-dimensional array) that preserves the structure of the original data.

A self-organizing network consists of a set of input nodes $V = \{v_1, v_2, ..., v_N\}$, a set of output nodes $C = \{c_1, c_2, ..., c_M\}$, a set of weight parameters $W = \{w_{11}, w_{12}, ..., w_{ij}, ..., w_{NM}\}$ $(1 \le i \le N, 1 \le$

Table 15.3 Summary of k-means iterations

	Cluster A		Cluster B		E
Step	Membership	Center	Membership	Center	
1	1,2,3,4,5	(2,2)	6	(4,3)	6
2	1,2,3,4	(7/4, 7/4)	5,6	(3.5,3)	4
3	1,2,3,4	(7/4, 7/4)	5,6	(3.5,3)	4

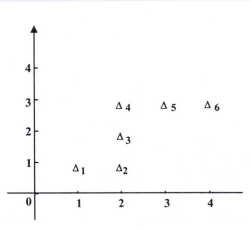

Figure 15.3

Sample data for K-Means

$j \leq M$, $0 \leq w_{ij} \leq 1$), and a map topology that defines the distances between any given two output nodes. The input nodes represent an N-dimensional vector, e.g. a single gene's expression level at N time points. *Figure 15.5* is an illustration of how the gene expression example in *Figure 15.1b* may be analyzed using SOM. In this case, each gene is expressed at six time points, therefore there are six input nodes. The output nodes in SOM are usually arranged in a 2-dimensional array to form a 'map'. In this illustrative example, there are 20 output nodes on the map although the most appropriate number of output nodes usually needs to be determined via experimentation.

Each input node is fully connected to every output node via a variable connection. A weight parameter is associated with each of these connections, and the weights between the input nodes and output nodes are iteratively changed during the learning phase until a termination criterion is satisfied. For each input vector v, there is one associated *winner node* on the output map. A winner node is an output node that has minimum distance to the input vector. This can be seen for the output map for winner node corresponding to gene 1 in *Plate 11*.

Here is the sketch of the SOM algorithm:

1. Initialize the topology and size of the output map (M).
2. Initialize the connection weights to random values over the interval [0, 1] and normalize both the input vectors and the connected weight vectors. Initialize the gain value η (the learning rate) and the neighborhood size r.
3. Repeat until convergence:
   - Present a new input vector **v**.
   - Calculate the Euclidean distance between the input vector and each node on the output map, and designate the node with the minimum distance as the winner node c.

$$c = \min \sqrt{\sum_{k=1}^{N} \left(v_k - w_{jk} \right)^2}, \, j = 1,2,...,N \qquad (15.6)$$

   - Update weights W, learning rate η and N_c, the neighborhood surrounding the winner node c, in such a way that the vectors represented by output nodes are similar if they are located in a small neighborhood. For each node $j \in N_c$, perform the following:

$$w_j^{(new)} = w_j^{(old)} + \eta \, [v_i - w_j^{(old)}] \qquad (15.7)$$

   - Decrease both the neighborhood size and the learning rate.

The neighborhood set N_c is a set of nodes that surrounds the winner node c. These nodes in N_c are affected with weight modifications, apart from those changes made to the winner node, as defined in the algorithm. These weight changes are made to increase the matching between the nodes in N_c and the corresponding input vectors. As the update of weight parameters proceeds, the size of the neighborhood set is slowly decreased to a predefined limit, for instance, a single node. This process leads to one of the most important properties of SOM that similar input vectors are mapped to geometrically close winner nodes on the output map. This is called *neighborhood preservations*, which has turned out to be very useful for clustering similar data patterns.

Going back to *Plate 10*, we can see that the output map status at time t and the final map when the computational process has ended. It becomes clear that there are three distinctive clusters on the output map: {1,2,3}, {4,5,6} and {7,8}.

It should be noted that for illustrative purposes, each gene profile in this example has been mapped onto a distinctive node on the output map. This is not always the case. As long as two gene profile vectors are sufficiently similar, they may be mapped onto the same node on the output map.

It is not always straightforward to visually inspect the projected data on the two-dimensional output map in order to decide the number and size of natural clusters. Therefore, careful analysis or post-processing of output maps is crucial to the partition of the original data set. Like the K-means algorithm, the SOM produces a sub-optimal partition if the initial weights are not chosen properly. Moreover, its convergence is controlled by various parameters such as the learning rate, the size and shape of the neighborhood in which learning takes place. Consequently, the algorithm may not be very *stable* in that a particular input pattern may produce different winner nodes on the output map at different iterations.

15.3.4 Discussion

Different clustering algorithms may produce different clusters from the same data set. Some may fail to detect obvious clusters, while many will always be able to produce clusters even when there is no natural grouping in the data. This is because each algorithm implicitly forces a structure on the given data. Thus it is extremely important to emphasize that cluster analysis is usually one of the first steps in an analysis. Accepting the results of a cluster algorithm without careful analysis and follow-up study can lead to misleading results. The validation of clustering results is difficult. There is no optimal strategy for cluster interpretation, although a few formal approaches have been proposed such as the use of 'measures of distortion' and the use of 'internal criteria measures' to determine the number of clusters (Webb, 1999). The application scope of these methods is rather limited. Relevant domain knowledge such as biological information about the genes and associated proteins in a cluster can help check the consistency between the clustering findings and what was already known and remains one of the most important ways of validating the results.

Hierarchical clustering, K-means and SOM are probably the most commonly used clustering methods for gene expression analysis. By no means should they be considered the only ones worth using. The global search and optimization methods such as genetic algorithms or simulated annealing can find the optimal solution to the square error criterion, and have already demonstrated certain advantages. Finally, fuzzy clustering may have much to offer as well. Since a gene may have multiple functional roles in various pathways, it can in effect be the member of more than one cluster.

15.4 Classification

The task of *classification* exists in a wide variety of domains. In medical diagnosis, there may exist many cases that correspond to several diseases, together with their associated symptoms. We would like to know whether there are any diagnostic rules underlying these cases that would allow us to distinguish between these diseases. In credit scoring, a company may use information regarding thousands of existing customers to decide whether any new customer should be given a particular credit status. This information can include a variety of financial and personal data. In the gene expression settings, we might want to predict the unknown functions of certain genes, given a group of genes whose functional classes are already known. In all these examples, there are the following common themes. First, we have a set of *cases*, e.g. patients, customers, or genes. Second, we are interested in assigning each new case to one of the pre-defined *classes*, e.g. different disease categories, credit status, or gene function classes. Third, there is a set of *features* on which the classification decision is based, e.g. a set of symptoms, personal or financial information, or gene expression level at different time points. Essentially we are interested in how to construct a classification procedure from a set of cases whose classes are known so that such a procedure can be used to classify new cases.

The objective of learning classification models from sample data is to be able to classify new data successfully, with the corresponding error rates as low as possible. Often one would want to split all the available data randomly into two groups: one for designing the classification model, and the other for testing the model accuracy. Traditionally for a single application of the training-and-testing method, the proportions of the data used for training and testing are approximately a 2/3, 1/3 split. However, if only a small number of cases are available, such a split would have problems since this will leave one with either insufficient training or testing cases. A more suitable way of dealing with such situations would be the use of *resampling* techniques such as *cross-validation* or *bootstrapping*. Take the k-fold cross-validation as an example; the cases are randomly partitioned into k mutually exclusive test partitions of roughly equal size. For each of the k test partitions, all cases not found in this partition are used for training, and the corresponding classifier is tested on this test partition. The average error rates over all k partitions are then the cross-validated error rate. In this way, we have made the full use of the limited number of cases, for testing as well as training. The 10-fold cross-validation method appears to be one of the most commonly used.

There have been many classification methods proposed in the statistical, artificial intelligence and pattern recognition communities. Commonly used methods include Bayesian classifiers, linear discriminant analysis, nearest neighbor classification, classification tree, regression tree, neural networks, genetic algorithms, and more recently, support vector machines. Detailed discussions of most of these topics may be found in Hand (2001) and Webb (1999).

15.4.1 Support vector machines (SVM)

Over the last few years, the SVM has been established as one of the preferred approaches to many problems in pattern recognition and regression estimation, including handwritten digit recognition, object recognition, speaker identification, face detection, text mining, time series predictions, and gene expression classification. In most cases, SVM generalization performances have been found to be either equal to or much better than that of the conventional methods. This is largely due to the SVM's ability to achieve the right balance between the accuracy attained on a particular training set, and the *capacity* of the learning machine, that is, the ability of the machine to learn any training set without error. Statistical learning theory (Vapnik, 1995) shows that it is crucial to restrict the class of functions that the learning machine can implement to one with a suitable capacity for the amount of training data available.

Suppose we are interested in finding out how to separate a set of training data vectors that belong to two distinct classes. If the data are separable in the input space, there may exist many hyperplanes that can do such a separation, but we are interested in finding the optimal hyperplane classifier – the one with the maximal margin of separation between the two classes. Without going into full details the maximal margin of separation can be uniquely constructed by solving a constrained quadratic optimization problem involving *support vectors*, a small subset of patterns that lie on the margin. The support vectors, often just a small percentage of the total number of training patterns, contain all relevant information about the classification problem. For the illustration of these concepts a simple partitioning of a data set into '+' and '−' classes is shown in *Figure 15.4*. The corresponding linear SVMs can be easily constructed to separate the two classes.

When the two classes are not separable in the input space, the support vector machine can be extended to derive non-linear decision rules using so-called *kernel functions*. Kernel functions in this context are meant to be those that map the input data into some feature space. The key idea is to map the training data nonlinearly into a high-dimensional feature space, and then construct a (linear) separating hyperplane with maximum margin there. This yields a non-linear decision boundary in input space the concept of which is shown in *Figure 15.5*. The development of SVM began with the consideration of two classes problems, but has recently been extended to multi-class problems.

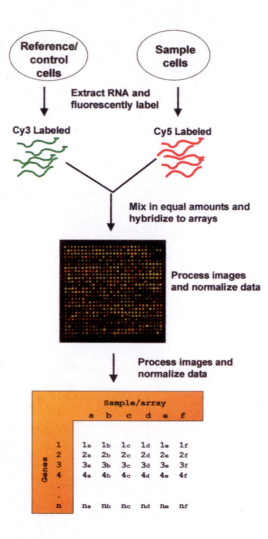

Plate 8

Microarray process from labeling of mRNA, through hybridization, to production of the gene expression matrix. The matrix is essentially a large table with genes listed in rows and samples/arrays listed in columns. The values correspond to each gene's expression level in each sample. (See Chapter 14, Section 14.3.1)

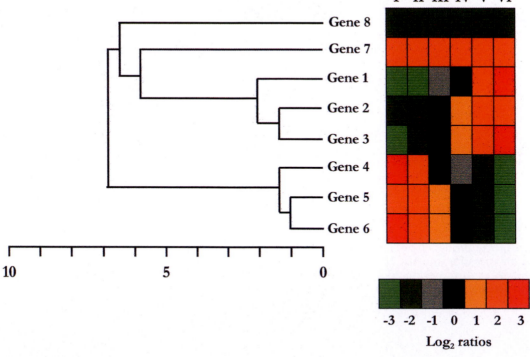

Expression values

Cy5	5000	10000	6000	20000	15000	10000	16000
Cy3	40000	40000	12000	20000	7500	2500	2000
Ratio	1/8	1/4	1/2	1	2/1	4/1	8/1
Log2 Ratio	-3	-2	-1	0	1	2	3

Plate 9

Representation of gene expression measurements from the 'false color scanner images, raw gene expression intensity measurements, ratio of Cy5/Cy3 and $\log_2$ ratios. (See Chapter 14, Section 14.4)

Plate 10

The Gene Expression dendrogram. (See Chapter 15, Section 15.3.1.3)

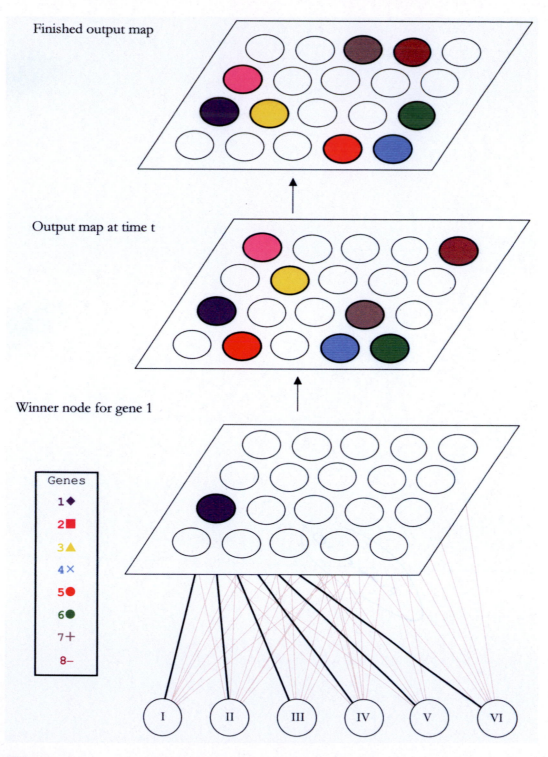

Finished output map

Output map at time t

Winner node for gene 1

Genes
1 ◆
2 ■
3 ▲
4 ✕
5 ●
6 ●
7 +
8 –

I   II   III   IV   V   VI

Plate 11

Self Organizing Map. (See Chapter 15, Section 15.3.3)

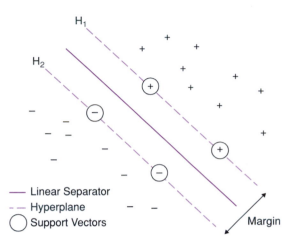

Figure 15.4

Hyperplanes and support vectors

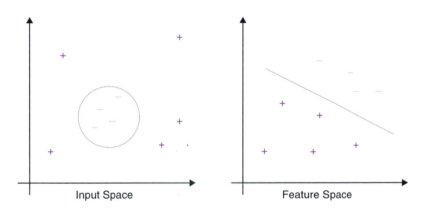

Figure 15.5

Non-linear mapping

Support vector machines offer a novel approach to classification. SVM training always finds a global minimum, while its neural network competitors may get stuck with a local minimum. It has simple geometric interpretation and it can be analyzed theoretically using concepts from statistical learning theory. It has also found much success with many practical problems. However, it should be noted that currently there is no established theory that can guarantee that a given family of SVMs will have high accuracy on a given problem. SVM is very much characterized by the selection of its kernel, and it is not always easy to choose a kernel function that is most suited to the classification problem in hand. Computational efficiency is also a concern for large-scale applications. A detailed introduction of the SVM can be found in Cristianini (2000).

15.4.2 Discussion

Classification, or discriminant analysis, is a well-researched subject with a vast literature. However, it is not always easy to decide the most suitable method for a classification task in hand. Given that most of these methods are included in many of the commonly used statistical or data mining packages, performance comparison of these methods on a given problem has been made easier. It is perhaps useful to point out that predictive accuracy on unseen test data is not the only, or even the most important, criterion for assessing the performance of a classification system. Other important factors in deciding that one method is preferable to another include computational efficiency, misclassification cost, and the interpretability of the classification models. Much more work is needed to study the relative strengths and weaknesses of the various methods.

15.5   Conclusion and future research

While there are many statistical, data mining, or new gene expression data analysis tools that can be relatively easily used for clustering, classification and visualization, we will witness very intensive research effort on the construction of networks for understanding gene interaction and regulation. Following the clustering of potentially co-expressed genes, it is desirable to develop a computational model that uses the expression data of these genes and relevant biological knowledge to understand interactions and regulation of these genes, for example, how the proteins of the expressed genes in a cluster interact in a functional module. Although the work in this relatively new area of functional genomics, *network modeling*, has begun more recently, a variety of approaches have already been proposed. These range from the Boolean network to Bayesian network, and from time series modeling to the learning of such networks from the literature. The development of most of these approaches is still at an early stage, and it is often very difficult to compare them (D'haeseleer, 2000).

Major breakthrough in this area will not come easily. This will require a truly interdisciplinary effort in which experimentalists, bioinformaticians, computer scientists and mathematicians work closely together. A much closer integration of network modeling with other data pre-processing activities as well as with the experimental design of microarrays will be necessary. It will also require much more experience in constructing and comparing gene networks using different methods for the same expression data. Last but not least, the timely and dynamic use of various kinds of relevant information such as genes, proteins, their relationships, diseases, patients, and clinical data, which may exist in various bioinformatics resources, medical databases, scientific literature, and a variety of web sites will greatly aid network modeling.

References and further reading

Cristianini, N., and Shawe-Taylor, J. (2000) *An Introduction to Support Vector Machines.* Cambridge University Press, Cambridge, UK.

D'haeseleer, Liang, S., and Somogyi, R. (2000) Genetic network inference: from co-expression clustering to reverse engineering. *Bioinformatics* **16**: 707–726.

Everitt, B.S. (1993) *Cluster Analysis.* 3rd edition. Arnold, London, UK.

Hand D.J., Mannila H., and Smyth P. (2001) *Principles of data mining.* MIT Press, Cambridge, MA.

Jain, A.K., and Dubes, R.C. (1999) Data clustering. *ACM Computing Survey* **31**: 264–323.

Vapnik, V.N. (1995) *The Nature of Statistical Learning Theory*, Springer-Verlag, New York.

Webb, A. (1999) *Statistical Pattern Recognition.* Arnold, London, UK.

Proteomics

16

Malcolm P. Weir,
Walter P. Blackstock and Richard M. Twyman

- Proteomics is an emerging scientific discipline encompassing any method for the large-scale analysis of proteins. Two main areas are recognized.
- *Expression proteomics* involves the separation, quantitation and characterization of large numbers of proteins. This is usually achieved by high-throughput 2-D-gel electrophoresis and mass spectrometry, but chip-based technology platforms may succeed these methods in the future. Expression proteomics is used to catalog the proteins expressed in a given cell type or tissue, and study differential protein expression.
- *Cell map proteomics* involves the large-scale analysis and characterization of protein interactions. Identifying the components and localization of protein complexes can help to determine functions and reconstruct cellular pathways and networks, providing a dynamic and holistic view of the cell.
- Bioinformatics is required to analyze and interpret the large datasets derived from proteomics experiments and is also important in the provision of databases. Specific algorithms and statistical tests are also needed for some types of proteomic analyses, including the imaging and comparison of 2-D gels and the use of mass spectrometry data to query sequence databases.

16.1   The proteome

The entire complement of proteins produced in a cell or organism is termed the *proteome* and this may comprise tens or even hundreds of thousands of different proteins. In order to study the proteome, it has therefore been necessary to develop large-scale and high-throughput protein analysis techniques, and this essentially defines the scientific discipline known as *proteomics* (Blackstock and Wier, 1999; Pandey and Mann, 2000; Lee, 2001).

The proteome is derived from the *genome*, which is the entire complement of genetic information in the cell. However, for several reasons, the proteome is much more complex than the genome. First, each gene can potentially produce a number of different mRNAs (mechanisms include alternative splicing, RNA editing and the alternative use of different promoters and terminators), and each of these mRNAs can give rise to a separate protein. The proteins can then be processed or modified in a number of different ways. At each stage of the process, further diversity can be introduced, resulting in many more proteins than genes.

The *comprehensive proteome* may be an order of magnitude more complex than the genome, but it exists only as a concept. In real cells, only a subset of the comprehensive proteome is ever present due to the regulation of gene expression, protein synthesis and post-translational processing. Together, these regulatory processes control the abundance (steady state level) and stability (turnover rate) of different proteins and therefore define a *cellular proteome*. The exact nature of the cellular proteome depends on the cell type and its environment. All cells contain a *core proteome* comprising *housekeeping proteins* that are essential for survival, plus an additional set of proteins conferring specialized functions. It should also be noted that the cellular proteome is in a state of dynamic flux and can be perturbed by changes in cell state or the environment. The analysis techniques described in this chapter cannot monitor the proteome in real time, but can provide snapshots of the proteome under different sets of conditions.

16.2   Proteomics

16.2.1   Why study the proteome?

Proteins, not genes or mRNAs, are the functional agents of the genome. However, even where a complete genome sequence is available, very little information about the encoded proteins can be extracted. In many cases it is possible to determine the primary structure of a protein, but scant information can be learnt about how that protein is expressed, processed and recycled, where it is localized in the cell, how it interacts with other proteins and small molecules, and its exact function. Therefore, to gain full insight into the functioning of a cell it is necessary to study proteins directly.

16.2.2   How to study the proteome

Proteomics can be divided into two major subdisciplines, which we term expression proteomics and cell-map proteomics. The analysis of protein expression profiles (*expression proteomics*) can be used for large-scale protein characterization, essentially a cataloging process to identify all the proteins in the proteome or a subset thereof, or differential expression analysis, where alternative samples are compared. The latter approach has numerous applications, such as the identification of disease markers, potential drug targets, virulence factors, polymorphisms for genetic mapping and species determinants.

  Cell-map proteomics is the large-scale characterization of protein interactions, and includes methods for studying protein–protein interactions, methods for studying the interaction between proteins and small molecules (ligands) and cellular localization studies. The goal of cell-map proteomics is an integrated view of cellular processes at the protein level. Given the role of proteins as the primary effectors of biological function, such *virtual cells* will allow biologists to begin to navigate the enormous complexity that confronts them and enable better hypothesis generation and experimental design.

16.2.3   The role of bioinformatics in proteomics

Bioinformatics is an integral part of proteomics. In a general sense, bioinformatics is required for analysis, interpretation and storage of the large datasets produced in many proteomic experiments. For example, protein expression data may be converted into a distance matrix, which is then clustered to identify proteins with similar expression profiles. This is very similar in principle to microarray analysis. Proteomics data encompass all levels of protein structure, from primary sequence to tertiary and quaternary conformations. This requires various forms of analysis including sequence alignment, structural modeling and docking algorithms that allow small molecules to be matched to protein structures (Chapter 13). Bioinformatics also provides tools that are unique to proteomics, such as the algorithms used to query protein and DNA sequence databases using mass spectrometry data. These applications, and others, are discussed in more detail below.

16.3   Technology platforms in proteomics

The cellular proteome contains tens of thousands of proteins differing in abundance over six orders of magnitude. Essential components of proteomic research therefore include the

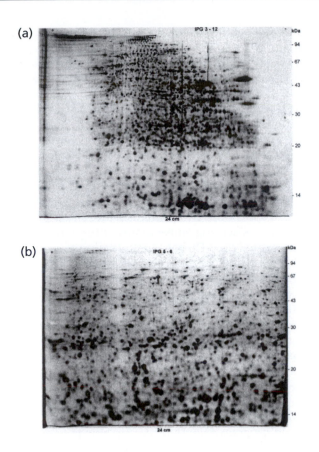

Figure 16.1

(a) Wide IPG with extended separation distance. Analytical IPG-DALT of mouse liver proteins (Lysis buffer extract). First dimension: IPG 3–12. Separation distance: 240 mm. Sample application: cup-loading near the anode. Running conditions: 6 h at 3500 V max. Second dimension: vertical SDS-PAGE (13% T constant). Silver stain. (b) Narrow IPG with extended separation distance. Analytical IPG-Dalt of mouse liver proteins (Lysis buffer extract). First dimension: IPG 5–6 (laboratory-made). Separation distance: 240 mm. Sample application by cup-loading at the anode. Running conditions: 40 h at 3500 V max. Second dimension: vertical SDS-PAGE (13% T constant). Silver stain. Reproduced from Proteomics: from protein sequence to function, Pennington and Dunn, BIOS Scientific Publishers, Oxford.

separation of complex protein mixtures (*protein separation technology*) and the characterization of individual proteins within such mixtures (*protein annotation technology*).

16.3.1 Protein separation technology

16.3.1.1 2-D gel electrophoresis

A suitable technology platform for protein separation was available as early as 1975 in the form of *2-D-gel electrophoresis* (2-DE) (Dunn and Gorg, 2001). The technique is simple in concept. A protein mixture is loaded onto a non-denaturing polyacrylamide gel and separated in the first dimension by *isoelectric focusing*, which means the proteins migrate in a pH gradient until they reach their *isoelectric point* (the position at which the pH is equal to their pI value and they have no charge). The gel is then equilibrated in the detergent sodium dodecylsulfate (which binds to the protein backbone and confers a uniform negative charge) and the proteins are separated in the second dimension according to their molecular mass. The gel is then stained and the proteins are revealed as a pattern of spots (*Figure 16.1*). Sensitive detection is required because there is no way to selectively amplify 'rare' proteins (rare mRNAs can be amplified using the polymerase chain reaction (PCR) but there is no analogous process in the protein world).

Although 2-DE is the most widely used method for protein separation in proteomics, it suffers from several limitations. These limitations involve resolution, representation, sensitivity and reproducibility. In the first experiments using 2-DE it proved possible to resolve over 1000 proteins from the bacterium *Escherichia coli*. The resolution of the technique has improved steadily, and it is now possible to separate 10,000 proteins in one experiment, although such skill is not readily transferable. However, even this may still only represent a fraction of the proteome. Individual spots often contain a mixture of proteins, which reflects the overlapping or merging of spots from two proteins with similar properties and large spots representing abundant proteins obscuring those representing rare proteins. Some classes of proteins tend not to be represented on 2-DE gels, including proteins with poor solubility and many membrane proteins, plus those present at less than 1000 copies per cell. A major disadvantage of 2-DE is that the technique is not easily reproducible, particularly across different laboratories, which makes it difficult to carry out comparative expression analysis and to replicate experiments in different settings.

Many of these gel-related technical difficulties have been overcome with the development of immobilized pH gradients for the first dimension, and reproducible commercial gels for the second dimension. Membrane proteins remain a challenge requiring case-by-case tuning of the detergents used. A variant of the standard 2-D-approach using different detergents in each dimension can be useful. Fluorescent dyes are supplanting colloidal Coomassie blue and silver staining and offer a more linear dynamic range for quantification. Powerful image analysis software capable of spot recognition and volumetric integration is now widely available (e.g. the freely available program MELANIE II) and this is one area in which bioinformatics plays an important role in proteomic research (Pleissner *et al.*, 2001). Where maximum resolution of proteins is required, narrow range pH gels can be used in the first dimension, and the overlapping images stitched together by computer.

With such a wide variation in protein expression level in tissue, cell lysate or serum, it is beneficial to remove abundant proteins. In serum, for example, albumin accounts for over half the protein content by weight, and many low-abundance proteins would not be quantified or identified unless this and other abundant proteins were removed. *Affinity depletion* methods are often used for serum. For cell lysate and tissue samples free flow electrophoresis and density gradient ultra-centrifugation may be used for the *prefractionation* of protein samples. In all such approaches, great care has to be taken to avoid artifacts.

16.3.1.2 Liquid chromatography methods

While 2-D-gels cannot yet be surpassed in terms of their ability to separate complex protein mixtures, a liquid-based method would offer easier automation and better quantification. Pre-fractionation also reduces sample complexity to a simpler mixture containing the target proteins, for which the separating power of 2-D-gels may not be necessary. Two orthogonal chromatography separation methods are typically employed, using combinations of size exclusion, ion exchange and reverse phase chromatography. These approaches can be combined into an integrated system controlled by software. Equally powerful, but less readily integrated, is the combination of free-flow electrophoresis and reverse phase liquid chromatography. Integrated systems are still in the development stage, particularly for limited sample amounts, but offer great promise, not least because they can be directly coupled to a mass spectrometer.

16.3.1.3 Affinity chromatography techniques for cell-map proteomics

The techniques described above are used in expression proteomics, where large numbers of proteins are studied in a single experiment. Cell-map proteomics typically involves the study of smaller groups of proteins, e.g. those that form a particular complex or those involved in a particular cellular pathway. The use of affinity chromatography or batch separation techniques, whether through attachment of *affinity tags* at the gene level, or through the use of specific antibodies directed towards particular proteins (*immunoprecipitation*), is the most powerful way of enriching samples in a directed fashion. Affinity purification enriches not only target proteins, but also any proteins associated with them in a non-covalent manner, and this can be exploited to isolate entire complexes and characterize interacting proteins. The method relies on a sequence determinant or post-translational modification (e.g. phosphotyrosine) in the protein, which forms a strong, specific interaction with an antibody or purification matrix. Such methods are capable of over 10,000-fold enrichment and can therefore be used to extract rare proteins in sufficient quantities for further analysis.

16.3.2 Protein annotation technology

Although technology for protein separation was available more than 25 years ago, a significant bottleneck was the subsequent process of protein identification (*protein annotation*) (Anderson and Anderson, 1998). The bottleneck existed because there were no high-throughput methods for characterizing proteins in 2-D gels and very few protein sequences available in databases for comparison. The database problem has been largely solved by the exponential growth in sequence information over the last 20 years, particularly data from genome sequencing projects and high-throughput single-pass sequencing of cDNA library clones to generate ESTs. The annotation problem itself was initially addressed by the development of methods for the direct sequencing of proteins immobilized on membranes, but the technique that has had the biggest impact on protein annotation in proteomics is mass spectrometry.

16.3.2.1 Protein annotation by mass spectrometry

Mass spectrometry (MS) is used to determine the accurate masses of molecules in a particular sample or analyte. A mass spectrometer has three components. The *ionizer* converts the analyte into gas phase ions and accelerates them towards the *mass analyzer*, which separates the ions according to their mass/charge ratio on their way to the *ion detector*, which records the impact of individual ions. Large molecules such as proteins and DNA are broken up by standard ionization procedures, but *soft ionization* has been developed to allow the ionization of such molecules without fragmentation. The soft ionization procedures used most commonly in proteomics – MALDI and ESI – are discussed briefly in *Box 16.1*. There are several different ways to annotate a protein by mass spectrometry (*Figure 16.2*). The most commonly applied procedures are:

- Peptide mass fingerprinting (PMF)
- Fragment ion searching
- *De novo* sequencing of peptide ladders.

In all cases, cleanliness is essential, as the background level of keratin in a typical laboratory is high. Clean-room working with laminar flow hoods is recommended.

Peptide mass fingerprinting is most successful with very simple mixtures of proteins of comparable abundance. In practice this means isolated spots from 2-D-gels or highly enriched protein samples separated on denaturing 1-D-gels. First, a protein sample is digested with trypsin, to generate a collection of *tryptic peptides*. These peptides are analyzed by MALDI mass spectrometry with a time-of-flight analyzer (*Box 16.1*) to determine accurate molecular

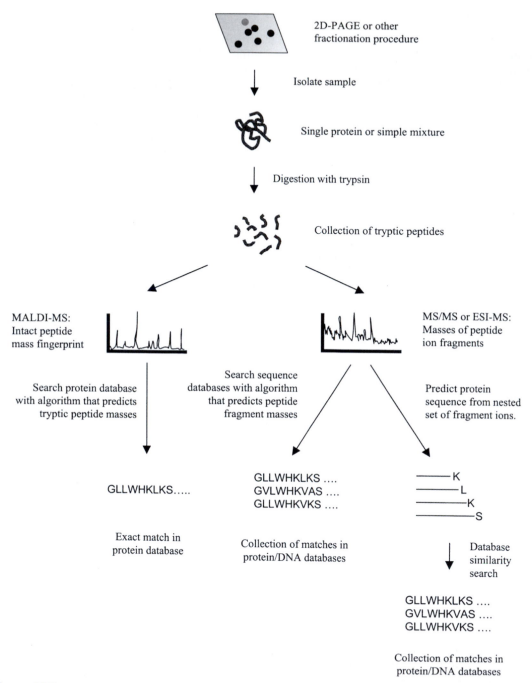

Figure 16.2

Strategies for protein annotation by mass spectrometry

masses. The results are then used as a search query against a protein database such as SWISS-PROT. An algorithm is required that can perform *virtual digests* on protein sequences *in silico* using the same cleavage pattern as trypsin. In this way, the tryptic peptide fragments of all known proteins can be predicted, their masses calculated and compared with the

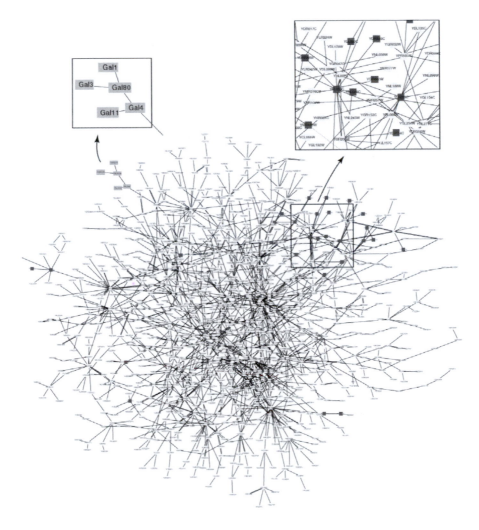

Figure 16.3

A highly complex interaction map representing 1500 yeast proteins. Reproduced from Tucker C.L. *et al.*, Towards an understanding of complex protein networks. *Trends Cell Biol* **11**: 102, 2001 with permission from Elsevier Science.

experimental masses. The molecular weight of a whole protein is insufficiently discriminating for its identification, which is why trypsin digestion is required. If a protein exists in the database, it can be identified from as few as two or three peptides. Excellent commercial and public domain software is available to carry out this task, some of which uses probability-based scoring to give quantitative results. Confidence is also improved by increased precision in peptide mass measurement, now resolved to better than 50 ppm, an order of magnitude improvement over the first generation of instruments. Many of the procedures from the cutting of gel spots, proteolysis, MALDI MS and database searching are being automated, and PMF is widely used for high-throughput studies.

Not all proteins can be identified by PMF because the technique relies on the availability of complete protein sequence data. PMF is therefore most successful with fully sequenced genomes of modest size, in which variable splicing and post-translational modification are

---

Box 16.1   Mass spectrometry in proteomics

Soft-ionization methods. The ionization of large molecules without fragmentation and degradation is known as soft ionization. Two soft ionization methods are widely used in proteomics. *Matrix-assisted laser desorption/ionization* (MALDI) involves mixing the analyte (the tryptic peptides derived from a particular protein sample) with a light-absorbing matrix compound in an organic solvent. Evaporation of the solvent produces analyte/matrix crystals, which are heated by a short pulse of laser energy. The desorption of laser energy as heat causes expansion of the matrix and analyte into the gas phase. The analyte is then ionized and accelerated towards the detector. In *electrospray ionization* (ESI), the analyte is dissolved and the solution is pushed through a narrow capillary. A potential difference, applied across the aperture, causes the analyte to emerge as a fine spray of charged particles. The droplets evaporate as the ions enter the mass analyzer.

Mass analyzers. The two basic types of mass analyzer used in proteomics are the quadrupole and time of flight (TOF) analyzers. A *quadrupole analyzer* comprises four metal rods, pairs of which are electrically connected and carry opposing voltages that can be controlled by the operator. Mass spectra are obtained by varying the potential difference applied across the ion stream, allowing ions of different mass/charge ratios to be directed towards the detector. A *time of flight analyzer* measures the time taken by ions to travel down a flight tube to the detector, a factor that depends on the mass/charge ratio.

Tandem mass spectrometry (MS/MS). This involves the use of two or more mass analyzers in series. Various MS/MS instruments have been described including triple quadrupole and hybrid quadrupole/time of flight instruments. The mass analyzers are separated by a collision cell that contains inert gas and causes ions to dissociate into fragments. The first analyzer selects a particular peptide and directs it into the collision cell, where it is fragmented. A mass spectrum for the fragments is then obtained by the second analyzer. These two functions may be combined in the case of an *ion trap analyzer*.

---

the exception. The proteome of the yeast *Saccharomyces cerevisiae*, with over 6000 potential gene products, has thus been widely studied using MALDI MS. For the human and mouse proteomes, where variable splicing and post-translational modification are the rule rather than the exception, PMF often fails due to the lack of correlation between peptide masses and sequences stored in databases. In such cases, an alternative technique called *fragment ion searching* can be used. As with PMF, protein samples are digested with trypsin and analyzed by mass spectrometry. However, *tandem mass spectrometry* (MS/MS) is used to fragment the peptides prior to mass determination. The tandem mass spectrometer has two or more mass analyzers in series, either of the same type or a mixture. The first analyzer is used to focus the ion stream into a collision cell, which causes ion fragmentation, and the resulting fragment ions are separated by the second analyzer. A widely adopted approach is electrospray ionization on an ion trap or quadrupole time-of-flight mass spectrometer (*Box 16.1*). The small fragments generated can be searched not only against known genes, but also against less robust sources of sequence information such as *expressed sequence tags* (ESTs). These are sequences obtained by the random, single-pass sequencing of clones from cDNA libraries. Although short (2–300 bp) and somewhat inaccurate, they are the most abundant form of sequence information focusing on genes, and therefore prove a rich source of data on protein sequences. Data from peptide mass fingerprinting cannot be searched against ESTs.

Even fragment ion searching may not be sufficient to annotate all proteins. This may be because a given protein simply is not represented in any of the sequence databases including the EST databases. However, more likely reasons include the presence of unanticipated post-translational modifications or protein polymorphisms. Most common post-translational modifications can be allowed for, at the expense of computing time, but polymorphisms remain a challenge. For example, the most abundant form of genetic variation appears in the form of single nucleotide polymorphisms (SNPs) which are single base variations at individual positions. In the human genome, a SNP occurs on average every kilobase, so there will be in the order of 3,000,000 potential base differences between any two people. Although SNPs are more abundant in non-coding compared to coding DNA, and many SNPs in coding sequences represent synonymous changes, there still remain up to 50,000 amino acid polymorphisms in the human proteome. If a deposited sequence represents one of those variants and another is detected by mass spectrometry, the masses of the peptides or fragment ions will not correspond.

One way to approach this problem is to use sequence candidates from tandem mass spectrometry data as the input for a homology search. A program called *CIDentify* is capable of identifying variants of known proteins (taking into account isobaric variants of amino acids, such as glutamine/lysine and leucine/isoleucine for example) but the computer time can be long. Another useful program is *MS-BLAST*, which is more suited to the numerous short sequences produced by tandem mass spectrometry. An alternative approach applicable in such situations is *de novo peptide sequencing*. In this technique, the peptide fragments generated by MS/MS are arranged into a nested set differing in length by a single amino acid. By comparing the masses of these fragments to standard tables of amino acids, it is possible to deduce the sequence of the peptide fragment *de novo*, even where a precise sequence match is not available in the database. This new sequence can be used in a BLAST or FASTA search to identify related sequences. In practice the *de novo* sequencing approach is complicated by the presence of two fragment series, one nested at the N-terminus and one nested at the C-terminus. The two series can be distinguished by attaching diagnostic mass tags to either end of the protein. The different mass spectrometry and database search techniques outlined in this section have been recently reviewed (Mann *et al.*, 2001).

16.3.2.2 Combined HPLC and MS

Combining low-flow rate HPLC with a tandem mass spectrometer is a very powerful analytical tool for protein identification. The data storage requirements from on-line techniques are challenging to most laboratories, with thousands of mass spectra from each run to be stored and analyzed. The data are valuable for re-searching against nightly updated databases, and so cannot be discarded after the first analysis. Increasingly, bigger laboratories are using tape silos or secure commercial archiving providers. Using an on-line HPLC-tandem mass spectrometer, mass spectra containing sequence information are generated continuously over several hours as each peptide enters the mass spectrometer. It is not possible to achieve complete separation of tens of thousands of peptides in a typical digest from a mixture of proteins, and invariably, mixtures of peptides are entering the mass spectrometer simultaneously. No human operator could keep pace with the decisions needed, and so data-dependent software has been developed. The incoming sample is ionized and the molecular weight and charge state of the peptides in the mixture determined by software while the peptides are still entering the mass spectrometer. Some peptides will be known contaminants (from, for example, keratin) or they may be peptides from proteins in a user-defined exclusion list, or they may have been analyzed in the previous scan. These can be ignored and the remaining peptide molecular ions fragmented. This is done very rapidly, with new mass spectra containing sequence pre-processed and stored for database searching, which in most cases identifies the proteins.

16.3.2.3   Protein quantification by mass spectrometry

Stable isotopes have been widely used for quantification in drug metabolism, as the mass spectrometer can easily distinguish and quantify two isotopically labeled forms of the same compound that are chemically identical and can be co-purified. Extending this idea, an isotope coded affinity tag (ICAT) method has been developed using a biotinylated iodo-acetamide derivative to selectively label protein mixtures at their cysteine residues. The biotin tag allows affinity purification of the tagged cysteine peptides after proteolysis with trypsin. The ICAT reagent is available in 'heavy' (d_8) and 'light' (d_0) isotopically labeled forms, and for example, may be used to differentially label cell pools before and after stimulation. This allows the proteome to be studied as a dynamic system, and the results of any perturbation to be characterized in terms of up-regulated and down-regulated proteins. Alternatively the samples could be healthy and diseased tissue.

After labeling, the cells are combined, lysed and the proteins isolated: purification losses occur equally on both samples. Even after reducing the peptide pool size to tagged cysteine-containing peptides, the mixture remains very complex and the amount of potential data overwhelming as both quantification and identification could be carried out for each peptide. It is therefore prudent that only the proteins showing changes in expression level are identified. Low flow rate HPLC tandem mass spectrometry has been used for studies so far, but other mass spectrometers may offer throughput advantages. Isotope intensities are compared for peptides as they enter the mass spectrometer. If they are equivalent, then no up-regulation or down-regulation has occurred, and the protein is of no immediate interest. If the intensities differ, then a change in protein expression has taken place, and the protein is of interest. The amount of the two forms is measured, and the peptide d_0-form is fragmented and identified by database searching. This approach can be extended as different chemical tags, including some for phosphorylation, are developed.

16.3.3   Protein chips

Protein chips are miniature devices on which proteins, or specific capture agents that interact with proteins, are arrayed. As such, protein chips can act both to separate proteins (on the basis of specific affinity) and characterize them (if the capture agent is highly specific, as in the case of antibodies) (Templin *et al.*, 2002). For *de novo* protein annotation, protein chips with broader capture agents can be integrated with mass spectrometry. Indeed several chip formats are available where the chip itself is used as a MALDI plate. The advantages of protein chips over other technology platforms used in proteomics are their size and their potential for ultra-high throughput analysis. The bioinformatic support required for protein chips is similar to that required for DNA chips, and involves areas such as image analysis, signal quantification and normalization, pairwise comparisons, clustering, classification and feature reduction. Different types of protein chip and their applications are discussed briefly in *Box 16.2*.

16.4   Case studies

16.4.1   Case studies in expression proteomics

As noted above, the foundation of proteomics as a field was the use of 2-DE for high-resolution separation of proteins from complex mixtures. This technique remains of great importance for the identification of protein disease markers from clinical samples, in which diseased and normal tissues are compared for differences in protein expression, and specific proteins of interest identified by mass spectrometry. An example of expression proteomics applied by the group of Julio Celis to the diagnosis of human bladder cancer is cited in Banks *et al.* (2000). Over 90% of bladder cancers in the Western world are transitional-cell

---

Box 16.2 Different types of protein chip

Antibody chips. These consist of arrayed antibodies and are used to detect and quantify specific proteins in a complex mixture, essentially representing a miniature, highly parallel series of immunoassays.

Antigen chips. The converse of antibody chips, these devices contain arrayed protein antigens and are used to detect and quantify antibodies in a complex mixture.

Universal protein arrays. These devices may contain any kind of protein arrayed on the surface and can be used as detection instruments or to characterize specific protein–protein and protein–ligand interactions. Various detection methods may be used, including labeling the proteins in solution or detecting changes in the surface properties of the chip, e.g. by surface plasmon resonance. Includes devices such as lectin arrays, which are used to detect and characterize glycoproteins.

Protein capture chips. These devices do not contain proteins, but other molecules that interact with proteins as broad or specific capture agents. Examples include oligonucleotide aptamers and chips containing molecular imprinted polymers as specific capture agents, or the proprietary protein chips produced by companies such as BIAcore Inc. and Ciphergen Biosystems Inc., which employ broad capture agents based on differing surface chemistries to simplify complex protein mixtures.

Solution arrays. New technologies based on coded microspheres or barcoded nanoparticles release the protein chip from its two-dimensional format and will probably emerge as the next generation of miniature devices used in proteomics.

---

carcinomas but in regions of Africa where the parasite *Schistosoma haematobium* is present about 80% are squamous cell carcinomas. It is hard to distinguish the two types simply from histological examination, but 2-D gel electrophoresis allowed six squamous cell cancers to be correctly identified from 150 bladder tumors on the basis of characteristic protein expression patterns. The bioinformatics techniques required in this analysis were clustering and classification of expression data to reveal commonly expressed proteins in each class of tumor. Several protein markers including psoriasin and keratins 10 and 14 were especially useful in assessing the degree of differentiation. Such studies as these will increasingly reveal markers of disease progression, enable diagnosis and reclassification of disease, and increase understanding of the molecular basis of pathophysiology.

There are numerous further applications of expression proteomics. Of great interest to the pharmaceutical industry is the use of these methods to understand the toxic effects of drugs (toxicology) and the cellular basis for drug action (efficacy). For example, the immunosuppressive drug cyclosporin, which is used to prevent rejection of tissue transplants, has side effects on the kidney due to calcification of the tubules. Aicher and colleagues (cited in Banks *et al.*, 2000) showed that renal expression of the calcium-binding protein calbindin is reduced in kidney transplant patients who suffer from such toxicity, providing a possible explanation for this problem. Many such proteomic studies are under way in pharmaceutical and academic laboratories, promising to considerably increase our understanding of the mechanistic basis of cellular toxicities.

16.4.2   Case studies in cell-map proteomics

Cell map proteomics takes as its starting point the observation that proteins rarely work alone, but as part of larger assemblies. In some cases, these are multi-protein assemblies, while in other cases different types of molecules may also be involved. For example, complexes may assemble on DNA or RNA. The complexes may be transient or stable, and vary in terms of their cellular location.

There are many different ways of identifying interacting proteins, including genetic methods (e.g. suppressor mutations, dominant negatives), biochemical methods (e.g. cross-linking, co-immunoprecipitation), cell biological methods (e.g. fluorescence resonance energy transfer, co-localization) and atomic methods (X-ray crystallography, nuclear magnetic resonance spectroscopy). Most of these are not high-throughput methods and are not, therefore, suitable for systematic proteome analysis. One large-scale approach is the yeast two-hybrid system, which measures binary interactions by assembling a functional transcription factor from interacting fusion proteins. This method has the advantages of scalability and detection of transient interactions. However, the context of the interaction is artificial, and complexes that require cooperative interactions to be stable may go undetected. Another generic approach, if a suitable cDNA is available, is to affinity tag the gene, and to express the protein under as near to physiological conditions as possible, preferably in a cell line of the proper lineage. For yeast, this can be done by homologous recombination under the control of the endogenous promoter. For mammalian systems, transfection with expression vectors has proved effective. In both cases, the bait protein is isolated in such a way as to retain any binding partners, and then the complex is analyzed by mass spectrometry (Vorm et al., 2000).

A number of protein complexes have been isolated and characterized in this manner (*Table 16.1*). More recently, a study in yeast involved the production of 1739 affinity tagged genes under the control of their natural promoters, leading to the isolation of 232 distinct multi-protein complexes. Ninety-eight of these were already known and present in yeast proteome databases (http://genome-www.stanford.edu/Saccharomyces/yeast_info.html). The remaining 134 complexes were novel. Complexes contained between two and 82 separate protein components, with an average of five. Since the novel complexes generally contained some proteins of known function, it was possible to propose tentative functional roles for 231 associated proteins based on circumstantial evidence ('guilt by association'). For example, the complex that polyadenylates mRNA comprised 21 proteins, four of which had no previous annotation of any kind.

The observed protein complexes frequently contained common components that suggested interconnections between them, providing a glimpse of the dynamic functional network of the cell. For example the protein phosphatase PP2A was found in separate complexes with cell-cycle regulators and proteins involved in cellular morphogenesis. Finally, the study showed the presence of *orthologous complexes* (not just orthologous proteins) between yeast and man, consistent with conservation of key functional units defining a core proteome in eukaryotes. There is already evidence for variations in the composition of complexes in metazoans depending on cellular context, effectively *paralogous complexes*, trends that are likely to be reinforced as studies such as this are extended.

Table 16.1 Some protein complexes that have been characterized by mass spectrometry

1996	Yeast anaphase promoting complex
1998	Yeast spindle pole body
1999	Human anaphase promoting complex
1999	Yeast ribosome
1999	Human interchromatin granule cluster
2000	Yeast nuclear pore complex
2000	EGF signaling complex
2001	NMDA receptor complex

Information on protein assemblies is typically rendered as a graph in which the nodes are individual proteins or protein complexes and the edges are physical connections between them (*Figure 16.3*). Bioinformatics methods will be needed to display the full complexity of the interacting network and demonstrate its plasticity and dynamism. Data from other proteomic studies (particularly yeast two-hybrid screens) and from alternative sources (e.g. DNA array profiling) will need to be integrated into databases of interacting proteins. A number of such databases have already been established and are discussed in a recent review (Xenarios and Eisenberg, 2001).

16.5 Summary

The focus of this chapter has been on large-scale identification of expressed or interacting proteins, two vital aspects of proteomics that are proving amenable to high-throughput techniques. Data analysis and management techniques have enabled such data to be integrated and viewed to good effect, although much work remains to be done in the representation and mining of such complex datasets. Proteomics however will not rest with expression and interaction. Post-translational modification is often a critical modulator of function, and methods for large-scale measurement of phosphorylation are already in train. The three-dimensional structure of individual proteins or whole complexes at various levels of resolution are increasingly available, and bioinformatics methods need to anticipate growth in such data which arguably ultimately subsume other proteomics data types. Finally, proteomic information must be placed in the context of the cell, and its relationship to the ligands that comprise the metabolome, or indeed the chemicals that are pharmaceutical medicines, must be clarified by annotation or preferably structure solution. Such multi-faceted combinations of data form the basis for the algorithmic analysis of biological pathways, which will enable evolutionary and functional relationships to be mined and simulated at increasing levels of biological complexity.

References and further reading

Anderson, N.G., and Anderson, N.L. (1998) Proteome and proteomics: New technologies, new concepts, new words. *Electrophoresis* **19**: 1853–1861.

Banks, R.E., Dunn, M.J., Hochstrasser, D.F., Sanchez, J.C., Blackstock, W., Pappin, D.J., and Selby, P.J. (2000) Proteomics: new perspectives, new biomedical opportunities. *Lancet* **356**: 1749–1756.

Blackstock, W., and Weir, M. (1999) Proteomics: quantitative and physical mapping of cellular proteins. *Trends Biotechnol* **17**: 121–127.

Dunn, M.J., and Gorg, A. (2001) Two-dimensional polyacrylamide gel electrophoresis for proteome analysis. In: *Proteomics – From Protein Sequence to Function.* BIOS Scientific Publishers Ltd., Oxford UK, pp 43–64.

Lee, K.H. (2001) Proteomics: a technology-driven and technology-limited discovery science. *Trends Biotechnol* **19**: 217–222.

Mann, M., Hendrickson, R.C., and Pandey, A. (2001) Analysis of proteins and proteomes by mass spectrometry. *Annu Rev Biochem* **70**: 437–473.

Pandey, A., and Mann, M. (2000) Proteomics to study genes and genomes. *Nature* **405**: 837–846.

Pleissner, K.-P., Oswald, H., and Wegner, S. (2001) Image analysis of two-dimensional gels. In: *Proteomics – From Protein Sequence to Function.* BIOS Scientific Publishers Ltd., Oxford, UK, pp 131–150.

Templin, M.F., Stoll, D., Schrenk, M., Traub, P.C., Vohringer, C.F., and Joos, T.O. (2002) Protein microarray technology. *Trends Biotechnol* **20**: 160–166.

Vorm, O., King, A., Bennett, K.L., Leber, T., and Mann, M. (2000) Protein-interaction mapping for functional proteomics. In: *Proteomics: A Trends Guide* **1**: 43–47.

Xenarios, I., and Eisenberg, D. (2001) Protein interaction databases. *Curr Opin Biotechnol* **12**: 334–339.

Data management of biological information

Nigel J. Martin

- Why managing biological data poses problems
- Databases and database software:
  - Advantages, disadvantages, architectures
- Data management techniques – key concepts:
  - Accessing a database
  - Designing a database
  - Overcoming performance problems
  - Accessing remote data
- Challenges arising from biological data

17.1 Concepts

Bioinformatics methods aim to extract information about biological functions and mechanisms from computer-held biological data. These data[1] range from genomic sequence data to the transcriptomics and proteomics data discussed in previous chapters. Data management techniques enable these increasingly huge amounts of data to be stored and extracted from computers reliably and efficiently. Most of these techniques were developed for business applications, where the data are structured and it is clear what object any particular piece of data relates to, such as in the case of recording a customer's name, address and account balance. However, managing biological data and extracting useful information from them poses particular problems for the following reasons.

- Biological data come in many different forms: structured, as in GenBank or PDB file formats; unstructured, as in biomedical literature abstracts; sequence, as in genomic or protein sequences; or more complex, as in biochemical pathway data.
- There is no single biological object to which the data refer: individual users of the data have their own perspective depending on the science they are carrying out, ranging from the atomic level to that of whole organisms.
- Similarly, the biological functions and mechanisms to which data refer are also dependent on the perspective of the user of the data.
- Even if the appropriate biological object can be identified, there are many incompatible formats and nomenclatures for the data about that object and its functions.
- The data about any given biological object are likely to be held at many different sites, each with a different approach to coping with the previous problems.

In this chapter we introduce the fundamental ideas underpinning data management techniques and their use. We then examine some of the techniques for dealing with the particular problems arising from the management of biological data.

---

[1] In computing, *data* is treated as a collective noun as in 'this data'. However, in this chapter *data* is treated as a plural noun following the usual scientific usage.

17.2   Data management concepts

17.2.1   Databases and database software

The term *database* can be used simply to mean any collection of data stored in a computer, perhaps even in a single file. Biological data have indeed frequently been held in simple files often with tags to indicate the meaning of a particular piece of data as, for example, in the case of a GenBank file. Files may also have a fixed record format, as in the case of a PDB file. As discussed below, problems often arise with file-based databases, and this has led to the development of database management techniques to overcome such problems. Using these techniques, a database is still a collection of data stored in a computer, but the storage of and access to the data is controlled by specialized software called a *database management system* (DBMS). In practice, even biological data which have traditionally been in a file format, such as GenBank or PDB data, are in reality now managed by a DBMS and the familiar file formats are retained as one way of making such data available to scientists.

17.2.2   Why are DBMSs useful?

Scientists have traditionally written programs in languages such as Perl, C or C++ to extract data relevant to their work from files, perform processing to generate new data of interest, and then write those data to a new file. This way of working leads to the following problems.

- Data are not shared. The data generated by each scientist are stored in different files, the format of which will be dependent on the precise programs written by the scientist. It will not be possible for other scientists to share those data without knowing of the existence of the files and their format.
- Data are not integrated. The new files generated by each scientist have no relationship to the files of data from which they were derived, nor with the many files being generated by other scientists. The information which would be available if the data stored in the separate files were integrated and inter-related is lost.
- Data are duplicated. The same or closely related data are generated many times by different programs, and stored in different files. This inevitably leads to problems when revisions are made to one file of generated data, perhaps to correct an error, but the other files of identical or closely related data are not revised. Multiple versions of data which should be consistent are then inconsistent.
- Programs are dependent on file formats. If the format of a file is changed, perhaps to add some new type of data, then the programs which access that file have to be modified to take account of the new format.

DBMSs were developed to overcome these problems. They do this in the following ways.

- Data are held in a single shared resource: the database. Each scientist still writes programs to access data, but the format of the data is defined centrally for the database rather than being determined by an individual scientist's program. Further, if the data generated by the scientist are themselves stored in the database, they can be shared by other scientists.
- Data are integrated in the database. When data are stored in a DBMS-managed database, the DBMS enables the relationships between all the data to be explicitly represented, and the information inherent in those relationships is available.
- Data need not be duplicated. In reality data *are* almost always duplicated in a DBMS-managed database, but the duplication is controlled by the DBMS. The reasons for this duplication are to enhance performance, and to enable recovery of lost data if some failure results in their loss.

- Programs are independent of file formats. The format of the data may be changed or new types of data added to a DBMS-managed database, but existing programs which have no requirement to access the new types of data need not be modified: the DBMS itself ensures existing programs still work as though there had been no change to the data formats.

The difference between file-based and DBMS-managed data is illustrated in *Figure 17.1*. The upper half of the figure shows file-based 'databases': many different programs access data in multiple files and write out new data to yet more files. It becomes an impossible task to maintain data as a shared, integrated, consistent resource. The lower half of the figure shows DBMS-managed data. Programs now access data in a shared, integrated database by using the services of a DBMS, which is able to ensure data are shared, integrated and consistent.

It might be imagined from the above that having data managed by a DBMS rather than using ordinary files is always advantageous. Indeed, many scientists use a DBMS as though the very fact of having data in a 'real' database gives benefits. In fact there are disadvantages in using a DBMS including the following.

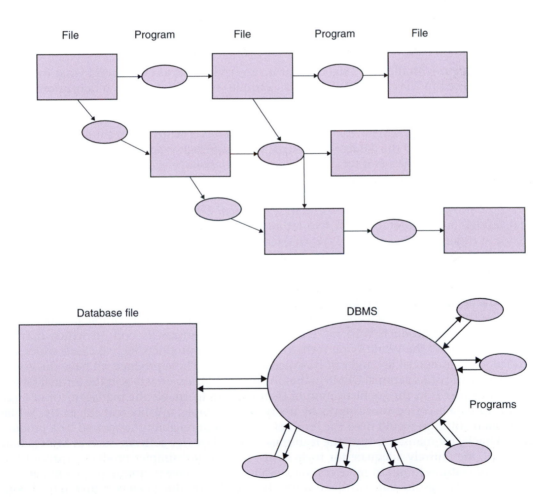

Figure 17.1

File-based v DBMS-based data

- A DBMS is expensive. This can include the cost of purchasing the DBMS software, but equally importantly a DBMS is a large piece of software which requires more computer resources such as memory than conventional programs accessing files. File-based access to data does not incur these costs.
- DBMSs are complex. If a scientist writing a Perl program to access a file finds the program is not working as expected, there is every chance that the source of the problem can be discovered and corrected. If the same program is using a DBMS, there are many areas of the DBMS's operation which are complex and can only be corrected with specialist knowledge of the DBMS.
- Access to data by a DBMS may be slow. The DBMS software is potentially carrying out a great deal of work to enable access to data reliably by many different programs concurrently and there is an inevitable overhead associated with this. A skilled programmer will always be able to write a program accessing file-based data more quickly.

In summary, the advantages of a DBMS are that it provides shared, integrated access to data, with the ability to modify the types and formats of stored data for the needs of new applications, while allowing existing applications to continue working without modification. Usually these are significant advantages to a scientist with data, but if the costs arising cannot be justified, then it makes sense to continue working with file-based data.

17.2.3 Relational and other databases

Another problem with file-based data is that it can be difficult to understand what information is contained in the file. A PDB file, for example, certainly contains much information about the 3D structure of a macromolecule, but it is no simple matter to understand from the file what are the biological and other entities and their relationships inherent in the data. An important feature of a DBMS is that it supports a *data model* which makes it easier to understand what information the database contains. Many different types of data model have been developed since the 1960s, but for over 10 years the *relational model* has been dominant, and DBMSs supporting the relational model are currently the most widely found. The essential characteristic of the relational model is that all data are represented as being stored in tables. *Figure 17.2* shows example tables stored in a *relational DBMS*, i.e. one supporting the relational model. Some columns from two tables, PDB_DOM and PDB_DOM_SEG are shown with some example rows in each. PDB_DOM records the CATH structural classification (see Chapter 7) for each identified domain in the PDB. A domain may encompass non-contiguous sections of a chain, each such section being known as a domain segment. PDB_DOM_SEG records the start and finish residue numbers of each domain segment.

Once the decision has been made to use a relational database, there are a large number of different relational DBMSs which are available. These range from relatively simple products such as *MySQL* through products with more features such as *PostgreSQL*, both of which may be obtained freely over the World Wide Web, to more complex products such as *Oracle* which are intended for commercial use and can be expensive to buy. These products all have the essential characteristic of a relational DBMS in that the representation of data is in the form of tables; where they differ is in the sophistication of their features to enable the management of large amounts of data accessed concurrently by many users securely, reliably and efficiently. While a financial institution would need the levels of security and reliability supported by a product such as Oracle, for many scientists the features supported by free software such as MySQL and PostgreSQL are entirely adequate for their needs. Indeed, the simpler products, without the overhead of features found in more complex products, may have performance advantages for straightforward access to data. Further reading about DBMS products can be found in the books by Reese *et al.* (MySQL), Worsley and Drake (PostgreSQL) and Abbey *et al.* (Oracle).

The relational model has the great advantage of being a very simple way of representing data, but when there are complex relationships between entities, as is the case for many bio-

PDB_DOM

PDBCODE	CHAIN	DOMAIN	C	A	T	H	S35	S95	S100
1lga	A	1	1	10	520	10	1	2	1
1lga	A	2	1	10	420	10	2	2	1
1lga	B	1	1	10	520	10	1	2	2
1lga	B	2	1	10	420	10	2	2	2

PDB_DOM_SEG

PDBCODE	CHAIN	DOMAIN	SEGMENT	RESNUM1	RESNUM2
1lga	A	1	1	11	144
1lga	A	1	2	269	299
1lga	A	2	1	145	268
1lga	A	2	2	300	339
1lga	B	1	1	11	144
1lga	B	1	2	269	296
1lga	B	2	1	145	268
1lga	B	2	2	297	339

Figure 17.2

Example relational table data

logical data, it can be difficult to understand how the tables represent those relationships. Taking the example of *Figure 17.2*, each PDB structure may have many domains, and each PDB structure may have many chains. The relationship between domains and chains is a many–many one: namely, each domain may be related to many chains, while each chain may be related to many domains. However, the nature of these relationships is not clearly represented in the tables shown. It is desirable that any data model makes it easy to understand what relationships are represented therein, and the relational model does not always achieve this. An alternative data model, the *object-oriented model*, is supported by some DBMSs. This represents objects and the relationships between objects directly, as well as controlling the way in which objects may be created and accessed in the database. So, for example, rather than having data about a protein domain spread across many tables as would be the case in a relational database, in an object-oriented database all the information about a protein domain, including what a domain is related to and the ways in which domain data may be created and accessed, would be brought together in the definition of a protein domain object.

Some DBMSs have been designed specifically for managing scientific data. One having features specifically for managing genetic map, physical map and sequence data is *AceDB*, originally designed for the *C. elegans* sequencing project. AceDB uses an object-oriented model with each object having a tree-like hierarchical structure. Graphical interfaces for displaying the map and sequence data come with the DBMS.

Although an object-oriented model can be well-suited to representing data with complex relationships, the simplicity of the relational model and the strengths of well-established relational DBMSs in supporting reliable, shared access to integrated data have ensured their continued wide use. Given this current dominance of relational DBMSs, object-oriented DBMSs are not considered further in this chapter.

17.3   Data management techniques

Whatever relational DBMS product is used, the fundamental requirement of any user is knowing how to access the data in the database. Those scientists developing their own databases additionally need to know how to design and create a relational database. Both of these tasks require an understanding of how a relational DBMS works, in order that the database may be used in an efficient way. The techniques underpinning such use of a relational database are now examined.

17.3.1 Accessing a database

The standard language for accessing a relational database is *SQL* (Structured Query Language): no matter which relational DBMS product is used, SQL will be supported. In fact SQL is more than just a language to access the data in a relational database, it is also the language used to create the tables in a database and insert, modify or delete rows of data in those tables. Here we only consider its use for accessing table data; discussion of SQL's wider features may be found in many textbooks such as that of Mata-Toledo and Cushman.

SQL has a SELECT statement to retrieve rows from tables. For example:

```
SELECT C, A, T, H
FROM PDB_DOM
WHERE PDBCODE = '1lga'
AND CHAIN = 'B'
```

This query has three parts or clauses, a SELECT clause, a FROM clause and a WHERE clause. Every SELECT statement has SELECT and FROM clauses, while the WHERE clause is optional. If this SELECT statement were executed with the example tables of *Figure 17.2*, the rows retrieved forming the result would be:

C	A	T	H
1	10	520	10
1	10	420	10

The SELECT clause specifies which values in particular appear in each row in the query result – in this case C, A, T and H values. The FROM clause of the query specifies that data are to be retrieved by accessing the PDB_DOM table. If the query had no WHERE clause, the C, A, T and H values from every row in PDB_DOM would appear in the result. If the query does have a WHERE clause, as in this example, only rows which satisfy the WHERE clause conditions produce values in the query result.

Normally queries will require access to data in more than one table. For example, suppose we wish to know the CATH classification of the domain containing residue 100 of PDB structure 1lga chain A. A possible SELECT statement is:

```
SELECT C, A, T, H
FROM PDB_DOM, PDB_DOM_SEG
WHERE PDB_DOM_SEG.PDBCODE = '1lga'
AND PDB_DOM_SEG.CHAIN = 'A'
AND PDB_DOM_SEG.RESNUM1 <= 100
AND PDB_DOM_SEG.RESNUM2 >= 100
AND PDB_DOM.PDBCODE = PDB_DOM_SEG.PDBCODE
AND PDB_DOM.CHAIN = PDB_DOM_SEG.CHAIN
AND PDB_DOM.DOMAIN = PDB_DOM_SEG.DOMAIN
```

producing result:

C	A	T	H
1	10	520	10

The query specifies both the PDB_DOM and PDB_DOM_SEG tables in the FROM clause: the query requires access to data in both tables. The first four conditions in the WHERE clause specify the conditions that a row in PDB_DOM_SEG must satisfy in order to figure in the result. For any row that does satisfy these conditions, the final three conditions in the WHERE clause specify the conditions which a row in PDB_DOM must itself satisfy in order for its C, A, T, H values to appear in the result, namely it must match the PDBCODE, CHAIN and DOMAIN values in the PDB_DOM_SEG row.

SQL SELECT statements have many other features not illustrated by this example, but these features do not extend the fundamental way in which SELECT statements work. In essence, any SQL SELECT statement will retrieve data from one or more tables by using:

- a SELECT clause specifying the particular values which we wish to retrieve,
- a FROM clause to specify the tables which must be accessed, and
- an optional WHERE clause to specify conditions restricting the rows which we wish to retrieve from tables, and the conditions which specify how rows in different tables must match each other.

While SQL statements enable data to be retrieved from tables, developing a full application will usually require programming in a language such as Perl, Java or C++ for all the non-database tasks of the application. Hence, mechanisms have been developed to enable such languages to use SQL to access a database.

The main difficulty in using SQL in an ordinary programming language is that an SQL SELECT statement potentially results in many rows of data being retrieved to form the result, but programming languages are usually designed to handle a single row or record at a time. This mismatch is solved by use of an *SQL cursor* which enables row-by-row retrieval of results so a program can process a row at a time.

The details of how to use SQL from a programming language depend on the particular language used. For Perl, a programming interface called *DBI* is widely used, as described by Descartes and Bunce. This is database independent, meaning it will work with most DBMSs including PostgreSQL, MySQL and Oracle. A similar database-independent programming interface *JDBC* is used from within Java programs; Fisher *et al*. provide a thorough tutorial and reference for JDBC. For C or C++ programs, an interface supporting *embedded SQL* is often used, in which SQL statements are interspersed with the ordinary C or C++ program statements. Embedded SQL is typically more dependent on the particular DBMS used than either Perl DBI or Java JDBC.

17.3.2 Designing a database

There are two distinct but related aspects to designing a database, *logical design* and *physical design*. The former is concerned with designing the table structures which users of a relational database will see. The latter is concerned with designing the actual storage structures which underlie those table structures. This raises a fundamental aspect of databases which many database users remain unaware of: namely, the data model presented by a DBMS – a series of tables in the case of a relational database – may differ enormously from the storage structures maintained in reality by the DBMS to hold the data.

17.3.2.1 Logical design

Considering logical design first, there are usually many possible table designs for any given collection of data. *Figure 17.2* showed one possible table design for some domain data. *Figure 17.3* shows an alternative design. This has introduced a single numeric identifer DOMAIN_ID to represent the combination of PDBCODE, CHAIN and DOMAIN in *Figure 17.2*. A new table PDB_DOM_ID has been introduced to record the relationship between the new DOMAIN_ID

PDB_DOM_ID

DOMAIN_ID	PDBCODE	CHAIN	DOMAIN
1	1lga	A	1
2	1lga	A	2
3	1lga	B	1
4	1lga	B	2

PDB_DOM

DOMAIN_ID	C	A	T	H	S35	S95	S100
1	1	10	520	10	1	2	1
2	1	10	420	10	2	2	1
3	1	10	520	10	1	2	2
4	1	10	420	10	2	2	2

PDB_DOM_SEG

DOMAIN_ID	SEGMENT	RESNUM1	RESNUM2
1	1	11	144
1	2	269	299
2	1	145	268
2	2	300	339
3	1	11	144
3	2	269	296
4	1	145	268
4	2	297	339

Figure 17.3

Table Design 2

and the existing identifiers. *Figure 17.4* gives a third possible design consisting of just one table.

The process of database logical design is that of choosing between different possible table designs such as seen here. One advantage of Design 2 is that domains are identified by a single numeric identifier. Hence, in queries such as the last one seen in section 17.3.1, only this

PDB_DOM_SEG

PDBCODE	CHAIN	DOMAIN	SEGMENT	RESNUM1	RESNUM2	C	A	T	H	S35	S95	S100
1lga	A	1	1	11	144	1	10	520	10	1	2	1
1lga	A	1	2	269	299	1	10	520	10	1	2	1
1lga	A	2	1	145	268	1	10	420	10	2	2	1
1lga	A	2	2	300	339	1	10	420	10	2	2	1
1lga	B	1	1	11	144	1	10	520	10	1	2	2
1lga	B	1	2	269	296	1	10	520	10	1	2	2
1lga	B	2	1	145	268	1	10	420	10	2	2	2
1lga	B	2	2	297	339	1	10	420	10	2	2	2

Figure 17.4

Table Design 3

single identifier needs to be matched when combining rows from more than one table, rather than the values of three columns as was needed for the original design. Conversely, a disadvantage of the Design 2 is that the query would involve three tables rather than the two or one of the alternative designs, and queries involving matching rows in multiple tables – *joins* – can be very slow.

17.3.2.2   Physical design

Turning to physical design, while a relational database appears to consist of tables, in reality the data are stored in a variety of storage structures designed to support quick retrieval of data. It is the aim of physical database design to choose the most appropriate storage structures for a database. Here we consider only the use of index storage structures.

Indexes in databases work in a similar fashion to indexes in books. A reader wanting to find pages in a book related to a particular topic can simply work through the book page by page examining each until one is reached mentioning the topic of interest. For short books this may be acceptable, but for long books it is simply too time-consuming. Similarly for a table in a database, if rows are sought containing a particular value or values, the DBMS can examine each row one-by-one, but this will be slow for large tables potentially holding millions of rows. Therefore, DBMSs support indexes on tables. An index identifies which rows contain particular values for one or more columns in a table. For example, consider again the design of *Figure 17.2* and the last query of section 17.3.1. The query suggests two indexes which would be beneficial if the tables involved contain many rows.

- An index on the combination of values of columns PDBCODE and CHAIN in PDB_DOM_SEG would enable a DBMS to retrieve rows very quickly in that table with particular values for the PDBCODE and CHAIN columns, as was needed for the query for the combination of values ('1lga', 'A').
- An index on the combination of values of columns PDBCODE, CHAIN and DOMAIN in PDB_DOM would mean that once a row in PDB_DOM_SEG had been retrieved, the DBMS could very quickly retrieve matching rows in PDB_DOM with the same combination of PDBCODE, CHAIN and DOMAIN values.

The example illustrates the two main ways indexes are used by a relational DBMS: to quickly retrieve rows in a table with a given combination of values therein, and additionally to retrieve rows in one table matching rows in a second table as required in a join of the tables. Although indexes may give improved query performance, they do result in slower update performance since as values in table columns are updated, associated indexes must also be updated by the DBMS. This cost, as well as the space required to store the indexes, must be borne in mind when considering a physical design.

17.3.3   Overcoming performance problems

Once a database has been designed and loaded with data, it can be a shock to find that the SQL SELECT statements written to query the tables run very slowly indeed. While this might be expected for large tables containing millions of rows, even with tables containing thousands of rows it is very easy to write SQL queries which take many minutes and possibly hours before results are produced. Conversely, it can be a surprise that a SQL SELECT statement which does query tables containing millions of rows nonetheless produces results almost instantaneously.

Imagine you are using the design of *Figure 17.2* and have written the last SQL query of section 17.3.1. You are disappointed to find performance is slow and you are wondering if there is any way you could make the query run faster. What should you consider?

- First, examine the SQL query itself. Has it been written correctly to return the results you intended? In particular, check that the query does not refer to more tables than are necessary in the FROM clause, or that the WHERE clause is missing a join condition specifying how the values in two tables should match.
- Second, make sure the physical design of the tables is appropriate. It was noted in section 17.3.2 that indexes enable a relational DBMS to retrieve rows quickly. Have the indexes been defined to support the problem query? Are there other physical storage structures supported by the DBMS which would help the query run faster?
- Third, if the physical design seems to be appropriate, is the DBMS choosing a sensible strategy to retrieve rows? When a DBMS is given an SQL query to execute, it considers alternative ways to produce the required results. If, for example, it needs to do a join retrieving rows in one table matching rows in a second table, it has a number of options. It could simply retrieve the rows in the first table one by one, and check each against every row in the second table to find matches. Alternatively, it could use an index to find matching rows in the second table. Yet another alternative would be for it to sort both tables on the values being matched, and then scan each sorted table in parallel looking for matches. The DBMS is carrying out *query optimization* in choosing between the many options it has for producing the results of a query, and sometimes it does not make the best choice. Many DBMS products have tools to enable a user to view the strategy used by the DBMS to execute a query – the *execution plan* – and provide ways enabling a user to change the execution plan so that the DBMS executes the query more quickly. This might involve writing the SQL query in a slightly different way, or running a utility which updates the information the query optimizer uses to choose an execution plan, or changing one of the parameters controlling the detailed way the DBMS works. The details will be specific to each DBMS.

If none of these results in the query running quickly enough, more drastic solutions must be considered:

- Should the computer be upgraded with more memory or processors? DBMS products are notoriously greedy in their use of computer resources.
- Should the logical design of the database be changed? For example, should the design of *Figure 17.2* be changed to Design 3 of *Figure 17.4*? The problem query would then have to retrieve data from only one table rather than two, so removing the table join which could be causing the performance problems. The implications of changing a table design are considered further below.
- If all else fails, then there is no alternative but to accept that it is not feasible to run the query against the database. In that situation, one would look for a simpler, more efficient query to extract data from the database, and then manipulate that data outside the database using a conventional language such as Perl or C++.

The option of changing the logical design of a database was noted above as one way of removing performance problems with some queries. Of course, it is highly desirable that the most appropriate logical design is chosen in the light of expected queries before the database is created. A change to the logical design of an existing database is not a task to be undertaken lightly, since that may then force changes to be made to all the SQL queries in applications retrieving data from the original tables. This is one form of the problem of programs being dependent on the format of data as discussed in section 17.2.2. DBMS products reduce this problem by allowing the definition of *view* tables, which look like the original tables but do not exist as tables stored in their own right. Instead, view tables are defined by SQL queries which are run by the DBMS when needed, to give the impression that the view tables are actually stored. In this way existing application programs may be able to continue with SQL queries as originally written, but those queries now reference view tables giving the appearance of the original table design.

Since what appears to be a single view table may in reality be produced by an SQL query joining a number of real tables, retrieving data from a view table may be slow given the joins involved. Some DBMS products support *materialized views* which, like ordinary view tables, are tables defined in terms of an SQL query, but in which the data for the resulting table are explicitly stored as a real table. As changes are made to the tables on which the materialized view is defined, the DBMS can then sometimes automatically make corresponding changes to the stored materialized view table.

17.3.4   Accessing data from remote sites

It was noted in section 17.1 that the data about any biological object are almost certainly held at many sites rather than in a single database. Hence, scientists typically need to be able to access data from many sites in order to carry out their work. Traditionally, this has meant downloading files from remote sites and writing programs to access those local copies of files, which has added to the problems of file-based data described in section 17.2.2.

With the advent of the World Wide Web, it has become common for web-based interfaces to be made available for browsing and searching data at remote sites, but in order for a scientist then to write programs to process such data, there has often still been no alternative other than to download the data to a local machine.

DBMS-based databases have a number of features to make it easier for programs to access data held remotely.

- First, DBMS software usually follows a *client–server* model. This means that the database and the DBMS software for managing it can be at one *server* site, while an application program and the DBMS software needed to enable that program to access the database can be at a separate *client* site. In terms of the bottom half of *Figure 17.1*, the database file is at a server site, each program may be at a different client site, and the DBMS software itself runs partly at the server site and partly at each client site.
- A further level of flexibility is provided by a *distributed database*. This enables the database data to be spread across a number of sites, so what appears to a program to be a single database may in fact be a series of databases distributed across multiple server sites.

The techniques just described may be sufficient if all the remote data are held in the databases managed by a particular DBMS product. Even if the remote data are held in databases managed by different DBMS products, there is generally no problem in writing a program to access those data. Both the Perl DBI and Java JDBC interfaces referred to in section 17.3.1 are designed to work with multiple databases managed by different DBMS products. The Java JDBC interface can also be used from a client program without any DBMS software being installed on the client machine: the DBMS software needed is downloaded automatically from the server site. Another useful feature of Java JDBC is that it enables data held in files outside any database to be accessed nonetheless as though they were stored in relational database tables.

The real problem with accessing remote data stored at multiple sites managed by different DBMS products or outside any database is that the data are no longer integrated and shared. Each database is completely separate and so, once more, the model of working becomes that of the top half of *Figure 17.1*, with each box being a database rather than an individual file. All the problems of that model then reappear.

One approach to dealing with such problems is to build specialized software to cross-link data in different biological data sources, together with facilities to enable data from those sources to be retrieved in response to a query. *SRS* (*S*equence *R*etrieval *S*ystem) was developed at EMBL to achieve this for molecular biology data, and provides facilities to link sequence data (nucleic acid, EST and protein) together with wider data sources such as protein structure and bibliographic data resources. SRS supports querying of those resources from both Web-based interfaces and a command line interface suitable for use from programs.

Some general technologies have been developed to enable data from multiple remote sites nonetheless to be treated as a single, integrated resource. In particular:

- *Data warehouses* are formed by extracting data from primary data sources and then holding the data in a database specifically designed to allow the relationships between the different data from the primary sources to be modeled and searched.
- *Mediator* software is designed to run between the server software of remote data server sites and the application programs of client sites. A client program is given the illusion that the data are all stored according to a single common data model. This is achieved by the mediator software intercepting client requests for data, translating those requests into the format required by a particular data server, and then transforming the data retrieved from a server site into a form consistent with the common data model for transmission back to the client program. The term *federated database* is sometimes used to refer to a system consisting of a number of server sites managing independent databases controlled by different DBMS products, but for which a single common data model is also supported to enable the databases to be accessed as a single resource.
- *XML technology* enables data from a remote data server to be retrieved with tags which explicitly represent the meaning of the data in a form suitable for processing by computer programs. XML has become particularly important for enabling data to be passed between computer programs over the World Wide Web. It is described in more detail together with related World Wide Web technologies in Chapter 18.

In recent years, DBMS products have become increasingly sophisticated in their support for these kinds of technologies. A DBMS no longer just manages data in tables in a single local database; increasingly it provides access to data stored locally and remotely, the data being stored in conventional databases, or in files external to the database, or on the World Wide Web.

17.4   Challenges arising from biological data

In section 17.1 some of the particular problems posed by biological data were highlighted. Section 17.3 introduced the data management techniques available with current DBMS products. To what extent do those techniques solve the problems identified?

The first problem was the many differing formats in which biological data are found, ranging from the regular structured data of GenBank or PDB files, to data of complex structure such as pathway data, or unstructured data such as literature abstracts. This is a problem for a DBMS which only supports a simple relational table model of data. However, many DBMSs have wider capabilities such as the following.

- Object-oriented DBMSs do have facilities for defining objects having a complex structure, together with the operators for storing and retrieving those objects. Hence, sequence or pathway objects could, for example, be defined and manipulated with such a system. Other DBMS products, known as *object-relational* DBMSs, have the ability to define such objects while retaining tables as the primary representation of data in the database.
- Some DBMS products have special facilities for holding and managing unstructured text data such as are found in biomedical literature abstracts.
- *Semi-structured database* techniques have been developed to handle data which, while not wholly unstructured, do not have the regular structure making them suitable to hold in table form. Currently, XML techniques are often used to handle such data.

The second problem was that there is often no single biological object to which data refer: individual users of the data will have their own perspectives. This is equally the case in considering data concerned with biological functions and mechanisms.

- This remains a problem with many DBMS products. The relational model provides a fixed table representation of the data although view tables can go some way to providing alternative views of the data. More flexible views can be obtained by holding the data in a semi-structured form, with table views being generated as required. However, this can result in a significant performance overhead unless materialized views, introduced in section 17.3.3, are used.

The third problem was the many incompatible formats and nomenclatures for biological data. The key to overcoming this is agreement on standards for representing biological data and their semantics. There are a number of initiatives in this area. For example:

- The *GO* (*Gene Ontology*) consortium has developed a controlled vocabulary for terms describing gene products from three perspectives: molecular function, biological process and cellular component. The ontology explicitly defines the relationship between more specific and more general terms in graph structures for each of these three perspectives. The ontology may be obtained in XML format, and also exists as a relational database which may be queried from a Web-based interface.
- A very large number of XML-based definitions and other resources have been published. These range from XML representations of specific data resources such as GO, InterPro and PIR, to proposals for XML representations of sequence data (BIOML, BSML), BLAST output (BlastXML) and expression data (MAGE-ML), as well as more general XML definitions for ontology specification (DAML+OIL). One example of XML being used in a practical way is DAS (Distributed Annotation System) which uses an XML representation of sequence annotation information to enable integration of annotations retrieved from multiple sites.
- In the object database world, the OMG LSR (Object Management Group Life Sciences Research) provides a focus for similarly defining standards and interfaces supporting a common format for biological data.

The final problem identified in section 17.1 was the fact that data about any given biological object are likely to be held at many different sites, each with a different approach to coping with the previous problems.

- Techniques for retrieving data from remote sites are well established, as outlined in section 17.3.4. Once again, the problems are those of being able to treat those data as a single integrated resource, and the key is agreement on standards for representing data and their semantics.

17.5 Conclusions

The data management techniques outlined in this chapter will undoubtedly be developed further to enable distributed biological data to be used as an integrated resource. Many of the most difficult outstanding problems arise from the semantic integration of such data rather than any technical problems in its storage and access as such.

Currently the World Wide Web enables distributed data to be accessed and displayed in a browser, but a program cannot extract the meaning of the data. The term *semantic web* is used to describe what data management techniques will increasingly support: ways of enabling distributed data to be accessed in a form whereby the semantics of the data are explicit and may be extracted and exploited by computer programs (see Chapter 18). In addition, the computer programs themselves will be distributed resources on the Web, rather than being written by individual scientists to run on local machines. This is the aim of *grid* computing: turning the Web into integrated data and computational resources which a scientist can plug into in the same way that the electricity grid can be plugged into when electrical power is required.

This chapter has focused on fundamental data management issues. The question remains how data may be turned into knowledge. Data mining and other data analysis techniques are used for extracting knowledge from data, with DBMSs themselves increasingly incorporating data mining and data analysis tools. These techniques are described in Chapter 15.

References and further reading

Abbey, M.C., Michael, J., and Abramson, I. (2004) *Oracle 10g: A Beginner's Guide.* Oracle Press, Osborne/McGraw-Hill, Berkeley, CA.

Descartes, A., and Bunce, T. (2000) *Programming the Perl DBI.* O'Reilly, Sebastapol, CA.

Fisher, M., Ellis, J., and Bruce, J. (2003) *JDBC API Tutorial and Reference,* 3rd Edn. Addison-Wesley, Reading, MA.

Mata-Toledo, R.A., and Cushman, P.K. (2000) *Fundamentals of SQL Programming.* Schaum's Outline Series, McGraw-Hill, New York, NY.

Reese, G., Yarger, R.J., King, T., with Williams, H.E. (2002) Managing and Using *MySQL,* 2nd Edn. O'Reilly, Sebastapol, CA.

Worsley, J.C., and Drake, J.D. (2002) *Practical PostgreSQL.* O'Reilly, Sebastapol, CA.

Internet technologies for bioinformatics

Andrew C.R. Martin

- The link between the Internet and Bioinformatics is demonstrated by showing how the two have evolved together. The chapter aims to give the reader an overview of modern Web technologies and the importance of these in Bioinformatics.
- Some of the major technologies of the Web are introduced – HTML and CSS used for creating web pages, XML used to exchange and store data, XSLT used to transform and reformat XML files and SOAP, a mechanism for accessing services on remote computers.
- The progression of the Web from a system for simple serving of documents to a distributed database using semantic markup is described.

18.1 Concepts

The *Internet* and the *World Wide Web* have revolutionized data and information handling. Access to the hundreds of gigabytes of genome and protein structure data by the worldwide community of scientists would simply not be possible without it. Indeed, it is likely that the genome projects themselves would have faced severe hurdles without the ability to share data across the Internet.

The rapid evolution of the Internet over the last 10 years has resulted in a huge number of web sites that present diverse biological data in a human-friendly format. The Web itself was developed during the early 1990s by Tim Berners-Lee at the European Laboratory for Particle Physics (CERN) and is just one application of the Internet. Web pages contain plain text with '*markup*' to indicate how text should be presented. Markup simply consists of plain text labels introduced by some 'special' character. The current markup language of the Web is *HyperText Markup Language* (HTML). Other markup languages include the printed text presentation languages *TeX* and *LaTeX*.

The current generation of Web pages which are marked up with HTML are all, by their very nature, computer readable, but are largely lacking in data that are also computer-understandable. HTML concentrates on describing the visual presentation of the data rather than its content. If we are to integrate and analyze these data in any meaningful automated way, it is essential that data be made more accessible with some type of *semantic markup* that is not only computer readable, but also computer understandable.

Currently we face a situation where data are frequently stored in some form of database. This may be a *relational database* system, or simply a plain text file using some agreed standard format (e.g. the columns of a *Protein Databank* file) or containing some form of tag-based markup (e.g. a *Swiss-Prot* or *Genbank* file). These data are frequently presented on a Web page using attractive layout and tables providing the scientist with a powerful visual image, which is easy to digest. However, when the data are restructured for attractive visual display, there is often a loss of information. The semantic meaning of the text is lost and perhaps only a subset of the available information is presented. Should a user of the Web page wish to extract information in an automated fashion for further processing, a computer program must be written that parses the HTML file and interprets the visual markup (tables, bold fonts, etc.) to regenerate the contextual meaning of the text. Inevitably this results in further information loss. If the web page designer decides to change the layout of the page, then the user's parser no longer works and has to be re-written.

The Web therefore provides a huge resource of data that, being computer readable, could be used directly by computers, but since these data are currently hidden behind visual markup, they are not computer understandable. The importance of this deficiency has led Tim Berners-Lee and the *World Wide Web Consortium* (W3C) to introduce the concept of the '*Semantic Web*'. Their vision is two-fold: first, rather than the Web simply presenting information from databases, by using semantic markup, the Web will itself become a huge database. Second, rather than browsing the Web for information using search engines, the user will be able to make a request of a *software agent* which will be able to browse the Web, compile and sort information and provide the user with only the relevant results. Tim Berners-Lee *et al.* describe their vision in an article in *Scientific American* (Berners-Lee *et al.*, 2001).

18.2   Methods and standards

The rapid evolution of the Web from a simple medium for sharing text documents using hyperlinks, through today's highly visual medium to the Semantic Web of the future has led to a plethora of new and evolving standards. This chapter does not pretend to be comprehensive, but simply attempts to describe the major concepts and outline the standards that are becoming available to encourage the bioinformatician to make use of what is currently available and to delve further into this area.

The Internet is littered with even more acronyms than bioinformatics itself! We will start with a simple overview of HTML, familiar to most people as the language that provides visual markup of today's Web and *CSS* (Cascading Style Sheets) which provide customized stylistic presentation. We will go on to *XML* (eXtensible Markup Language), which is similar in concept to HTML but provides semantic markup, and *XSL* (eXtensible Stylesheet Language) and *XSLT* (eXtensible Stylesheet Language Transformations), which allow visualization and restructuring of XML data. We will briefly describe methods for accessing software and data on remote computers and finally give an overview of necessary supporting standards which describe available resources and the meanings of words together with the importance of restricted vocabularies and ontologies.

18.2.1   HTML and CSS

HTML (HyperText Markup Language) is well known to most as the language of the Web. It was originally designed simply to support scientific text documents and to provide *hyperlinks* for linking documents between computers. The original HTML design was inspired by the ideas of *SGML* (Standard Generalized Markup Language). SGML, standardized in 1986 (ISO:8879, 1986) and based on an earlier effort by IBM to produce '*Generalized Markup Language*' (GML), is a complex standard for document markup used mostly by a few large companies and research organizations.

HTML contains two types of markup. The first, major type is the '*tag*' consisting of a label contained between angle brackets (< and >) followed by some text. The end of the tagged region is marked by an 'end-tag', identical in form to the opening tag, but started by </ rather than <. For example, the following would indicate a piece of text to be set in a bold font:

```
<b>This text is in bold</b>
```

The second type of markup is the '*attribute*' which consists of an attribute type and value pair contained within an opening tag. For example, the key feature of the World Wide Web, the hyperlink, consists of an 'anchor' (<a>) tag with which the text displayed as a hyperlink is indicated and an 'href' attribute containing the URL of the page to which the link is made. This is illustrated in the following simple example of a Web page written in HTML which might be used to display the content of this chapter:

```html
<html>
  <head>
    <title>
        Internet Technologies
    </title>
  </head>
  <body>
    <h1>
        Chapter 18: Internet Technologies for Bioinformatics
    </h1>
    <h2>
        Andrew C.R. Martin
    </h2>
    <p align="right">
        andrew@bioinf.org.uk <br>
        <a href="http://www.bioinf.org.uk">
           http://www.bioinf.org.uk
        </a>
    </p>
    <h2>
      Concepts
    </h2>
    <p>
        . . .
    </p>
    <h2>
        Methods and standards
    </h2>
    <p>
    . . .
    </p>
    <h3>
      HTML and CSS
    </h3>
    <p>
        . . .
    </p>
    . . .
    </body>
</html>
```

The resulting web page is illustrated in *Figure 18.1*.

All spacing and indentation are provided only to show the structure of the document. Most web browsers are very forgiving and will allow much of the overall structure (such as the <html>, <head> and <body> tags) to be left out. Many closing tags such as </p> can also be left out.

It was quickly realized that simple text markup was insufficient and HTML was extended to provide images, tables, fill-in forms and frames. As the Web developed into a commer-

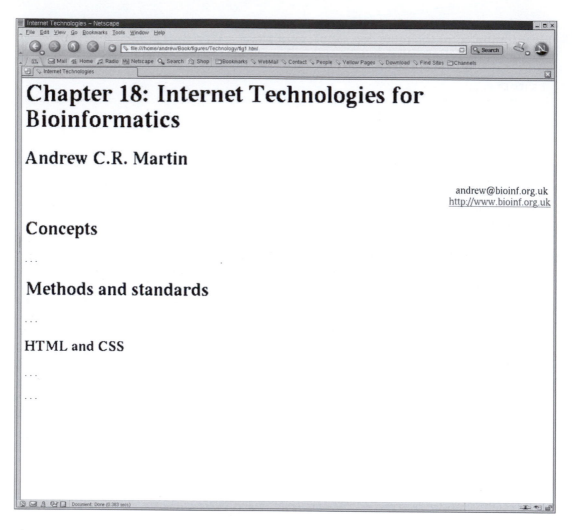

Figure 18.1

HTML example shown in the text rendered using Internet Explorer

cial force, browser plug-ins were developed to support dozens of additional multi-media concepts including complex animations, streaming video and audio. Browsers also began to support *JavaScript*, a simple object-oriented scripting language and *Java*, a full-scale platform independent programming language which can run as an *'applet'* within a Web browser.

Initially, HTML provided simple in-line control over presentation (fonts, colors, etc.). To provide a small degree of separation between content and presentation, and to make it easier to produce consistent documents, *Cascading Style Sheets* (CSS) were introduced. These allow the font, color and other rendering styles for elements such as headings to be defined once and then applied automatically to all occurrences of that element.

The following example describes styles for <h1>, <h2> and <h3> header elements and for paragraph text. All are rendered in a sans-serif font (Helvetica or Arial if available otherwise the default system sans-serif font). Level-1 headings are rendered at 18pt bold in white on a black background while level-2 are also 18pt bold, but in black on a white background while level-3 headings are in 14pt bold italic:

```
h1 { margin: 0em;
border: none; background: black; color: white; font: bold 18pt Helvetica,
Arial, sans-serif; padding: 0.25em;
  }
   h2 { font: bold 18pt Helvetica, Arial, sans-serif;}
   h3 { font: bold italic 14pt Helvetica, Arial, sans-serif; }
   p { font: 12pt Helvetica, Arial, sans-serif;}
```

Most people are familiar with the 'point' (pt) as a unit of measure of text size (1 pt = $\frac{1}{72}$ inch). In rendering a document on screen, such measures only give a relative idea of size, since the displayed size depends on the dimensions and resolution of the monitor (although typically Microsoft Windows uses a resolution of 96 pixels per inch). Another measure is used in this example as well, the 'em'. One em is defined as being the point size of the current font, so in this example, the padding around the level-1 heading will be 0.25×18 pt or 4.5 pt. The advantage of using em units is that if the point size for the font is changed, then the amount of padding around the heading will also change to give the correct appearance to the document automatically. The same document, rendered using this style sheet appears in *Figure 18.2*.

Unfortunately, as HTML has evolved and new tags have been introduced, it has become somewhat unwieldy. The situation has been made worse by the vendors of different browsers (primarily Netscape and Microsoft) adding vendor-specific extensions leading to incompatibilities between Web pages some of which render properly only on a given browser. The W3C has now started to address this problem by introducing the latest HTML standard, HTML4.0, also known as *XHTML*. XHTML is now an application of XML (see below) and has been neatly segregated into core functionality and additional packages to support more advanced features such as tables, forms, vector graphics, multimedia, maths, music and chemistry. Thus all browsers can implement the core functions and be selective in a modular fashion about what other support they provide. As the Web becomes more pervasive and appliances as diverse as computers, hand-held devices, mobile phones and even refrigerators provide Web browsers, this modular design becomes more and more important. A document can provide a profile describing the minimum set of features a browser must support in order to render it correctly and rendering devices similarly are able to provide 'device profiles' describing which tag modules they support.

XHTML is much stricter than earlier versions of HTML. All opening tags must be paired with closing tags and tags must be correctly nested. All tag names must be provided in lower case and attribute values (such as the URL in the example above) must be enclosed in inverted commas.

18.2.2 XML

HTML is still a presentation language. It provides a set of tags around pieces of text that describe how the text should be rendered. It does provide a minimal level of content markup by, for example, providing 'heading' tags. For example <h1> indicates a top-level heading, <h2> a second level heading, etc. CSS then allows rendering styles to be applied to these tags. However this level of content markup is extremely limited.

XML (eXtensible Markup Language) was developed to overcome this problem and has rapidly become the accepted standard for data interchange. In the past, many proprietary formats have arisen and software will frequently only understand one format. For example,

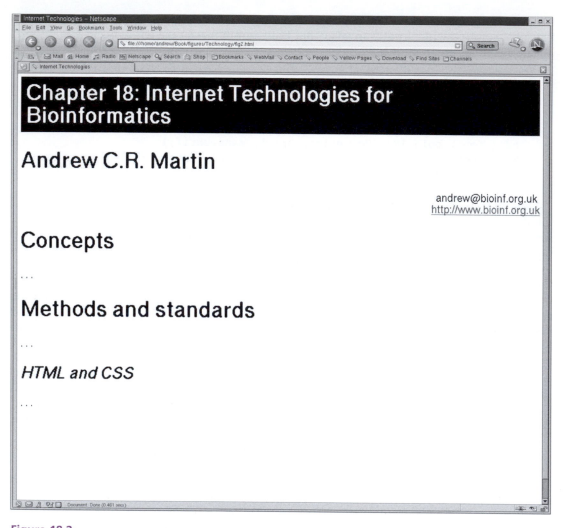

Figure 18.2

Example HTML document rendered using the example CSS stylesheet

sequence data may be stored in FASTA, PIR, Swiss-Prot and Genbank formats amongst others. Some of these formats are richer than others, i.e. they may contain more annotation information. The extensibility of XML means that a simple format can be designed and new data types can be added for richer markup as necessary. Programs requiring only the basic types of data can extract just those types of data that they need.

Conceptually and syntactically, XML is very similar to HTML, but rather than providing a fixed set of tags which describe rendering, the document author can design a *data type definition* (DTD) that provides *any* desired tags which have some sort of syntactic meaning. Semantic markup, in which tags indicate the meaning of the data rather than what should be done with those data, allows data to be exchanged between different applications which can use the data in different ways. For example, the same HTML example might be represented by the following fragment of XML:

```
<chapter number="18">
  <title>Internet Technologies for Bioinformatics</title>
  <author>
    <name>Andrew C.R. Martin</name>
    <email>andrew@bioinf.org.uk</email>
    <url>http://www.bioinf.org.uk</url>
  </author>
  <section title="Concepts">
    <p>
      . . .
    </p>
  </section>
  <section title="Methods and standards">
    <p>
      . . .
    </p>
    <subsection title="HTML and CSS">
      <p>
        . . .
      </p>
    </subsection>
  </section>
  . . .
</chapter>
```

Thus XML describes only the content of a document; it says nothing about what should be done with the data – in this case how the document should be rendered. In the case of this XML document, given that we would probably want to render it on a Web browser, or on paper, it is then necessary to translate the XML into HTML or the printer control language *PostScript* (perhaps via a page markup language such as *LaTeX*). However, the important point is that every piece of text is associated with a tag which labels it and from this label, meaning can be derived. There is thus a clear and total separation between the content of the document and the way in which it is presented. Having marked up a document in XML it becomes possible to render it for any output device in any way necessary. For example, the same piece of XML shown above could be converted into a number of Web pages and printed forms using different stylistic layouts. Rather than render the document, one might write a program to extract a table of contents, index keywords, or to extract code fragments and examples and place them in separate files.

A number of parsers are available for reading XML data and making it available within a computer program. Such parsers are available in most of the common programming languages including C, C++, Java and Perl. By using such a tool, the data and its semantic meaning are immediately available to any piece of software and the problems of trying to parse and extract semantic meaning from HTML or a variety of flat file formats are removed.

18.2.3 XSL and XSLT

When XML is used to create a document designed for electronic viewing or printing, *XSL* (eXtensible Stylesheet Language) may be used to describe how the document should be rendered on screen or on paper. It provides for a direct translation from XML markup to printed or electronic presentation. The first XSL standard was issued in October 2001 and a number of pieces of software are available to provide rendering from XML files with XSL stylesheets (see http://www.w3c.org/style/XSL/).

Part of the XSL standard provides for 'transformations' of an XML document. This part of XSL, known as *XSLT* (eXtensible Stylesheet Language Transformations), is more mature than the rest of XSL and a number of XSLT translators are available. XSLT style sheets are themselves written in XML and take input in XML and generate output which may be in XML, HTML or plain text. XSLT incorporates the *XPATH* standard for selecting parts of an XML document. Thus an XSLT style sheet can be used to parse an XML document, select parts relevant for a particular purpose and generate an HTML document containing only these data for rendering in a Web browser. Different style sheets can provide different on-screen layouts or perhaps a simplified *WML* (Wireless Markup Language) version for rendering on a mobile phone. Alternatively, the same XML document could be processed with a different XSLT style sheet to generate plain text output for rendering the text on paper using LaTeX. Goosens and Rahtz (1999) provide a very practical introduction to such techniques.

For the *Semantic Web* to become a reality, it is necessary for Web browsers to become XML/XSL capable. Microsoft Explorer already contains an XSLT translator, which can be used to convert an XML page to HTML, which can then be rendered using CSS style sheets. An alternative route to using XML on the Web is to run the XSLT translator on the Web server to generate HTML pages that can be rendered with any Web browser. The considerate Web page author will also make the XML available!

18.2.4   Remote procedure invocation

Web pages are generally accessed using a web browser via *'Hyper Text Transfer Protocol'* (HTTP). The browser sends a simple request in plain text in the form `GET /path/page.html` and the Web server returns the page. Parameters may also be passed tacked onto the end of the GET request, or the alternative POST method may be used to pass more complex data. This is the way in which data are sent from forms on Web pages to a Web server. While this system works well for normal pages and occasional access to Web-based servers, the use of a Web-browser is inconvenient when one wishes to access a Web-based service for analysis of many hundreds or thousands of protein sequences.

For example, suppose someone has implemented a Web server for secondary structure prediction. There are of course many of these accessible over the Web. The server requires the user to paste an amino acid sequence into the Web page and returns a page (in HTML, or if you are very lucky in XML) which displays the sequence together with the secondary structure assignments for each amino acid. Now suppose you have 1000 sequences for which you want to perform secondary structure predictions. One option is to spend a week or two sitting in front of the computer cutting and pasting in the sequences and extracting the results – not exactly an appealing proposition! Alternatively, by analyzing the content of the Web page used to access the server, one could write a small Web client that emulates the Web browser, sends each amino acid sequence in turn to the server and waits for the results. If these are returned in XML, then extracting the secondary structure assignments should be relatively straightforward, but more likely the data will come back in HTML and this must be parsed to extract the secondary structure assignment.

Remote Procedure Calling (RPC) is a general term for a protocol that wraps up all these operations in a manner that makes it possible to write software which calls a server on a remote computer to perform an operation as if it were just part of the local program. When such servers are made available, it becomes easy to write a program that reads through 1000 or more sequences and dispatches each in turn for secondary structure prediction collecting the results for output or further processing.

Many RPC protocols are available. *CORBA* (Common Object Request Broker Architecture) is one such system. Another system gaining in popularity is *SOAP* (Simple Object Access Protocol). SOAP has the advantage that all communication is done using the HTTP protocol and all data that are passed backwards and forwards between the computers are formatted in

XML. Since *firewalls*, used to protect computer networks from outside access, are configured to allow access to Web pages, use of standard HTTP ensures that SOAP services can be accessed with no changes to the firewall.

Thus providers of Web-based services can now easily set up an RPC server such as a SOAP server to support the same functions in a manner more suitable for automated access. In the case of SOAP, the request is packaged together with data in XML and the results are also returned in XML. Numerous libraries (such as the *SOAP::Lite* package for use with the Perl programming language) are available that look after the XML packaging and network access automatically such that invoking a method on a remote machine is just as easy as on the local machine.

18.2.5 Supporting standards

Standards also exist to add explanations to semantic markup. *RDF* (Resource Description Framework) is one such standard allowing more verbose descriptions of the definitions of markup tags used in XML. Ideally, definitions and selected markup tags should be defined using an *ontology*. At the very least, a standardized restricted dictionary together with definitions of terms must be used such that the markup is always used by different people to mean the same thing. There are also standards that can be used to annotate services available as remote procedures. *WSDL* (Web Services Description Language) is one such standard. Both RDF and WSDL descriptions are themselves created using XML, providing an integrated XML-based system.

18.3 Insights and conclusions

The World Wide Web started as a very simple idea. The aim was simply to share scientific text documents and provide links from one document to another across the Internet. HTML rapidly evolved from a simple text-based medium to a highly graphical presentation medium integrating animation, streaming video and audio as well as dynamic content and pages generated on-the-fly. However, this is all based around the basic idea of HTML pages containing visual rather than semantic markup. Some 10 years after the beginnings of the Web, its explosion in popularity has revealed the weaknesses of the initial approaches and the Web is starting to be driven by semantic markup.

Similarly, bioinformatics started as a few small specialized databases. With the huge explosion in available biological data, the deficiencies of having a multitude of small databases with no way to integrate them has been realized and many efforts are underway to integrate data using semantic markup and ontologies to allow databases to be linked. Bioinformatics and the Internet have grown up together – so much so that some of the developments in Internet and Web technology have been driven by the needs of bioinformatics.

The Internet and the Web are constantly evolving. As well as the Web, there is a move towards more distributed computing networks with access to supercomputer power over the 'Grid' – a concept for distributing computer power across a faster version of the Internet (see Chapter 17). While the 'first generation' of the Web used HTML as the basis for all pages, possibly with style sheets written with CSS to customize the visual appearance, the 'second generation' uses XML. XML can be translated to HTML for rendering by the Web server using XSLT or by an XSLT engine embedded in the Web-browser (either provided as part of the browser itself or in a Java applet). Alternatively, when the XSL standard is adopted within Web browsers, XML may be rendered directly using XSL style sheets. Critically, the XML should always be available to the client application such that the semantic markup is not lost.

In the 'third generation' of the Web as envisioned by Tim Berners-Lee, semantic markup will rule. This moves away from Web browsing in order to find information and moves to a 'data-push' technology where Web agents interpret our queries and seek out only the information we require. The whole of the World Wide Web becomes a single integrated database of immense power.

However, a major part of the success and popularity of the Web has been the ease with which anyone with a connection to the Internet can put up Web pages using HTML. The future of the Web will see it diversifying into two streams – the popular Web where Joe Bloggs can create his home page in simple HTML and the semantic information Web which will provide an unimaginably huge information resource for everyone's use.

References and further reading

Berners-Lee, T., Hendler, J., and Lassila, O. (2001) The Semantic Web. *Scientific American*, May 2001. Also available free online at http://www.sciam.com/

Goosens, M., and Rahtz, S. (1999) *The LaTeX Web Companion*. Addison Wesley. ISBN: 0201433117.

Glossary

Ab initio

Means 'from first principles' or 'from the beginning.' This implies that a method requires only knowledge of fundamental physical constants. However, in practice it has come to mean methods that require no prior constraints or parameterization.

Algorithm

Any well-defined computational procedure that takes some value, or set of values, as input and produces some value, or sets of values, as output.

Analogs

Non-homologous gene products that have acquired shared features, such as a similar folding architecture or similar functional site, by convergent evolution from unrelated ancestors.

Bit score

In sequence searching programs, the bit score is a measure of sequence similarity that is independent of database size. Because of this independence, bit scores from searches on different databases (using the same query and search program) are comparable in a way that *E-values*, which are scaled by database size, are not. Typically, the higher the bit score, the better the match. Bit scores reported by BLAST are discussed further under *E-values in BLAST*.

Clustering

Unsupervised classification of patterns or observations into groups (clusters).

Correlation

A measure of the strength of association between two variables.

Data mining

Extracting valuable information from large volumes of data.

Distance metric

A dissimilarity measure between two patterns.

Distance plot

A distance plot is the 2-D matrix used to visualize the intramolecular distances between residue positions in a protein. The residue positions of the protein are used to label both axes of the matrix and cells are shaded depending on whether residues are within a threshold distance from each other. Distances between Cα positions are typically exploited, though distances to any of the atoms in a residue can be used. Characteristic patterns of lines are observed in the matrix depending on the fold of the protein structure. See Chapter 6 for details.

Domain

A protein domain is a well-defined modular unit within a protein that either performs a specific biochemical function (functional domain), or constitutes a stable structural component within a protein structure (structural domain). Domains can be classified into domain (super)families that arose through duplication events (*evolutionary domain*).

Dot plot

A dot plot is the 2-D matrix used to visualize the similarity between two protein sequences. The sequence of one protein is used to label the horizontal axis and the other sequence the vertical axis. Cells in the matrix are then shaded depending on whether the associated residue pairs are identical or possess similar properties. See Chapter 3 for details.

Dynamic programming

Dynamic programming is an optimization technique frequently used to obtain the alignment of protein sequences or structures. It is a computational method which efficiently explores a 2-D matrix scoring the similarity between two proteins to find a path through that matrix and determines where insertions and deletions are occurring relative to each protein. See Chapter 3 for details.

Empirical

Derived from experiment. Relates to a set of values based on experiment rather than on theory.

Entropy

In physics, entropy is a measure of the 'disorder' in a system. In mathematics, computer science and bioinformatics, entropy refers to the Shannon entropy (also known as the 'information theoretic entropy' or the 'information entropy'), which has a more abstract definition. The (Shannon) entropy measures the diversity among a set of symbols. For example, if set X comprises equal amounts of the symbol 'A' and 'B' whereas set Y comprises only one type of symbol 'A', then X is more diverse than Y. Correspondingly, the entropy of X is higher than the entropy of Y.

Entropy is traditionally measured in binary information digits (bits). This is because it was originally used in signal processing to measure the information content of a transmitted signal. A signal is deemed informative if it resolves a lot of uncertainty. For example, the set of symbols X (described above) might be our initial knowledge about some value of interest. X tells us there is a 50/50 chance of it being either 'A' or 'B', and so has an uncertainty of 1 bit, because in order to resolve the true value we must correctly make 1 binary decision ('A' or 'B'). A signal Y is received. Y contains only 'A' and so reveals the true value. The information content of the signal Y is the amount of uncertainty we lost on its receipt, i.e., entropy(X) − entropy(Y) = 1 bit.

E-value

The E-value (also known as the 'expected value' or the 'expectation') for an observed score x is the number of times you would expect to get a score of at least x by chance.

E-values are closely related to *P-values*. In fact, it is usual for an observation's E-value to be calculated directly from its *P*-value. In BLAST literature, however, the calculation is laid out differently and *P*-values are rarely mentioned. See the entry *E-values in BLAST* for information on why this is.

The E-value depends on two quantities: N, the number of times you repeat the experiment (e.g., the number of database sequences you try to match your query against), and p, the probability of getting such a result in a single experiment (i.e., the *P*-value). Multiplying one by the other gives the E-value, i.e., $E = Np$. For example, the probability of throwing a die and getting 5 or more dots is $p = 2 \times 1/6 = 1/3$. If you throw a die 10 times, the number of times you expect to get 5 or more dots is $E = Np = 10 \times 1/3 = 3.3$.

E-values in BLAST

BLAST compares one query sequence with many match sequences, looking for stretches of high similarity. These stretches are known as high scoring pairs (HSPs). Each has a raw match score of S, calculated by summing matrix similarities of aligned residues and deducting gap penalties over its length.

A traditional type of *P*-value might measure the probability of the query sequence matching a particular database sequence with a score of at least S; the corresponding E-value in this case would be the expected number of times that happened in a database of size N (hence $E = pN$). That is what FASTA does; BLAST is different.

In BLAST, the E-value and *P*-value do not refer to the whole query sequence; rather they refer to the fragment of the sequence that is in a given match (i.e., an HSP). The *P*-value is the probability that an HSP will score at least S is given by

$$P = Ke^{-\lambda S}$$

In a comparison between a sequence of n residues and a database m residues (i.e., m is the length all the database sequences laid end-to-end) there are $m \times n$ different places to start aligning from. Therefore, the E-value is

$$E = mnP$$

$$E = Kmne^{-\lambda S}$$

The *P*-value is usually a very small number and, in the BLAST statistical model, is not particularly intuitive. So rather than reporting the *P*-value for a hit, BLAST reports the 'bit score', denoted by S', which is a simple transformation of the *P*-value:

$$S' = -\log_2 P$$

$$S' = \frac{\lambda S - \ln K}{-\ln 2}$$

The bit score is just as useful as the *P*-value because, unlike the E-value, it is independent of sequence and database length. But it is more intuitive because matches that are highly significant yield high bit scores.

Exon

A segment of a gene which codes for all or a specific portion of a protein. Exons are spliced together into a complete protein-coding mRNA sequence by excision of introns where present.

Compare *intron*.

Fold

Fold describes the three-dimensional structure adopted by a polypeptide chain. The fold description details the topology of the structure. That is both the arrangement of secondary structures in 3-D space and also the order in which the secondary structures are connected along the polypeptide chain. See Chapter 7 for details.

Genome

The sum total of all the genetic material (usually DNA) in a given organism. For higher organisms, the genome is segregated into separate chromosomes.

Heuristics

Methods of solving problems by using past experience and moving by trial and error to a solution.

Homologs

Homologous sequences or structures are related by evolutionary divergence from a common ancestor. Homology cannot be directly observed, but must be inferred from calculated levels of sequence or structural similarity.

Indel

Indel is the abbreviated term used to describe the residue insertions or deletions which have occurred between two related protein sequences during evolution.

Internet

A worldwide system of computer networks following agreed standards for identification of computers and communication between them. It enables computers around the world to be electronically linked and supports applications such as electronic mail (email) and the *World Wide Web*.

Intron

An intervening sequence in a gene that may interrupt the coding region and divide it into exons. Introns are transcribed but are excised from the initial transcript in order to produce the processed mRNA. Introns can also occur in the untranslated region of the gene.
   Compare *exon*.

Knowledge-based

Derived from an organized body of information. Has different meanings in different areas of bioinformatics. <computer science> A knowledge-based system represents a program for developing or querying a collection of knowledge expressed using some formal knowledge representation language. <molecular modeling> A knowledge-based method derives rules from observed data. This applies both to using known parent structures to build models of proteins and to create scoring functions (e.g. derived from atomic contacts in a structural database).

Mean

The average of a set of values obtained by summing the values and dividing by the number of values there are. Hence, the mean $\bar{x}$ of n values x is given by the formula:

$$\bar{x} = \sum x/n$$

Alternative average values are given by the *median* and *mode*.

Median

The average of a set of values obtained by placing the values in increasing order and then choosing the value at the midway point. Alternative average values are given by the *mean* and *mode*.

Metabolome

The full complement of small molecular weight metabolites which interact with the proteins expressed in a cell (i.e. the proteome).

Mode

The average of a set of values obtained by choosing the value which occurs most frequently. Alternative average values are given by the *mean* and *median*.

Monte Carlo method

Monte Carlo is a *stochastic* method providing solutions to a variety of mathematical problems by performing statistical sampling experiments on a computer. The method uses a random number generator, essentially a roulette wheel. Hence it is named after the city in Monaco famous for gambling.

Motif

Motifs are typically small segments of a protein sequence or fragments of a protein structure which are well conserved and which may occur in different contexts. They are often important for the stability or function of the protein. For example the αβ-plait motif refers to a structural subdomain comprised of three β-strands and one α-helix, which occurs frequently in a number of different folds in the α–β class and also recurs within some folds. Motifs are also used to describe small highly conserved sequence patterns often associated with the functional properties of a protein. For example, a GxGGxG motif often found in nucleotide-binding proteins, where G is glycine and x is any residue, is the sequence pattern or motif of the residues involved in co-factor binding.

Normalize

To normalize a variable (such as a score or some other measurement) is to apply a mathematical operation to it that then allows it to be compared with other variables in a meaningful way.

Ontology

A defined vocabulary describing concepts and the relationships between concepts for a given knowledge domain, e.g., terminology used in a medical speciality or terms used to describe functions of biomolecules.

Open reading frame (ORF)

A stretch of DNA or RNA located between the initiation codon (usually ATG) and the stop codon. In prokaryotes, an ORF often corresponds to a single complete gene.

Orthologs

Orthologs are equivalent genes in different species that evolved from a common ancestor by speciation. They usually have the same, or a highly similar, function.

Paralogs

Homologous genes or gene products that are descendants of an ancestral sequence that has undergone one or more gene duplications. Paralogs typically perform a different but related function in the same or a different species.

Pattern

A qualitative *motif* description based on a regular expression-like syntax.

Profile

A quantitative *motif* description, assigns a degree of similarity to a potential match. A profile contains a matrix that lists the frequencies of finding each of the 20 amino acids at each position in a multiple alignment of a protein domain family.

Protein family

A protein family comprises a set of closely related proteins which have derived from a common ancestral gene and which have been demonstrated to be evolutionarily related by significant similarities in their sequences (≥35% identity), structures and functional properties. Relatives within a protein family are described as homologs.

Protein superfamily

A protein superfamily comprises a set of related proteins which have derived from a common ancestral gene, some of which may be quite distantly related. Identification of evolutionary relatedness is typically based on significant similarities in sequence (≥35%) or in distant relatives possessing low-sequence identities, by significant structural and functional similarity. Relatives within a protein superfamily are described as *homologs*. These can be further distinguished as *orthologs* or *paralogs*. See Chapters 3, 5 and 6 for more details. Within superfamilies, more closely related proteins can be grouped into families (see above).

Proteome

In bioinformatics, the proteome sometimes refers to the protein translations derived from the genome of an organism. However, more generally, the proteome is defined as the complement of proteins expressed by a cell or organ at a particular time and under specific conditions. The total number of proteins in the proteome can exceed the number of genes in the genome due to differential splicing for example.

P-value

The *p*-value for an observed score x is the probability that a score equal to or more extreme than x could have occurred by chance. In bioinformatics, *p*-values are often used to tell us how significant a measured similarity is or how unlikely some observed data are to have come from a particular null model or distribution.

For instance, suppose you have a protein sequence and are looking for homologs in a database of other protein sequences. (Equally, you could have a protein structure, or any other type of object, and be looking for significant matches in a database of structures, or of comparable objects.) Now suppose the query matches sequence *A* with a similarity score *x* of 34 units, where 'units' denotes some arbitrary similarity measure. Is sequence *A* a significant match? This depends on whether 34 units is unusually high or not, which in turn depends on the distribution of scores produced when comparing the query with all the other sequences in the database. The *p*-value associated with *x* is more informative than *x* by itself. It states the probability that a sequence randomly chosen from the database will give a score greater than or equal to *x*. For instance, if only 5 out of 100 of the sequences match the query with 34 or more similarity units, then the *p*-value for the match score, and hence the *p*-value for the similarity between sequence *A* and the query, is 5/100 = 0.05.

More generally, the smaller the *p*-value, the more unlikely the score associated with it is due to chance, and the more the match, data point or whatever is being measured, can be considered statistically significant.

RMSD

Root Mean Square Deviation (RMSD) is used as a measure of similarity between two protein structures. Mathematically, this is defined as:

$$R = \sqrt{\sum_{i=1}^{N} (d_i)^2 / N}$$

Given a set of N equivalenced atom pairs, the distance between each pair is calculated, squared and summed over the N atom pairs and divided by N to calculate the mean. Finally the square root of this value is taken:

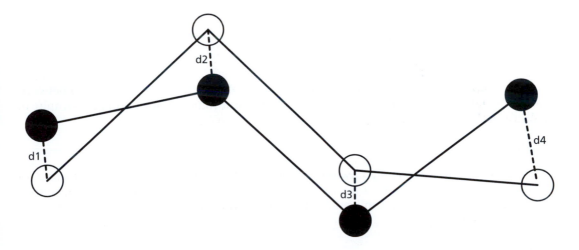

RMSD may be quoted over 'all atoms', Cα atoms, or 'backbone' atoms. This final term usually refers to N,Cα,C,O, but may also refer to N,Cα,C or to N,Cα,C,O,Cβ. Care should therefore be taken in comparing RMSDs quoted over backbone atoms.

Sequence homology

Sequence homology is commonly misused to mean *sequence similarity*. A 'percentage homology' is often quoted, but in reality a pair of sequences either are, or are not, *homologs*.

Sequence identity

The sequence identity is the percentage of identical residues in the optimal alignment between sequences X and Y. A number of variations in the choice of divisor in calculating the percentage is possible depending on the context. Typically, the divisor is the total length of the smallest sequence. As a variation, unaligned tails in the alignment may be excluded. Alternatively, if one is interested in one of the sequences compared with a number of other sequences (in, for example, comparative modeling), then the divisor will be the length of the sequence of interest.

Sequence similarity

The sequence similarity between two sequences is expressed as a percentage. Given a similarity scoring matrix such as the Dayhoff or Blosum matrices, the percentage similarity between sequences X and Y is generally calculated as the sum of the scores from the matrix for the optimum alignment divided by the lower of the two scores achieved by scoring sequence X against itself and sequence Y against itself. Optionally the score for the alignment between X and Y may take into account gap penalties used in generating the alignment in which case the numerator is the maximum score obtained in the dynamic programming matrix. For the term to have any real meaning, the scoring matrix and gap penalty scheme used should be quoted alongside the similarity value, but this is often not the case.

Standard deviation

A measure of the degree to which a set of values vary from the *mean* average value. The more that values differ from the mean, the larger will be the standard deviation. It is obtained by adding the squares of the deviation of each value from the mean, dividing by the number of values there are, and finally taking the square root. Hence, the standard deviation s of n values x with mean is given by the formula:

$$s = \sqrt{\left(\frac{\sum (x - \bar{x})^2}{n} \right)}$$

Some definitions of standard deviation and statistical software packages distinguish between standard deviation of a population, defined as above, and standard deviation of a sample, in which the divisor n is replaced by n-1.

Standardize

To standardize a variable usually means to *normalize* it such that it has mean 0 and standard deviation 1, that is, to convert it into a Z-score. Standardize may also refer to a more general normalization, such as the multiplication of a variable by a constant or the addition of a constant to all scores.

Stochastic

Has a specific meaning in statistics. Stochastic processes contain a random variable or variables. Stochastic search techniques (e.g. *Monte Carlo methods*) use randomized decisions while searching for solutions to a given problem.

Symbol: $\sum$

Pronounced 'sigma', this symbol denotes a sum of numbers. Symbols above, below and to the right of sigma indicate how the sum is to be formed. The number below indicates the first value, the number above the last value and the right-hand side indicates the operation to be performed on that value before it is added to the running total. The examples below show the sigma notation, with the meaning on the right-hand side:

$$\sum_{i=1}^{4} i = 1 + 2 + 3 + 4$$

$$\sum_{i=1}^{N} i = 1 + 2 + \cdots + N$$

$$\sum_{i} i = \sum_{i=0}^{N} i \quad \text{or} \quad \sum_{i=1}^{N} i \quad \text{depending on whether } i \text{ could reasonably be zero}$$

$$\sum_{i=1}^{4} f(i) = f(1) + f(2) + f(3) + f(4)$$

$$\sum_{i=3}^{5} f(x_i) = f(x_3) + f(x_4) + f(x_5)$$

$$\sum_{a \in BranchedChainAminoAcids} f_a = f_{leucine} + f_{isoleucine} + f_{valine}$$

Symbol: $\in$

Denotes membership of a set; reads as 'is a member of'. For example, consider the set A where $A = \{1,6,7\}$. If $x \in A$, then x must be one of 1, 6 or 7. Similarly, if $BCAA = \{$leucine, isoleucine, valine$\}$ and $a \in BCAA$, then a always refers to one of the branched chain amino acids.

Transcriptome

The full complement of activated genes, mRNAs, or transcripts in a particular cell or tissue at a specific time. Note that unlike the genome, the transcriptome is a dynamic entity, which changes over time and under different cellular conditions.

Tuple

A k-tuple refers to a row of k residues within a protein sequence. For example a tuple of 2 means two residues in a row. Tuples are often used when shading *dot plots* whereby only those cells associated with matching tuples between two protein sequences are shaded.

World Wide Web (abbreviated to WWW or W3)

An application of the *Internet* providing a network of information resources such as documents. Each resource may contain hypertext links which enable easy navigation by computer from one resource to related resources. The World Wide Web is often referred to as simply the Web.

Z-score

The Z-score for an item indicates how far and in what direction that item deviates from its distribution's *mean*. It is expressed in units of its distribution's *standard deviation*. You can convert an observed measurement, x_i, to its Z-score, z_i, with the following transformation:

$$z_i = \frac{x_i - \bar{x}}{\sigma},$$

where $\bar{x}$ is the mean and σ is the standard deviation of all observations.

The Z-score transformation is often used to compare the relative standings of items from distributions with different means and/or different standard deviations. However, this is only really meaningful when the observed measurements consistently fall into normal (i.e. Gaussian, or bell-shaped curve) distributions. In this case, the transformation acts to scale different normal distributions (i.e., normal distributions with different means and standard deviations) to the standard normal (i.e., a normal distribution with mean zero and standard deviation one). This means comparing Z-scores of observations from different normal distributions is equivalent to comparing raw scores from identical distributions. Moreover, because the transformation scales everything to a single distribution, there is a one-to-one mapping between an observation's Z-score and its *p-value*. This in turn means that observations with the same Z-score will be equally likely.

Index